THE FRONTIERS COLLECTION

THE FRONTIERS COLLECTION

Series Editors:
A.C. Elitzur L. Mersini-Houghton M.A. Schlosshauer M.P. Silverman
J.A. Tuszynski R. Vaas H.D. Zeh

The books in this collection are devoted to challenging and open problems at the forefront of modern science, including related philosophical debates. In contrast to typical research monographs, however, they strive to present their topics in a manner accessible also to scientifically literate non-specialists wishing to gain insight into the deeper implications and fascinating questions involved. Taken as a whole, the series reflects the need for a fundamental and interdisciplinary approach to modern science. Furthermore, it is intended to encourage active scientists in all areas to ponder over important and perhaps controversial issues beyond their own speciality. Extending from quantum physics and relativity to entropy, consciousness and complex systems – the Frontiers Collection will inspire readers to push back the frontiers of their own knowledge.

Other Recent Titles

For all volumes see back matter of the book

**Ulrich J. Frey · Charlotte Störmer ·
Kai P. Willführ**
Editors

HOMO NOVUS –
A HUMAN WITHOUT
ILLUSIONS

Springer

Editors

Ulrich J. Frey
Universität Gießen
Zentrum für Philosophie und
Grundlagen der Wissenschaft
Otto-Behaghel-Str. 10
35394 Gießen
Germany
ulrich.frey@phil.uni-giessen.de

Charlotte Störmer
Universität Gießen
Zentrum für Philosophie und
Grundlagen der Wissenschaft
Otto-Behaghel-Str. 10
35394 Gießen
Germany
Charlotte.Stoermer@phil.
uni-giessen.de

Kai P. Willführ
Universität Gießen
Zentrum für Philosophie und
Grundlagen der Wissenschaft
Otto-Behaghel-Str. 10
35394 Gießen
Germany
kai.p.willfuehr@phil.
uni-giessen.de

Series Editors:

Avshalom C. Elitzur
Bar-Ilan University, Unit of Interdisciplinary Studies, 52900 Ramat-Gan, Israel
email: avshalom.elitzur@weizmann.ac.il

Laura Mersini-Houghton
Dept. Physics, University of North Carolina, Chapel Hill, NC 27599-3255, USA
email: mersini@physics.unc.edu

Maximilian A. Schlosshauer
Niels Bohr Institute, Blegdamsvej 17, 2100 Copenhagen, Denmark
email: schlosshauer@nbi.dk

Mark P. Silverman
Trinity College, Dept. Physics, Hartford CT 06106, USA
email: mark.silverman@trincoll.edu

Jack A. Tuszynski
University of Alberta, Dept. Physics, Edmonton AB T6G 1Z2, Canada
email: jtus@phys.ualberta.ca

Rüdiger Vaas
University of Giessen, Center for Philosophy and Foundations of Science, 35394 Giessen, Germany
email: ruediger.vaas@t-online.de

H. Dieter Zeh
Gaiberger Straße 38, 69151 Waldhilsbach, Germany
email: zeh@uni-heidelberg.de

ISSN 1612-3018
ISBN 978-3-642-43952-0   ISBN 978-3-642-12142-5 (eBook)
DOI 10.1007/978-3-642-12142-5
Springer Heidelberg Dordrecht London New York

Cover design: KuenkelLopka GmbH, Heidelberg

Printed on acid-free paper

Springer is part of Springer Science+Business Media (www.springer.com)

To Eckart Voland

Preface

This book is intended to fill a longstanding gap and growing need for information and discussion. Although many interdisciplinary books consider aspects of human nature and human society in a more-or-less naturalistic framework, we know of none that concentrates—as this one does—on findings that emerge from evolution-inspired investigations. The chapters of this book illuminate these findings and, in particular, discuss their consequences and implications for the way we view ourselves and humankind's relation to the world-at-large.

The new insights are applied here to evolutionary psychology, evolutionary ethics, and evolutionary ecology, to name but a few. Exactly these disciplines are at the core of today's most controversial debates between traditional and naturalistic interpretations of human nature. In the various chapters, leading authors provide evidence for the need to switch to the new view of humans, in spite of the understandable resistance to giving up some of our most cherished illusions.

The concept for this book owes much to the inspiration provided by Eckart Voland of the University of Gießen. In fact it mirrors the multifaceted character of his work, his naturalistic way of thinking, and his commitment to spreading a new, scientifically grounded image of humankind. Throughout his scientific career as an anthropologist and primatologist, Eckart Voland has dispelled many illusions, frequently questioning the traditional view that humans are exceptional. With his doctoral advisor, Christian Vogel, he pioneered the brand-new discipline of sociobiology in Germany. Based mainly on the *Krummhörn project*, which would shape his scientific life for the following decades, he has contributed major findings to the grandmother hypothesis and the area of parental investment. His research reveals that many decisions in human life are not as independent of the biological imperative as our intuition leads us to believe—unmasking yet another illusion.

Our first thanks therefore go to Eckart Voland for his scientific inspiration. We are grateful, in addition, to Ingrid Weil for triggering the initial idea, to Angela Lahee at Springer-Verlag for the enthusiastic support that helped to get this project started, and to the authors for their uncomplicated willingness to contribute.

It is to be hoped that the insights inspired by evolution will—as they become better recognized—lead to a new understanding of man's place in nature, free of at least some of the illusions that encumber us today. This book should serve as a stepping stone along the way.

Gießen, Germany   Ulrich J. Frey
January 2010   Charlotte Störmer
                                                                                    Kai P. Willführ

Contents

Contributors

Kurt Bayertz Philosophisches Seminar, Westfälische Wilhelms-Universität, 48143 Münster, Germany, bayertz@uni-muenster.de

Athanasios Chasiotis Tilburg University Faculty of Social and Behavioural Sciences, 5000 LE Tilburg, Netherlands, achasiot@uvt.nl

Robin Dunbar Institute of Cognitive & Evolutionary Anthropology, University of Oxford, OX2 6PN Oxford, UK, robin.dunbar@anthro.ox.ac.uk

Harald A. Euler Institut für Psychologie, Universität Kassel, 34109 Kassel, Germany, euler@uni-kassel.de

Detlef Fetchenhauer Department of Social and Organizational Psychology, University of Groningen, 9712 TS Groningen, Netherlands; Institut für Wirtschafts- und Sozialpsychologie, Universität zu Köln, 50931 Köln, Germany, detlef.fetchenhauer@uni-koeln.de

Julia Fischer Cognitive Ethology Laboratory, German Primate Center, 37077 Goettingen, Germany, fischer@cog-ethol.de

Ulrich J. Frey Zentrum für Philosophie und Grundlagen der Wissenschaft, Universität Gießen, 35394 Gießen, Germany, ulrich.frey@phil.uni-giessen.de

Bernulf Kanitscheider Zentrum für Philosophie und Grundlagen der Wissenschaft, Universität Gießen, 35394 Gießen, Germany, bernulf.kanitscheider@phil.uni-giessen.de

Chris Knight Comenius University, Bratislava, Slovakia, Chris.Knight@live.com

Virpi Lummaa Department of Animal and Plant Sciences, University of Sheffield, Sheffield S10 2TN, UK, V.Lummaa@sheffield.ac.uk

Amy R. Parish Departments of Anthropology and Gender Studies, University of Southern California, Los Angeles CA 90089, USA, amyrparish@aol.com

Julia Pradel Institut für Wirtschafts- und Sozialpsychologie, Universität zu Köln, 50931 Köln, Germany, julia.pradel@uni-koeln.de

Gerhard Roth Institut für Hirnforschung, Universität Bremen, 28334 Bremen, Germany, Gerhard.roth@uni-bremen.de

Wulf Schiefenhövel Max-Planck-Institut für Verhaltensphysiologie, 82346 Andechs, Germany, schiefen@orn.mpg.de

Rebecca Sear Department of Social Policy, London School of Economics, London WC2A 2AE, UK, r.sear@lse.ac.uk

Volker Sommer Department of Anthropology, University College London, London WC1H 0BW, UK, v.sommer@ucl.ac.uk

Charlotte Störmer Zentrum für Philosophie und Grundlagen der Wissenschaft, Universität Gießen, 35394 Gießen, Germany, Charlotte.Stoermer@phil. uni-giessen.de

Christian Vogel (1933–1994) Institut für Anthropologie der Universität Göttingen, 37073 Göttingen, Germany

Gerhard Vollmer Seminar für Philosophie, Technische Universität zu Braunschweig, 38106 Braunschweig, Germany, g.vollmer@tu-bs.de

Kai P. Willführ Zentrum für Philosophie und Grundlagen der Wissenschaft, Universität Gießen, 35394 Gießen, Germany, kai.p.willfuehr@phil.uni-giessen.de

Chapter 1
Introduction

Ulrich J. Frey, Charlotte Störmer, and Kai P. Willführ

The title of this volume—*Homo Novus*—might need an explanation. In ancient Rome a "homo novus" was the first man in a family to serve in the senate. A second meaning attached to it is that of a parvenu or upstart. Neither meaning is intended here. Instead, we consider the third and most recent meaning, that of a freshman or a new beginning. The exploration of a new scientific field would be an example of such a new beginning. Often, such a new scientific beginning requires a revision of established theories and beliefs. To integrate or even replace old theories with new and fresh ones is difficult and cumbersome, and thus needs ambassadors and advocates with bold arguments and a new perspective. In particular theories and concepts that go against our intuition need perpetual support from keen thinkers.

Such new thinkers are endowed with a new self-perception and consequently a new conception of the world. They are willing to embrace not only one scientific revolution, but several—and see them as a chance and not a burden. They are able to follow a revolutionary insight such as the Darwinian theory of evolution into its diverging disciplines and explore its consequences.

Different scientific disciplines such as philosophy, psychology, ethnology, neurobiology, anthropology, and primatology are united by the interface of Darwinian evolutionary theory. Humans and their behaviors are a product of evolution. Contributors coming from these different areas of scientific research make this book a truly interdisciplinary work, throwing new light on human abilities and characteristics. In this way they are the pathfinders of future homines novi and at the same time building blocks of this enterprise.

There are six main illusions that are addressed in this volume:

1. Humans are exceptional
2. We are independent of our sociobiological roots
3. The biological imperative does not matter to us
4. The past does not echo in our heads

U.J. Frey (✉)
Zentrum für Philosophie und Grundlagen der Wissenschaft, Universität Gießen,
35394 Gießen, Germany
e-mail: ulrich.frey@phil.uni-giessen.de

U.J. Frey et al. (eds.), *Homo Novus – A Human Without Illusions*,
The Frontiers Collection, DOI 10.1007/978-3-642-12142-5_1,
© Springer-Verlag Berlin Heidelberg 2010

5. Moral, Religion and Culture are social constructions
6. We are free in what we want

In the first part, Bernulf Kanitscheider, Volker Sommer, and Julia Fischer try to answer the question of whether "humans are truly exceptional", the answer being that we are just one species among many, not the "crown of creation".

Kanitscheider analyses the transition in perception that displaced the earth from the cosmic center to a relatively peripheral place in space-time. The view today is: We are probably living in an infinite multiverse or in one of "a countable infinity of duplicates". We have to cope with the fact that there are no margins, no borders, and no distinguished places at all. All that is left is a slight coherence between the universes, as they share a common origin.

Volker Sommer and Amy R. Parish look into the question: What is culture? Do apes and other animals have it as well? They explore whether or not there is "culture in nature". This topic is particularly fascinating, because from antiquity onward culture has always been regarded as a distinctive attribute of humans.

Julia Fischer's contribution is about major differences in the language abilities of humans and animals, particularly apes. She is interested in how communication works in animals compared to how humans use it. Are there elements in language use that are unique to human language?

The next part is about the illusion that we like to think that "we are independent of our sociobiological roots". Christian Vogel, Robin Dunbar, Detlef Fetchenhauer, and Julia Pradel dispel this myth quite thoroughly. Vogel discusses how moral codes and ethical systems are based on "behavioral tendencies which we have acquired though natural selection from our genome".

Robin Dunbar makes it clear that our brains are at their core "social brains". They developed according to certain selection pressures in the mammal group. Culture is able to build upon this, but not without constraints. This means that we have to recognize certain biological limitations concerning human social behavior.

Julia Pradel and Detlef Fetchenhauer try to reconcile the "paradox" of altruism with evolutionary theory. They argue that there are stable altruistic intentions, which are recognizable by others. Based on these abilities, altruists cooperate electively with each other.

The third part emphasizes the point made by Illusion Number 2: We have to heed the "biological imperative". What does this mean? Charlotte Störmer, Kai Willführ, Virpi Lummaa, and Rebecca Sear explain this by pointing out that our lives are shaped by many different external influences. Working with life-history theory, which asserts that the allocation of resources is a permanent trade-off in life, these three chapters deal consecutively with trade-offs occurring between maintenance, reproduction, and growth. The high costs of investment in maintenance effort are the main subject in the chapter by Charlotte Störmer and Kai Willführ. They argue that all types of mortality crises (social, epidemiological, and subsistence crises) that we might be exposed to in early life have long-term consequences, and point out that all these consequences might be mediated by impairment of the immune system.

Virpi Lummaa focuses on the costs of reproduction for women and the impact of reproduction on senescence. Depending on the environment, either fast or slow life histories are favored to increase individual fitness.

In the third chapter Rebecca Sear discusses the question of whether taller is always better in terms of reproductive success. She outlines that the relationship between height and reproductive success is heavily dependent on the environmental context.

Ultimately, all authors in this section conclude that we are far from leaving the biological imperative behind. It has always been easy to see that we do not have any control over our birth, but it is far more difficult to recognize that the same is true for other stages in our life (e.g. the decision when to marry or to have children).

Since the "past echoes in our heads", Athanasios Chasiotis and Harald Euler both argue for a "human evolutionary behavioral science". Above all, psychology should be informed by evolutionary theory. Chasiotis first warns of the danger of dualistic psychology, then goes on to explain how developmental histories are the result of strategic decisions of the organism in response to the complex interplay of environment. Moreover, it is surprising to what extent "cultural" differences can be explained instead by childhood context, e.g. the number of siblings.

Euler concentrates on two illusions within psychology: The gender illusion, which states that men and women do not differ much, and the family socialization illusion, which claims the "power of parents to shape their child's personality permanently". Sex differences are shaped by evolution and formative influences on children outside the family are far stronger than those within the family. This is corroborated by behavioral genetic studies accounting for the exact opposite of these illusions: "Particularly life-history theory and parent–child conflict deliver plausible reasons why parents are not able to mold permanently their offspring's personality."

Another prominent illusion is discussed in Illusion Number 5, namely that morals, religion, and culture are purely social constructs, which is a core assertion of the Standard Social Science Model, attacked, for example, by John Tooby and Leda Cosmides.

Kurt Bayertz argues for a natural explanation of "the moral ought". Its phylogeny can be traced back to the realm of animals, but this does not mean that these biological roots of normativity support the claim that norms are "nothing but biology".

Chris Knight tries to show how the problem of human symbolic culture might be solved. For this reason he looks at possible models and how they might parsimoniously fit together. He concludes that there is not one superior model, but all explain a part of the riddle. For him, a Darwinian explanation is essential for understanding symbolic culture.

Wulf Schiefenhövel links religion of the Eipo in Melanesia to health in a vivid analysis. Above all, he describes the fascinating relation of the punishing power of spirits, that is, "black magic", to advantages in health. Religion is functional in reducing stress, having placebo and other psychosomatic effects.

Current debates about free will are taken up in Illusion Number 6: "We are free in what we want".

Gerhard Roth discusses the concept of free will using recent experimental evidence from neurobiology. He concludes that this concept of free will is not acceptable in light of this new evidence. This has, according to Roth, implications for our criminal justice system: How do we deal, as a society, with retribution, revenge, and punishment.

Gerhard Vollmer argues for a naturalistic concept of free will that must be compatible with determinism. He clarifies what this means for responsibility, punishment, and law generally. Surprisingly, it is perfectly possible in a determined world to talk of free will, punishment, and responsibility.

Finally, Ulrich Frey asks in the epilogue why the illusions discussed in this volume are so powerful. Why do they persist in spite of convincing counterevidence? He argues that they do not disappear because the consequences of giving them up are negative in historical and costly in psychological and biological respects.

Thus, authors from biology, anthropology, psychology, philosophy, neurobiology, and physics in this volume try to expose many illusions in our perception of ourselves and the world. These chapters deal with both historical and current illusions. Each era in scientific history is able to reveal some illusions of the scientists before them—but remains blind to own illusions. This task seems to always slip out of reach. We simply have to live with the remaining illusions, whatever they may be. (Un-)Fortunately, we will never know what we have missed.

Illusion Number 1
Humans Are Exceptional

Chapter 2
The Position of Man in the Cosmos

Bernulf Kanitscheider

Abstract The centrality of man in the universe has a long history, stretching from ancient times to modern cosmology. Stepwise man's homestead has been displaced to a typical, inconspicuous location in space. Today there is not the slightest evidence for the existence of distinguished places at all within an isotropic and homogeneous universe.

2.1 Historical Precursors

The place of man in the universe has been a matter of debate since Stoic times, but as late as the period of Copernicus it grew to be a really terrifying subject of anthropological thinking (Rossi 1972). The geocentric Aristotelian world possessed a natural place for dynamic movements; accordingly everybody tends in a natural way to reach the center of the physical space, which is at the same time the center of the universe, which in turn includes the core of the earth. But due to the growing discoveries of renaissance astronomy the earth-centered approach could no longer be upheld. The displacement of the cosmic center occurred step by step, but before the advent of relativistic cosmology it seemed self-evident to everybody that the universe ought to have a focal point somewhere and a peripheral region. In view of that, Copernicus shifted the cosmic center to the sun as the origin of the reference system of astronomical description, thereby simplifying enormously the older system of epicycle–excenter–equant geometry of Ptolemy. The idea of a central place and a spatial boundary remained an unavoidable detail of the "constructions of the heavens". Astronomers and cosmologists tried to identify the central system of the world, even if it were only a mathematical point in the depth of cosmic space. At the end of the eighteenth century, Sir William Herschel moved the imaginary center

B. Kanitscheider (✉)
Zentrum für Philosophie und Grundlagen der Wissenschaft, Universität Gießen,
35394 Gießen, Germany
e-mail: bernulf.kanitscheider@phil.uni-giessen.de

U.J. Frey et al. (eds.), *Homo Novus – A Human Without Illusions*,
The Frontiers Collection, DOI 10.1007/978-3-642-12142-5_2,
© Springer-Verlag Berlin Heidelberg 2010

to the core of our galaxy. But it couldn't stay there very long. The fundamental question at his time was the position of the entire Milky Way. Does it comprise the whole matter content of the universe or did Kant's speculation point in the right direction, that the universe contains many nebulae like our own? This was difficult to decide in those days. The uniqueness and the centrality of our mother galaxy grounded on the deceptive evidence that the nebulae were distributed well away from the disk of our galaxy on account of a fog that veils the innermost part of the disk. The extragalactic position was proved first for the Andromeda nebula by Heber Curtis, who showed that this system of stars is situated well outside our Milky Way by exploiting variable stars such as novae that change their intrinsic brightness in a well known way. In establishing that Andromeda is of comparable size to our galaxy he undermined the idea of the central position of the Milky Way. In the following era it was Edwin Hubble who corroborated by the method of Cepheid variable stars, using their period—luminosity relation, that in accord with Kant's supposition, our galaxy is only one medium-sized system within a huge entirety without any sign of specialness. Up to 1922 our sun was thought to be in a central position within the Milky Way but this location was overthrown by Harlow Shapley, who demonstrated that our Star is situated at a typical eccentric point a distance of 30,000 light years from the innermost part of our galaxy. The ultimate remainder of the former central position was the atypically large extension of the Milky Way, measuring tenfold that of a typical system of stars. The mistake was discovered by Walter Baade in 1952. He revealed that when using the method of measuring distances by Cepheid variables by the period—luminosity relation one has to distinguish between two types of variable stars and that therefore the intergalactic distance scale should be enlarged by a factor of 10. As a consequence our Milky Way ceased to be a giant system among its neighbors. Baade's discovery erased the ultimate vestiges of a privileged status of the cosmic home of man. To put the historical development in a nutshell: "Copernicus dethroned the earth, Shapley the sun and Baade the Milky Way. Since the local group of galaxies is a comparatively small cluster, the geocentric picture of the universe is completely discredited" (Sciama 1959, p. 62).

2.2 The Standard Concordance Model

Modern cosmology is based on Einstein's theory of gravitation, a theory in which gravity is treated as a trait of the metrical geometry of space-time. The field equations of gravity cannot be solved without the introduction of initial and boundary conditions. The most straightforward subclass of all space-times that are ruled by the Einstein equations is the set of isotropic and homogeneous world-models. In memory of their discoverers they are called Friedmann–Lemaitre–Robertson–Walker space-times or FLRW worlds. FLRW space-times comply with the constraints of isotropy and homogeneity; there are no distinguished directions or specially favored locations in these space-times. Most of them are expanding time-varying models in which curvature depends on the cosmic time parameter. Observation leads unequivocally to an isotropic distribution of matter and radiation. Local irregularities set aside, beyond 100 Mpc (megaparsec) galaxies are scattered evenly throughout 3-space up to the horizon. Radio astronomers have found that the very distant radio sources are distributed isotropically around us. The same is true for radiation, for example, the X-ray background and, foremost, the cosmic microwave relic radiation—the remnant of the fireball state, which today, after having expanded adiabatically for 15×10^9 years, has a measured temperature T of 2.9 K—show the same feature. Recent measurements reveal that the isotropy of the 3 K radiation amounts to about 1 part in 100,000, $\Delta T/T \leq 10^{-5}$. Although local isotropy (the rotational symmetry of light coming from distant galaxies around our special point of observation) is now well established, we cannot test empirically the homogeneity of space, because as observers we are bound to our terrestrial location. We cannot explore the vastness of space in order to assess the homogeneity of 3-space. It is a well-known theorem of differential geometry (Walker 1944) that exact spherical symmetry around any point entails that the universe is spatially homogeneous. Such a space-time admits a six-parameter group of isometries. Its surfaces of transitivity are space-like three-surfaces of constant curvature (Hawking and Ellis 1973, p. 135). In more colloquial terms: any point of 3-space in a homogeneous universe is physically equivalent to any other point on the same surface. Or, in phenomenological language, it is the impossibility of telling where one lives in a homogeneous universe. Since a uniform space has neither a boundary nor focal point, the very concept of place is inapplicable. A voyage to a far distant spiral galaxy would bring about no change in the overall appearance of the sky. Therefore it is prohibited to state that we look up to a world of stars from nowhere.

In order to fill the gap between local and global isotropy we need a bridging law. The customary procedure is to make use of the so-called Copernican principle. There are several formulations; an outstanding one has been stated in terms of the likelihood that man has a special privileged location in the universe. Hermann Bondi (Bondi 1968, p. 13) put it like this: "The Earth is not in a central, specially favoured position." The name of this principle is certainly a misnomer, since Copernicus, as mentioned above, believed that the sun occupies the central place in the universe. Nevertheless he initiated the tendency to displace the apparent focal point of the whole system of stars to the rim of the arrangement. Accordingly we have to realize that we are living on a medium-sized planet revolving around a normal main sequence star that is located on the edge of an average spiral galaxy, which in turn is a member of a local group of galaxies. If we recognize that there is not the slightest indication of a special position of our homestead, we can use without further ado the Copernican principle in order to pass from local to global isotropy, which in turn entails homogeneity. This is the received view of current cosmology. We should remark, however, that in 1978, now almost forgotten, Ellis, Maartens and Nel (Ellis et al. 1978) showed that the observed galactic redshift and the cosmic microwave background radiation (CBR) can be explained by a static spherically symmetric (SSS) model that contains two centers, a naked singularity, which in distinction to the big bang singularity continually interacts with the universe, and a cool center, in whose vicinity we are living, embedded in our galaxy. In this inhomogeneous SSS model the cosmological redshifts of the galaxies are interpreted as gravitational redshifts and the CBR as originating from hot gas near the naked singularity located at the second center of the universe. Although the inhomogeneous SSS model should not be taken as a realistic one, since it violates the cosmic censorship hypothesis (the conjecture, surmised by Roger Penrose, which states that black holes are always surrounded by an event horizon so that no outgoing causal effects can leave the singularity) and moreover does not provide a good fit to the magnitude—redshift relation (m, z), it points plainly to the necessity to take cognizance of the hidden selection effects. There is no need to assume that somebody has centered the universe to our advantage but it seems obvious that life would most favorably exist near the cool center of the universe and not in the vicinity of the hot radiating singularity. This corresponds to the obvious selection effect in a FLRW world, where nobody surmises the existence of life a short time after the big bang singularity. Temporal and spatial problems are mirror images of each other, in a FLRW world the initial singularity generates, probably by the process of inflation, the homogeneity of our observable universe; in an inhomogeneous SSS world the condition of static governs the overall structure.

The many advantages of dealing with time-like homogeneous hypersurfaces with constant curvature should not be overlooked. This high symmetry makes cosmology a much easier task in comparison with a highly inhomogeneous and irregular distribution of matter, which in turn would engender a complicated space time of variable curvature. A universe containing one or many special locations each with distinct properties of the pertinent matter could not be dealt with in a comprehensive way. The customary inference from a significant sample to the global matter distribution in space-time would be an invalid conclusion. Today we are convinced that the uniform distribution of galaxies stretches even beyond the event horizon and even if the global topology of 3-space is noncompact. It is not a matter of speculative physics that space is infinite and rather uniformly filled with galactic matter, it is rather the other way round, flat infinite models fit the astrophysical data much better than models with spatial curvature, hierarchical self-similar structure, or multiply connected topologies. The cosmic microwave background depends in a sensitive way on the model assumptions and this fireball remnant is only concordant with an infinite homogeneous flat expanding universe. Unsurprisingly matter distribution is quite irregular on a local scale, as we can observe in our planetary system and even on a galactic level. If, however, we move outwards towards the edge of the observable universe ($R \approx 10^{27}$ m) then the relative fluctuations shrink as $\Delta M/M \approx 10^{-5}$. What is more, there is no doubt that the homogeneity of the universe extends even beyond the observable realm (Tegmark 2003). From a theoretical point of view this uniformity seems to be startling, because Einstein's field equations do not demand anything like homogeneity in the distribution of matter and radiation, not even constant curvature for 3-space is required, and least of all flatness. Besides the mentioned Robertson–Walker spaces there exists a large class of solutions in which the requirement of isotropy is dropped but spatial homogeneity is retained. Even absolute rotation and shear of matter could be included in the field equations of gravitation. The more general Raychaudhury equations replace in this case the Friedmann equations. It might be of some interest that recently José Senovilla from the Universidad de Salamanca has discovered an exact solution of an inhomogeneous world that avoids the initial singularity, specifying a space-time that is regular from infinity to infinity (Senovilla 1990). In the meantime it has been shown that the solution is really geodesically complete and singularity-free. It satisfies the stronger energy and causality conditions such as global hyperbolicity, causal symmetry, and causal stability (Cinea et al. 2004). Anyway it seems to be clear that within the boundary of classical relativity there is ample scope for a cosmic past without breakdown of physically regular conditions.

2.3 Speculative Hypotheses on the High Energy Realm

Recently the entire debate on privileged locations received a new twist in relation to the concept of a *multiverse* that grows out of the high-level unified field theories. Unification has been a major aim of theoretical physics since the misty past. The Ionian dream has been a major attraction of theoretical physics since J.C. Maxwell unified electricity and magnetism with the stunning result of the wave theory of light. The replacement of Galilei symmetry group by the Lorentz group, which led to special relativity, amalgamated space and time in an intricate manner. With the advent of quantum mechanics the gauge group took up the role of a unifying instrument to bring about the standard model of elementary particle interaction. Electroweak theory in combination with quantum chromodynamics (QCD) gave rise to the standard model of elementary particle physics. The combination of electroweak theory and QCD was based mainly on similarity of mathematical structure; a lot of parameter values, especially the masses of leptons and quarks, were left as undetermined contingent empirical data. It could not be deduced from first principles that there have to be exactly three generations of leptons and quarks. On the other hand there were hints that such a grand unified theory (GUT) is not based on mere speculation. Although of very different strength on ordinary scales, these two forces seem to converge to a common power at higher energies. Concerning this meeting point of the running coupling constants of strong, weak, and electromagnetic interactions there is a powerful argument for superstring unification, because only with the inclusion of supersymmetry do the three coupling constants meet at exactly one point, with temperatures of 10^{28} K that prevailed 10^{-39} s after the Big Bang (Penrose 2005, p. 876).

Physical progress is propagated by the free parameters that remain unexplained by simpler theories; these contingent elements call for a justification of their numerical values. This was the leading motivation behind the construction of string theories. In the end, physicists discovered a growing number of solutions expressing local minima of energy, vacuum states that correspond to possible stable or metastable universes. Estimates of the number of these vacua reached the hair-raising number of 10^{500}. L. Susskind coined the name "string landscape" with a glance at biochemistry with its vast number of configurations. In some sense the original intention to minimize the contingent parameters had not been fulfilled, since although the theory was able to include gravity in a unified description it led to a proliferation of worlds instead of explaining the contingent parameter values of our distinctive world. With some repugnance physicists concede that we have to include environmental parameters in the scientific approach in order to come to grips with the gigantic number of vacua. Steven Weinberg (Weinberg 2005), by no means a promulgator of anthropocentric epistemology, concedes:

> The larger the possible values of physical parameters provided by the string landscape, the more string theory legitimates anthropic reasoning as a new basis for physical theories: Any scientists who study nature must live in a part of the landscape where physical parameters take values suitable for the appearance of life and its evolution into scientists.

What can now be concluded for the position of man within this huge array of worlds, most of which will not contain any life similar to ourselves? The first observation will be that the concept of position within an ensemble of worlds does not have any well-defined meaning, since the members of this set are not aligned in a spatial arrangement. As we have already seen, within an infinite single universe the very concept of position collapses; this is even more the case within an ensemble of worlds that do not have any common space-time or any causal connections whatsoever between them.

2.4 The Challenge of Quantum Cosmology

The very notion of a multiverse is a concept laden with enigmas and conundrums. It has a smell of metaphysics of those hoary days of yore. A deeply convinced logical empiricist of the days of the Vienna Circle would well-nigh reject any physical significance of an assertion belonging to a world besides our own. Nevertheless, the criterion of meaning has been liberated since the times of the founding fathers of scientific philosophy and science itself reintroduced this ancient idea into current discussion. With the advent of quantum cosmology it sounds less strange than before that our universe grew out of a primordial quantum configuration that engendered many other space-time regions as well. Andrei Linde's scenario of chaotic inflation gave rise to a somewhat more concrete picture, which delivered a dynamic underpinning of the idea of many worlds. Historically the proposal goes back to Leibniz, who introduced the expression of "possible world" in the metaphysical context of the theodicy problem. However, physics is not interested in the set of logically possible, that is noncontradictory, worlds but in the narrower concept of the really existing worlds. Conceptual problems arise at the very beginning in regard to the size of the class: Is the set infinite, which characters and peculiarities do the others have, and how will we gain knowledge of these other systems of total reality? Once more we are confronted with the mind-boggling concept of infinity but this time on a higher level. The very enticement to include this ancient, strange-sounding notion in the realm of scientific thinking had been that it seems to be the only logically coherent way to avoid a supernatural explanation of the fine-tuning of the cosmic parameters, elementary particle masses, and coupling constants anyway. There is no doubt that a cognizable universe permits only a rather narrow margin on these control parameters; the universe could have been only slightly different without denying the existence of observers. But within science there is no place for goal-directed teleological thinking. Why should a material universe be tailor-made for habitation, why are the initial conditions obliged to arrange themselves in a deliberate way that points to the evolution of man? (For a more elaborate criticism of the various types of anthropic principles see: Kanitscheider 1993.) The very occurrence of prescriptive terms in formulations of the strong anthropic principle disqualifies it as an assertion of a physical lawful.

At the moment there are only two ways to break the deadlock of explaining the contingencies of a welcoming and habitable universe: A deduction of all fundamental parameters and life-sustaining constants from first law-like principles or to treat these casual environmental elements as pointing to a hidden selection effect generated by our special membership within the ultimate ensemble of worlds, the multiverse. The first option has its shining example in the inflationary scenario, because it offers answers to a number of serious open questions of the standard FLRW big bang model. Instead of stipulating the constraints on the field equation of gravitation in order to get the homogeneity and isotropy of space-time, with only one further assumption of a scalar field at very early times it is possible to get a causal explanation as to *why* physical space has this peculiar feature. The dynamical process of exponential expansion led to a smoothing out of all possible earlier irregularities. The striking advantage of the inflation assumption, and what takes away the impression of an ad hoc postulation, lies in the simultaneous explanation of the horizon, the flatness, and the monopole problem.

2.5 The Intricacies of Infinity

The core of the problem of the position of man's homestead within the whole of reality is connected with the question of a physical infinity. The concept of infinity is laden with a host of harsh prejudices; the ancients were skeptical about anything that has no boundary. In medieval times infinity always had metaphysical overtones, since it was connected with the numinous and the supernatural. Scientists were suspicious because it transcends the empirically accessible. Ever since it has been argued that there are two types of infinity, a qualitative one that comprises perfection and self-sufficiency and which is more or less appropriate in the realm of metaphysics, and a quantitative one to be applied in mathematics. It must be recognized, however, that the defenders of the woolly concept of infinite qualities never came to grips with the vague semantics of that term. Therefore it can be argued that mathematics has for the first time cleared up the fuzzy connotations of the intuitive concept of infinity. But even within the territory of mathematics there had been age long quarrels on two types of infinity, the potential und the actual one. Aristotle tries to clarify in which sense an infinity can exist and in which not.

> On the other hand it is clear that, if an infinite does not exist at all, many impossibilities arise: time will have some beginning and end, magnitudes will not be divisible into magnitudes, and number will not be infinite. If therefore, when the case has been set out as above, neither

view appears to be admissible, we need an arbitrator; clearly there is a sense in which the infinite exists and another sense in which it does not.

Being means either being potentially, or being actually, and the infinite is possible by way of addition as well as by way of division. Now [...] magnitude is never actually infinite, but by the way of division [...] the alternative that remains therefore is that the infinite exists potentially. (Aristotle, Phys. III 6. 206a9) (Heath 1970)

Aristotle's conceptual distinction and argumentation became the received view on the two kinds of infinity, the potential or conceptual infinite as a necessary tool for analyzing mathematical problems, and the actual or concrete completed infinity in physical reality. The main objection to the latter was that it cannot have a determinate size, an infinite body or system of objects being a contradictory predication. In modern times David Hilbert reiterated this position. In his seminal paper on infinity (Hilbert 1925) Hilbert referred to the then prevailing opinion in cosmology of a curved elliptical metric that led to an unlimited but finite 3-space without boundary. But since his times the customary view in cosmology has changed definitely. Hilbert's main endeavor didn't concern cosmology but was intended to cope with the critics of the intuitionism of the School of Brouwer, who aimed to destroy the transfinite set theory of Cantor and his followers. Hilbert's movement to metamathematics in order to establish finite axiomatization of the whole edifice of mathematics takes the actual infinite as an idea without semantic reference. The doubts concerning a realized actual infinity have prevailed up to now and seem to be defended with the same empirically narrow minded arguments. In the influential paper of Stöger, Ellis, and Kirchner we encounter, for example, the opinion that Euclidean geometry will never be descriptive in physical cosmology, because these spatial sections are of indefinite extension.

In geometry we assume space extends forever in Euclidean geometry and [the same is true] in many cosmological models, but we can never *prove* that any realised 3-space in the real universe continues in this way—it is an untestable concept, and the real spatial geometry of the universe is almost certainly not Euclidean. Thus infinite Euclidean space as such is an abstraction that is almost certainly never realised in physical practice. (Stöger et al. 2004)

But here we encounter the old misunderstandings of logical empiricism, according to which every logical consequence of a physical theory has to be directly testable. It remains to be seen whether a theory that refers to an actual infinite space or a time coordinate that extends without limits to the past and the future can be tested, but it does not need to be the very consequence of infinity: It has to be controlled in regard to any of its empirical predictions. Only if the theory does not offer any contact with empirical reality at all can it be judged as metaphysical, superfluous, and cognitively worthless. As far as it concerns the position of man within an infinite Euclidean geometry, there is surely not the slightest possibility of a central position and even the rim is not defined because there are no distinguished directions. Jacque Monod judged the position of man with the famous words, in his book *Le hazard et la nécessité*: "L'homme sait maintenant que, comme un Tzigane, il est en marge de l'univers où il doit vivre. Univers sourd à sa musique, indifférent à ses espoirs comme à ses souffrances ou à ses crimes."[1] We can however strengthen these words, noticing that within an infinite universe there is no "marge" because there is no boundary and therefore we have neither a distinguished nor a peripheral position at all in a 3-space of indefinite extension.

2.6 Taming the Unfathomable

The problem iterates if we ascend from a single infinite universe to an ensemble of worlds, each of which can be thought of as finite or even infinite, but there is a difference: A universe can be thought of as a single causally connected space-time, where every subsystem can be conjoined by a time-like or null curve. A realized infinite ensemble of worlds does not necessarily have to be a causally linked system, although it might have a common origin, as e.g. within the chaotic inflationary scenario. Therefore the question of testability arises in an even more maddening manner. There can be neither direct nor indirect currents of information in order to gain some knowledge of these hidden worlds. The only way left is to specify the explanatory power of the many worlds assumption against some concurrent hypotheses. Stöger, Ellis, and Kirchner point out that some unsymmetrical generic worlds might need an unlimited amount of information in order to specify their initial and boundary conditions due to their field-like matter content, not every world being symmetric like FRW models. Within the ensemble there will be highly unsymmetrical worlds whose information content cannot be regarded as algorithmically compressible. But surely there is no need to postulate an ensemble of worlds that is so large that it comprises everything that can possibly happen, which means it is not logically contradictory. It suffices to hypothesize that the set is big enough to fulfill its explanatory function with regard to the otherwise unfathomable contingent traits of our visible universe. Unquestionably, within the approach of many worlds around the corner there lurks the problem of a slippery slope or runaway ontology, especially if we think of the continuum problem. How can we restrict the possible uncountable infinite set that pops up if we do not put a ceiling on the huge manifold of other worlds? Above that there is some quarrel over the size of infinity in physics. It is rather striking that none of our physical theories trespasses the cardinality of the real-number system. As Roger Penrose has argued, although the family of all real-number-valued functions on a space with points of the complex continuum C is indeed 2^C and therefore larger than C, the set of the continuous functions on C are only C in number (Penrose 2005, p. 378). A feasible procedure may be that we start with a set somewhat larger than surmised and then restrict it until we arrive at a minimum stage that suffices for our aim of explanation. Anyway the smallest set of possible worlds will do in accordance with Occam's principle of parsimony. But how large is the indispensable set of worlds compatible with the

---

[1] Man now knows that he lives, like a gypsy, at the edge of the universe. A universe that is deaf to his music and as indifferent to his hopes as it is to his sufferings or his crimes.

astrophysical data today? According to current methodology we have to accept the existence of objects that cannot be seen but deliver a coherent conceptual scheme for understanding visible things. Ships that disappear beyond the horizon do not vanish from existence, even for those who cannot follow them. Galaxies that drifted over the lookout limit did not change as material systems; they have the same onto-logical standing despite the fact that they do not have the same epistemological status. The most distant objects we can be aware of are at a distance of 14 billion light years away, it is the light from the hot initial state of our universe; the objects that sent us their radiation from the rim of the visible universe have in the mean-time reached a distance of 40 billion light years, far beyond the cosmic horizon. Therefore, even if we only acknowledge our unique world with its matter content we have to admit a part of space beyond any possibility of visibility. With respect to immediate observability there is no difference in principle between the far away parts of our Euclidean universe and any member of a set of worlds in a multiverse. The only question is what will be gained for our cognition by introducing such an ensemble. How can we avoid too much metaphysical excess baggage? One of the strongest defenders of various levels of reality, each consisting of many worlds, is Max Tegmark (Tegmark 2003). He correctly argues that even at the basic stage the concordance model that fits every type of empirical facts decomposes into a host of different causally detached parts. What is more, spatial infinity is a generic pre-diction of the inflationary scenario that is able to explain certain contingent riddles of the standard model, such as the absence of monopoles, the causal puzzle, and the flatness puzzle (Weinberg 2005, p. 202). Tegmark's strongest argument con-sists in the fact that finite alternatives to infinite Euclidean spaces are inconsistent with the cosmic microwave background; compact, hierarchical, or otherwise mul-tiply connected topologies cannot be integrated into the temperature distribution of the CBR. Furthermore, CBR and galaxy distribution suggest strongly a tendency to uniformity as we approach the edge of our observable universe. Within this level of existence the generic laws of physics will be the same, but there will be most likely a random distribution of initial conditions, since quantum processes were responsi-ble for generating density fluctuations within the primordial inflation epoch. One of the staggering consequences of this multiple-world structure in which almost every-thing that is possible will occur somewhere in a remote corner of the whole being is the existence of a double. At a distance of $10^{10^{29}}$ m we have to admit an identi-cal copy of ourselves, which has the same experiences, observing a Hubble volume with the same stars and galaxies (Tegmark 2003, p. 4).

2.7 Conclusion

All these cosmological facts having been considered, which conclusion can we arrive at concerning the place of man in an ensemble of worlds, be it countably infinite or uncountably large? What does it mean for our self-esteem, already pretty shaken by the offences of men's vanity from Copernicus to Freud? Should we be enjoyed or annoyed by our twin in another Hubble volume of our infinite universe at $10^{10^{29}}$ m distance? How do we rate infinity itself as the global environment?

Most people I asked, in order to get some psychological reaction, were not amused, either by an infinite multiversum or by a countable infinity of duplicates thinking the same ideas, playing the same music, or climbing the same peak over there. That means obviously that identity and unique existence seem to be a highly estimated value for mankind. But obviously the huge extension destroys the illusion of the uniqueness of existence too. Spatial infinity, be it on the level of a single space-time or an ensemble of worlds, reinforces the impression of being lost somewhere in the middle of nowhere that horrified the poets of the sixteenth century when they acknowledged that the medieval celestial spheres could not be upheld any longer within a heliocentric planetary system surrounded by an unfathomable array of stars. Already at the dawn of cosmological infinity John Donne confesses in his *Anatomy of the World* (1611):

> So did the world from the first hour decay,
> That evening was beginning of the day,
> And now the springs and summers which we see,
> Like sons of women after fifty be.
> And new philosophy calls all in doubt,
> The element of fire is quite put out,
> The sun is lost, and th'earth, and no man's wit
> Can well direct him where to look for it.
> And freely men confess that this world's spent,
> When in the planets and the firmament
> They seek so many new; they see that this
> Is crumbled out again to his atomies.
> Tis all in pieces, all coherence gone.
> All just supply and all Relation.

The case has strengthened since the time of Renaissance astronomy. What about the "coherence" John Donne alluded to? While there are possible space-time routes that could eventually lead from one Hubble volume to another, e.g., if cosmic expansion decelerates, there is no causal connection on the higher levels of cosmic manifoldness. According to Andrei Linde's scenario of chaotic inflation we have to surmise an infinite number of worlds with the same fundamental equations but different physical constants and dissimilar dimensionality. Evidently there remains some feeble law such as coherence, since there is a common origin of the space-times that afterwards are separated by superluminal velocities and aren't connectible by causal time-like lines. Apparently coherence crumbles in a piecemeal way, the causal bonds of the pertaining parts of reality getting weaker with every level of the multiverse. Shall we therefore break down in tears, being contrite due to the disintegration of reality? I surmise this to be a matter of personal temperament.

References

Bondi H (1968) *Cosmology*, 2nd ed. Cambridge University Press, Cambridge, MA

Cinea FJ, Fernández-Jambrina L, Senovilla JMM (2004) A singularity-free space-time. http://arxiv.org/abs/gr-qc/0403075. Accessed 10 August 2009

Ellis GFR, Maartens R, Nel SD (1978) The expansion of the universe. Monthly Notices of the Royal Astronomical Society 184:439–465

Hawking SW, Ellis GFR (1973) *The Large Scale Structure of Space-Time*. Cambridge University Press, Cambridge, MA

Heath T (1970) *Mathematics in Aristotle*. Oxford University Press, Oxford

Hilbert D (1925) Über das Unendliche. Mathematische Annalen 95:161–190

Kanitscheider B (1993) Anthropic arguments—Are they really explanations? In: Bertola F, Curi U (eds) *The Anthropic Principle*. Cambridge University Press, Cambridge, MA

Penrose R (2005) *The Road to Reality*. Random House, New York, NY

Rossi P (1972) Nobility of man and plurality of worlds. In: Debus AG (ed) *Science, Medicine and Society in the Renaissance*. Heinemann, London

Sciama DW (1959) *The Unity of the Universe*. Faber and Faber, London

Senovilla JMM (1990) New class of inhomogeneous cosmological perfect-fluid solutions without big-bang singularity. Physical Review Letters 64:2219–2221

Stöger WR, Ellis GFR, Kirchner U (2004) Multiverses and cosmology: philosophical issues. http://arxiv.org/abs/astro-ph/0407329. Accessed 10 August 2009

Tegmark M (2003) Parallel universes. http://arxiv.org/abs/astro-ph/0302131. Accessed 10 August 2009

Walker AG (1944) Completely symmetric spaces. Journal of the London Mathematical Society 19:219–226

Weinberg S (2005) Living in the multiverse. http://arxiv.org/abs/hep-th/0511037. Accessed 10 August 2009

Chapter 3
Living Differences

The Paradigm of Animal Cultures

Volker Sommer and Amy R. Parish

Abstract Variation in behavioural patterns between populations is a trademark characteristic of *Homo sapiens*. Indeed, it constitutes the basis of "cultural diversity". Social anthropologists tend to resort to a "humanist" stance, reserving the label "culture" for our own species, whereas biological anthropologists tend to be "universalists", assuming an evolutionary continuum of traits that constitute culture. A definition that aims to be species-inclusive would view culture simply as "socially transmitted behaviour". Field researchers have uncovered numerous examples of non-human animal cultures, in which members of different populations of the same species were found to possess dissimilar behavioural portfolios. The greatest degree of behavioural diversity amongst non-humans is perhaps exhibited by chimpanzees. Studies across Africa revealed that each chimpanzee community exhibits a unique combination of traits related to social customs, communication, territorial aggression, war-like raiding, hunting strategies, food processing and consumption, and ingestion of plant matter for self-medication. However, local traditions have also been described for other taxa as diverse as triggerfish, bottlenose dolphins, killer whales, capuchin monkeys, Hanuman langurs, Japanese macaques and orangutans. At least in primates, population-typical behaviours are not always due to different local ecologies, but may be idiosyncratic expressions of "social identity" that allow individuals to distinguish "us" from "them". Cultural diversity can thus be understood as a reflection of inter-group competition and strategies of resource acquisition. Human and animal cultures are also linked in a rather sad way: the current mass-extinctions of other animals caused by humans lead not only to a loss of genetic diversity, but likewise cultural diversity.

V. Sommer (✉)
Department of Anthropology, University College London, London WC1H 0BW, UK
e-mail: v.sommer@ucl.ac.uk

U.J. Frey et al. (eds.), *Homo Novus – A Human Without Illusions*,
The Frontiers Collection, DOI 10.1007/978-3-642-12142-5_3,
© Springer-Verlag Berlin Heidelberg 2010

3.1 What We Do and Monkeys Don't

Today's world is more complicated then it was for FitzRoy Richard Somerset. One of his brash aphorisms proclaims culture as "roughly anything we do and the monkeys don't". To be sure, most people would still agree with the British writer and 4th Baron Raglan: There are animals, and there are humans, and the former don't do what the latter can.

But—and not even to mention a certain Charles Darwin—there have always been those who were willing to look for similarities instead of differences. For example, in a nonfiction account of the Warsaw Zoo during World War II, the wife of the director, Antonina Zabinski, ruminates about our connection to the rest of the animal kingdom: "That night she lay awake thinking about the thin veil between human and other animals, only the faintest border, which people nonetheless drew as a 'symbolic Chinese Wall', one that she, on the other hand, saw as shimmery, nearly invisible. 'If not, why do we humanize animals and animalize humans?' " (Ackerman 2007, p. 239).

Wishful thinking such as this is not necessarily based on illusions about how the world should be. For those who study animals as professionals (as the Zabinskis did), it has become almost commonplace to zoomorphize people and to anthropomorphize animals, at least to a certain degree, given the fact that organisms share common ancestors and parts of their history. This conviction is increasingly nourished by the extraordinary accumulation of knowledge about the natural world and the organisms that have inhabited it during the last half a century since the opinionated Baron's death. A telling example relates to the pioneering 1960 discovery by the British primatologist Jane Goodall, who documented tool use in wild apes. Upon which her mentor, the paleoanthropologist Louis Leakey, reportedly declared: "Now we must redefine 'tool,' redefine 'man' or accept chimpanzees as humans" (Goodall 1971; Peterson 2006, p. 212).

Indeed, the task of toiling with definitions becomes ever more complicated. One of the latest exercises relates to one of the most enduring assertions employed to justify the dichotomy between "man" and "animal": the very claim that only humans possess *culture*. In this essay, we will join other evolutionary anthropologists and primatologists who dispute that this is a valid assumption.

Our view is, of course, not unbiased. As researchers at the interface of sciences and humanities, we have long gotten used to talk about "nonhuman animals" and "humans and other animals". Similarly, we do, right from the onset, not see nature and culture as opposites, definitional problems not withstanding. Nature and culture, if anything, are intertwined, and complement each other up to the point that it becomes pointless to try to distinguish them. For example, cultural practices such as marriage pattern influence the genetic make-up of future generations (Oota et al. 2001). Vice versa, our achievements and shortcomings in the cultural arena are only possible because our natural features—our physiology and our brain in particular—enable us to enact them.

Talking about culture, as we will elaborate upon, means to a large degree talking about differences between populations. It is probably no coincidence that the animal culture topic gained speed with the end of the cold war—at the same time that increasingly more human groups claimed to be somehow distinct from others while asserting their right to be different (Antweiler 2009). And thus, as we will also point out, the topic of animal cultures is, in its last consequence, likewise entangled in scenarios of genocide and culturecide.

3.2 How to Join the Culture Club

An exploration of whether or not there is "culture in nature" should best start with an attempt to define culture, because the answer will largely depend on what we are looking for in the first place. Unfortunately, there is no shortage of definitions and little agreement. One is tempted to quip that there are as many definitions of culture as there are cultures. . . .

The Latin roots of the word *cultura* are associated with cultivation or tending. In the early 1800s, it became associated with "high culture" in describing refined tastes and manners. Victorian Zeitgeist reserved the attribution to Western societies. The influential definition of the English anthropologist Edward Tylor follows this line of thought: "Culture or civilization [. . .] is that complex whole which includes knowledge, belief, art, law, morals, customs, and any other capabilities and habits acquired by man as a member of society" (cited in Miller 2005, p. 9). The definition, written in 1871, at the height of British colonialism, appears to be inclusive, but nevertheless lends itself to the exclusion of large segments of the human race from the culture club by allowing one to distinguish the "civilized" from the "primitive".

Times have since moved on, and we perceive the world around us differently. Not even a contemporary animal behaviorist will be anymore deterred by Tylor's definition, because none of the criteria listed in his catalog can easily serve as an a priori exclusion for other animals. Bees, lions, and gorillas—don't they form "societies"? Doesn't their behavior follow certain rules? Why can't these count as "laws"? Animals might also very well hold "beliefs". How can we know that an elephant doesn't have deep thoughts as if she touches the bleached skeleton of a deceased group member, or while looking into a colorful sunset?

Numerous other test criteria have been formulated, aimed to demonstrate that only humans fulfill them. The list includes traits such as "tool-use", "music", "political aptitude", or "story-telling". But ongoing research has shown that animals can very well hold their own in these arenas, be they woodpecker finches, humpback whales, or primates. The birds use small sticks to extract grubs from plants (Tebbich and Bshary 2004); the whales communicate through elaborate songs over thousands of kilometers (Rendell and Whitehead 2001); and two weaker monkeys may form an alliance to dethrone a stronger (Perry and Manson 2008). Animals may also "tell stories" and thus instill illusions in the heads of others; for example, orangutans hold leaves or their hands to their mouths with the effect of making their voices resonate more deeply and aggressively—a ploy that can fool attackers (Hardus et al. 2009). And bonobos trained in an artificial language "gossip" about events that have happened earlier in the day (Savage-Rumbaugh and Lewin 1994).

Of course, we could argue that only a Mozart mass is true music; that baboon politics are a far cry from what Bismarck did when he forged the German Reich; that story telling has to be of the Brother Grimm standard; that an implement has to have the sophistication of a Swiss-army knife to count as a tool. Or even better: That only those who make tools to make other tools are tool-makers. In which case yet another bonobo genie will have to be freed from the bottle, one who uses a stone to knap flakes from another stone, so that a rope can be cut (Schick et al. 1999).

At such points, the bargaining inevitably starts anew—over what a term really means or should mean, and whether or not an observation supports or falsifies a certain intellectual position (see McGrew 2004 for a comprehensive review of such "checklists"). This process is complicated by the fact that data do not speak for themselves—somebody has to interpret them. Thus, our willingness to credit animals with certain abilities or not will strongly depend on the conviction that we have in the first place.

Indeed, some academics will happily welcome the closer connection with other animals suggested by such discoveries. Others will play the game of raising the bar and re-defining the criteria: "Let's not get too excited and think there are no differences between them and us. [. . .] Chimpanzee tools are no more complex than a stick for termiting or ant dipping [. . .], whereas humans have built space ships to travel to the moon" (Rice and Maloney 2005, p. 194).

A complex term such as "culture" is particularly prone to be used as a pawn in these debates. As always, definitions are clearly influenced by the underlying agenda: either to establish what makes humans different from other organisms, or to reveal similarities. Consider the following definitions of culture collected from the glossaries in a dozen anthropology textbooks commonly used in undergraduate education throughout the USA:

- That which is transmitted through learning, behavior patterns, and modes of thought acquired by *humans* as members of society. Technology, language, patterns of group organization, and ideology are all aspects of culture (Kottak 1982, p. 490);
- The ways *humans* discover, invent and develop in order to survive. Culture is the human strategy of adaptation (Nelson and Jurmain 1991, p. 603);
- *Humans'* systems of learned behavior, symbols, customs, beliefs, institutions, artifacts, and technology, characteristic of a group and transmitted by its members to their offspring (Campbell and Loy 2000, p. 635);
- All aspects of *human* adaptation, including technology, traditions, language and social roles. Culture is learned and transmitted from one generation to the next by nonbiological means (Jurmain et al. 2001, p. 405);
- Ideas and behaviors that are learned and transmitted. Also, the system made up of the sum total of these ideas and behaviors that is unique to a particular society of *people*. Nonbiological means of adaptation (Park 2002, p. 449);
- Information stored in *human* brains that is acquired by imitation, teaching, or some other form of social learning (Boyd and Silk 2003, p. A6);
- All aspects of *human* adaptation, including technology, traditions, language, religion, marriage patterns, and social roles. Culture is a set of learned behaviors that is transmitted from one generation to the next by nonbiological means (Jurmain et al. 2003, p. 463);
- Learned, nonrandom, systematic behavior and knowledge that can be transmitted from generation to generation (Stein and Rowe 2003, p. 533);
- Behavior that is shared, learned and socially transmitted (Relethford 2005, p. G-3);

- A shared and negotiated system of meaning informed by knowledge that *people* learn and put into practice by interpreting experience and generating behavior. Interdependent with society (Lassiter 2006, p. 195);
- The sum total of learned traditions, values, and beliefs that groups of *people*, and a few species of highly intelligent *animals*, possess (Stanford et al. 2006, p. 557);
- Learned behavior that is transmitted from *person* to person (Larsen 2008, p. A12).

Clearly, all these definitions are driven by the underlying agenda to *not* do what the Victorians did, i.e., the definitions strive to allow for the finding that *all* humans have culture, independent of their ethnic or geographic background. On the other hand—as indicated by the added emphasis—ten of the twelve definitions are explicitly speciesist in that they describe culture as specifically human. Surprisingly, of these ten, only two come from cultural anthropology textbooks (Kottak 1982; Lassiter 2006), whereas eight are extracted from texts that focus on biological anthropology. Of these, only two (Stein and Rowe 2003; Relethford 2005) provide definitions that are not specifically human, and only one other (Stanford et al. 2006) specifically includes "highly intelligent" animals.

It is probably no coincidence that inclusive definitions all stem from post-1999 texts—the publication year of the benchmark article that put the idea of animal culture firmly on the map. This article identified dozens of behaviors in which populations of wild chimpanzees differ from each other (Whiten et al. 1999). Human cultures, this much can probably be agreed by everybody, exemplify a wide range of intra-specific differences in behavior. The same has now been shown for chimpanzees, as the study subjects live in various regions across Africa, but are all members of the same species.

Providing a post-1999 definition of culture that specifically excludes nonhuman animals thus probably reflects a conscious decision to reserve that label for *Homo sapiens*. Such authors stick to the traditional speciesism of cultural anthropologists that draws the line a priori: "Humans are animals with a difference, and that difference is culture" (Kottak 1982, p. 5).

Hardly any other term but culture is so central to the research and teaching conducted in anthropology departments, which, in Anglo-Saxon academia, historically encompasses four fields (Miller 2005):

- *biological* or *physical anthropology*: studying human variation, adaptation and change through, for example, the fossil record, comparative socio-ecology of primates (prosimians, monkeys, apes), ecology and nutrition;
- *archaeology* or *prehistory*: reconstructing life-styles of past cultures through examination of material remains;
- *linguistic anthropology*: documenting geographical distribution and development of native languages;
- *cultural* or *social anthropology*: exploring how and why human cultures are similar or different.

The new paradigm of "animal cultures" could, at least in principle, generate various synergies between these subdisciplines. For example, biological anthropologists could work with archaeologists to analyze tools of stone and wood abandoned by apes; or with linguists to detail intra-specific variations of sounds made by gibbons;

or with cultural anthropology to explore the degrees of behavior that differ between populations of the same species.

However, such cooperation is rare, as anthropology departments harbor academics that think in the "human–animal dichotomy" alongside those for whom the category "humans and other animals" comes naturally. In fact, departments have at times split along these lines. For example, the biological anthropologists of Duke University, USA, joined the Anatomy Department, and anthropology at Stanford University, USA, separated into the Department of Anthropological Sciences and the Department of Cultural and Social Anthropology. Part of the tension stems from the perennial nature/nurture debate over whether biology applies to people, reflecting dissent over the formula "nature in culture". The new twist of applying culture to nonhumans now allows for dissent over a new buzzword: "culture in nature".

The anthropologist and zoologist William McGrew is one of the lead researchers in the area of animal culture. He labels those who argue that nonhuman animals don't belong to the "culture club" as *humanists*, and distinguishes them from those with a more gradualist view, whom he labels *universalists* (McGrew 2004).

The primatologist and ethologist Frans de Waal was, from early on, an outspoken proponent of the latter position: "The 'culture' label befits any species, such as the chimpanzee, in which one community can readily be distinguished from another by its unique suite of behavioral characteristics. Biologically speaking, humans have never been alone—now the same can be said of culture" (de Waal 1999, p. 636). He is not shy about turning the tables on those who display a tiring propensity to deny the traits that we have in common with other species—by diagnosing the offenders as suffering from "anthropodenial" (de Waal 1997).

And as much as we would like to be called "humanists" in the traditional philosophical meaning of the word-being forced to choose a camp we would readily settle with the universalists.

3.3 Culture as the Way We Do Things

The textbook definitions listed above are not only different from Tylor's in that they try to be all-inclusive with respect to humans. They have also shifted from providing a catalog of criteria to instead focusing on an underlying mechanism: that of learned behavior.

A checklist of specific traits for what counts as culture is in some ways always arbitrary and one can easily add or identify traits that are not included. Tylor, perhaps for that reason, threw in the all-catching, albeit rather nebulous, expression of "that complex whole". Centering the definition on a mechanism is a more principled way of unifying phenomena.

The simple characterization of culture as "socially transmitted behavior" has the advantage that it can accommodate the considerable variation in behavioral patterns between human populations. Inhabitants even of neighboring villages may differ in the way they speak, how they greet each other, what they consider acceptable conduct and what counts as offensive, how and what they like to eat, and which implements they employ. Such "cultural diversity" is, without any doubt, a trademark of *Homo sapiens*.

At this stage, we have to back-pedal slightly, because we cannot necessarily assume that all behavior is based on learning. Instead, it could be caused by underlying variation in genetic make-up. For example, many adult humans are unable to drink milk without adverse affect. They do not have the genotype to produce the enzyme lactase beyond infancy, a trait that evolved rather recently among certain pastoralist populations (Ingram et al. 2009). The consumption of milk or the absence of this practice is therefore strongly influenced by genetic factors.

Then there are those learned behaviors, which—while not channeled by a particular genetic make-up—are nevertheless "constrained" by the environment (Parish and Voland 2001). An example would be how people eat rice—as varied customs reflect to a large degree the consistency of this food. Thus, chopsticks are the implement of choice when rice is sticky, whereas forks are more feasible when loose long-corn rice is consumed, while rice cooked into a mush is often eaten by hand. Even simpler, rice eating can be absent altogether in a population if rice was never available.

Finally, there are traits that are quite likely neither molded by genetic nor environmental influences. For example, people greet each other by bowing (Thailand), shaking hands (Germany), kissing on one cheek (Argentina) or both (Switzerland), or by moving the right hand towards the heart (Nigeria), etc. Of course, subtle environmental influences are always difficult to discount. Hygienic conditions could be a consideration, so that direct body contact is suppressed in Thailand but not Germany. However, such explanations border on silliness. It is more likely that such customs are arbitrary, thus representing pure *cultural variants*.

A definition of culture as socially transmitted behavior would not include traits that are genetically determined. Nevertheless, whether or not adults drink milk as mitigated by the absence or presence of lactase persistence would still contribute towards intra-specific variability—just that this patterning does not count as a "cultural trait" in the way we have defined culture. On the other hand, traits such as rice eating, which are more or less influenced by the ecology, *can* be socially transmitted. These customs, together with the arbitrary variants, would constitute the *cultural profile* of a population.

Differences in behavior can serve as important amplifiers of between-group differences and play a crucial role in competition over resources. However, environmentally constrained behaviors have a certain likelihood of appearing in neighboring groups, due to convergent evolution in similar surroundings and the likelihood of common ancestry. Behaviors that can serve as badges of social belonging therefore develop most efficiently if they do not serve any practical use or rational cause, but if they are simply arbitrary.

A fictitious tale illustrates how perfectly idiosyncratic do's and don'ts can become entrenched into a societal narrative: "There are five monkeys in a cage. String a rope to the ceiling, tie a banana to it, and push a stair-ladder under the bait. Once the first monkey climbs the ladder, hose all the others down with cold water. If a second one tries, do the same. A subsequent attempt by yet another monkey to get to the banana will be met by forceful intervention from his colleagues. Move the hose out of sight, remove one monkey from the cage and substitute him with a new one. When the newcomer goes for to the banana, he will be in for a nasty surprise, as four others will immediately start to beat him up. Exchange one more of the initial monkeys for a new group member. He is attacked as soon as he climbs the stairs—with the previous newcomer as part of the gang. Change a third monkey, and a fourth. As soon as the newcomers approach the ladder, they get whacked. Most of the aggressors have no idea why they beat up a fellow primate. Once you changed the last and fifth original monkey, there is nobody left who experienced the displeasure of being hosed down. Still, all monkeys will simply ignore the banana. Why? Well, nobody knows."

The story is a fitting illustration of a definition of culture developed by William McGrew as "the way we do things" (2004, p. 24f).

3.4 Multiculturalism Amongst Animals

The generous definition "socially transmitted behavior" has allowed zoologists to interpret numerous cases of the way things are done in one animal population as opposed to another as expressions of culture (reviews in Galef and Heyes 1996; Fragaszy and Perry 2003).

The potential cognitive mechanisms that allow the perpetuation of acquired behavioral pattern are the subject of intense research programs. Candidate processes include simple learning events such as conditioning, social facilitation, local enhancement, as well as more complex mental machinations related to insight, theory of mind, and teaching (Byrne 1995; Tomasello and Call 1997; Hurley and Nudds 2006).

Contemporary cognition research draws on methods from diverse disciplines such as developmental psychology, meme theory and neurobiology. Still, naturalistic studies of behavior are clearly as important, as a full understanding of cognitive mechanisms requires that they are tied back to selective processes under which they evolved. For this, we need to unearth what nonhuman animals actually do as well as the bandwidth of what they are capable of achieving. The following examples are meant to provide a glimpse into the variety of thoughtscapes.

Some descriptions highlight the existence of a natural barrier such as a river that may have prevented the diffusion of a certain locally made "invention". Others refer to subsistence techniques or thermoregulation. However, still others seem to be perfectly "useless" behaviors, and some resemble human "fashions", that appear just to die out soon after.

The case studies can serve as illustrations of the phrase "the way we do things". It implies characteristics that need to be present in cultural beings—chiefly that the behaviors in question have to be somehow standardized, collective, and that they can serve as a source of "identity" (McGrew 2004, p. 25). The following accounts also tell that identity can well form around something that is *not* done in a particular population.

- *Triggerfish.* Members of this taxon prey on sea urchins protected by spikes. The fish turn them over by "blowing" a jet stream of water against the urchins, exposing their unprotected body parts, into which the fish can bite. A different technique is observed in the Red Sea—but nowhere else. Here, fish first bite off the spines of the urchins, which allows them to drag their prey towards the surface. They then let go, and start feeding on the unprotected underneath of the prey while the urchins tumble slowly back towards the bottom (Bshary et al. 2002).
- *Bottlenose dolphins.* A common foraging technique sees these whales plow through the bottom of the ocean with their pointed beaks to unearth prey. But, doing this, they risk stings from bottom dwellers. Dolphins in Shark Bay, Australia, overcome this hazard by wrapping their sensitive beaks with a sea sponge that acts like a glove. The inventive covering probably protects against scrapes and stings (Krützen et al. 2005).
- *Orcas.* North America's west coast is a favorite hunting ground of these predators. Pods of killer whales can be distinguished on the basis of characteristic vocalizations as well as hunting styles. One group will drive schools of salmon into a spiral before gorging on them; another is specialized to attack seals that rest on land. The orcas, in a rather dangerous maneuver, will beach themselves, try to catch a seal, and then rock back into the sea (Baird 2000).
- *Otter.* Californian sea otters will float belly up, holding an anvil stone between their paws. They then smash mollusks that they balance on their chests. Otters further up the coast will not use this modus operandi (Hall and Schaller 1964).
- *Capuchin monkeys.* Some rather strange fads develop from time to time in certain populations of these New World primates, and die out again after a while. Pairs of monkeys in a Costa Rican population will suck each others' toes; they will stick a finger up each other's nose; or, most bizarre, poke a fingertip into the eye-socket of a partner, noticeably dislodging the eye-ball. These penetrations are probably associated with some discomfort and danger, as capuchin monkeys neither cut their fingernails nor clean under them. The practices perhaps serve to build trust between the partners. And those with a proven track record of not hurting their partner—despite the fact that they could—will make reliable allies in aggressive encounters with third parties (Perry and Manson 2008).
- *Hanuman langurs.* The famous temple-monkeys of India seem to follow different rules about social etiquette. Those who live around Jaipur in the northwestern state of Rajasthan will huddle with each other, rain or shine, in wintry chilly air as well as in the humidity of the rainy season. Just 150 km or so towards the west, around the city of Jodhpur, one will hardly ever see huddling between adult monkeys; they keep their distance, even when temperatures approach freezing point (Sommer 1996).
- *Japanese macaques.* The red-faced monkeys of Japan have been studied for more than 50 years. We thus know about numerous behaviors that reflect local traditions restricted to a particular group or their neighbors. Some mothers will "wash" their babies in the ocean; some will dive to hunt for seafood; some will sit around human-made fire to warm their hands. Not all behaviors have to do with subsistence or thermoregulation. Females in some groups will reject the advances of males and prefer homosexual mounts instead. Elsewhere the monkeys handle pebbles, pile them, roll, rub or clack them together (Huffman 1996; Vasey 2006).
- *Orangutans.* Several dozen local traditions are described from the Indonesian island of Sumatra. This includes a peculiar way of greeting each other by making a "raspberry" sound, and the habit of using vegetation as a cover to escape from the rain. Some apes utilize small wooden picks, held in the mouth, to extract fatty seeds from the very prickly Neesia fruit. These trees grow at both sides of the Alas river. The forest floor on one side of the river, in the Singkil swamp, is littered with wooden tools. Those orangutans inhabiting the Batu-Batu swamp on the other side of the waterway do not use the implements (van Schaik et al. 2003).

The above examples illustrate the emergence of *cultural zoology* as a multi-taxonomic discipline. Nevertheless, the topic of animal cultures is still largely primatocentric—simply because we are anthropocentric, and a disproportionate amount of attention is heaved upon primates as our closest living relatives. As the field moves on (see, e.g., Emery and Clayton 2004), it will certainly be increasingly difficult to get away with a new definition of culture as "roughly anything that we do and the monkeys do". Still, it can be justified to single out studies of chimpanzees as they provide not only more primate examples, but also stand for the prime examples which served as a catalyst for a fresh look at other animal taxa.

3.5 Panthropology

The professionals who explore human societies and their varied local customs of how to obtain, prepare and ingest food, or the regional expressions of rituals and social conventions are called anthropologists. According to Greek etymology, the diversity of humans (anthropos) is the subject of their words and wisdom (logos). However, "anthropology" has spawned one new subdiscipline, aptly nicknamed "panthropology" (Whiten et al. 2003): the exploration of the diversity of behavior in our closest living relatives, i.e., apes of the genus *Pan*.

Members of this genus, the chimpanzee (*Pan troglodytes*) and the bonobo (*P. paniscus*), display the most astounding degree of behavioral diversity in non-human animals (McGrew 1992; Wrangham et al. 1994; McGrew et al. 1996; Parish and de Waal 2000; Boesch et al. 2002; Hohmann and Fruth 2003; McGrew 2004). Research across Africa revealed for each study community a unique combination of the presence or absence of traits related to social customs, communication, territorial aggression, war-like raiding, hunting strategies, tool-kits, food processing and consumption, and ingestion of plant matter for self-medication. This degree of plasticity in behavioral patterns is perhaps not surprising given that *Pan* and the likewise very flexible *Homo* shared a common ancestor until about 5–7 million years ago. However, the relative richness of documented cultural diversity is also at least partly the result of an observation bias, given that the ecology and behavior of chimpanzees is far better explored than that of other apes.

The classic study already mentioned above (Whiten et al. 1999, 2001) compiled the behavioral patterns at nine long-term chimpanzee research sites. Behaviors for which ecological explanations reflecting environmental constraints seemed plausible were carefully discerned from a couple of dozen traits customary or habitual among some groups, but absent in others, which could count as purely cultural variants.

Well-known arbitrary traits are the styles of mutual grooming amongst chimpanzees. At some sites the apes will face each other while sitting on the ground, and each will fully extend the free hand overhead and clasp the partner's hand. It is hard to conceive an environmental pressure that would induce the apes to attain this peculiar A-frame style of grooming.

Another famous example is nut cracking with stone or wooden hammers against an anvil. This technique is restricted to West Africa, despite an abundance of nuts and potential tools elsewhere. The practice is in all likelihood neither genetically determined nor a reflection of particular environmental conditions because some communities exhibit the behavior while others within a closely related population do not—although they are separated only by the banks of the N'Zo-Sassandra River in Ivory Coast and exposed to virtually identical environmental conditions. Interestingly, it was recently discovered that chimpanzees in the Ebo forest, Cameroon, more than 1,700 km east of the previously proposed riverine "information barrier", also crack *Coula* nuts with stone hammers, while sitting in trees and using thick branches as anvils (Morgan and Abwe 2006). The observation does not necessarily challenge the "cultural variant" explanation, but questions the existing model of the cultural diffusion of nut-cracking behavior. Instead, nut cracking could have been invented multiple times, or perhaps it became extinct in the region between the N'Zo-Sassandra River and the Ebo forest.

The attitude that chimpanzees display towards water can count as another cultural variant. The apes will normally avoid coming into any contact with the wet element, carefully circumventing puddles, and treading carefully along the edges of rivers or lakes. However, in Senegal, they enjoy a good soak and splash in shallow ponds. Explanations that seek to correlate this cautiousness towards water with factors such as climate or parasites have failed so far (McGrew 2004).

Similarly, it is quite surprising that chimpanzees in Gashaka, Nigeria, will eat ants virtually every single day, using stick tools. However, they will never ever consume a termite—despite the fact that these insects are available to them and more nutritious in the first place (Fowler and Sommer 2007).

The above examples lend themselves to question some deeply ingrained assumptions about human uniqueness. For example, an experienced chimpanzee researcher can, by just looking at a catalog of absent or present behaviors, discern a West African culture from the culture in East Africa. She could even differentiate Ugandan chimpanzees into those that live in Budongo and those that live in Kibale— in very much the same way that human ethnographers could tell a lifestyle that is predominant in Japan apart from the ways of day-to-day life conducted in France, with further differentiation between Breton and Alsatian customs.

If chimpanzees were humans, we would probably also not hesitate to assume that they respect certain "taboos". Nigerian villagers at Gashaka abhor the thought of eating cats or dogs—whereas those across the Cameroon border, just 2 days' walk away, have no such inhibitions. The psychology of apes may well correspond to such ways of defining one's ethnic group: "You want to be a Gashaka chimpanzee? Well, then eat ants as often as you like. But don't dare to devour termites, and never ever disturb the spirits in the water. We don't do this here...."

In-depth research of an animal population can easily be likened to the cultural anthropological practice of *ethnography*, which produces a descriptive account of behavior observed within the particular study population. A further step is *ethnology*, the analysis across populations, which aims to detect patterns and causes of them. Only through investigations of "as many groups of chimpanzees in as many parts of Africa as possible" (Goodall 1994, p. 397) can universal behaviors be discerned from variants and whether these differences reflect genetics, environment, or arbitrary customs. Major tools for cross-cultural comparisons of human populations are the HRAF ("Human Relations Area Files"). The dynamic development of the paradigm of "cultural primatology" thus comes with the explicit aim of creating CRAF: "Chimpanzee Relations Area Files" (McGrew 2004).

We are thus living in an era brimming with surprising discoveries about our closest living relatives, an era of *Pan novus* that can and will lead to significant re-formulations of what it means to be human. This meaning will add to a paradigm of *Homo novus*, properly informed by evolutionary biology.

Sadly, however, the era of *Pan novus* is coming to an end, just when it was about to begin. The field of "cultural panthropology" has already become another "urgent anthropology", given that many wild populations of apes are threatened by extinction due to habitat destruction, disease, and the trade in bush-meat (Sommer and Ammann 1998; Peterson and Ammann 2003; Caldecott and Miles 2005; Sommer 2008). Correspondingly, not only genetic diversity is lost, but also cultural diversity—similar to how globalization increasingly destroys the inherited multiplicity of human ways of life. It is already clear that "we can never know the true extent of cultural diversity in chimpanzees because so many communities, along with their cultures, are already gone" (Goodall 1994, p. 397).

3.6 Cultural Capacity: Blessing and Curse

The swan song that can be sung for chimpanzees can also be sung for other animal forms, which are, as we are only about to discover, often surprisingly akin to us in their cultural diversity—whether they count as birds, primates, or whales.

This realization has practical consequences for efforts to conserve biodiversity. Cultural zoology, like modern molecular genetics, amplifies the differences between organisms that we might have previously just lumped into a single category, a single type. The new and more nuanced picture complicates straightforward appeals to save "the chimpanzee", "the orca", or "the otter". Because, who is going to decide which clusters of cultural (or genetic) markers should have priority over others doomed to go extinct in the near future?

And from where is this impeding mass extinction garnering its momentum in the first place? Ironically, it seems as if cultures come with an inbuilt tendency to gravitate towards their own annihilation.. . .

Human groups have always competed, be it over mating partners, water, land or grazing grounds. Traditions are a major tool in these conflicts, as they allow members of one's own group to set themselves aside from those of others—who speak in an ugly dialect (such as Bayerisch), wear strange clothes (such as Lederhosen), and have despicable customs (such as eating Weisswurst).

Inter-group competition does, however, also lead to the need to foster ever-larger alliances, because those who cooperate across prior borders of in-group vs. out-group will be able to outcompete those who stick to their aged ways of thinking small. Thus, a well-known spiral is set in motion: "I against you; we against our neighbors; our village against your village; our tribe against your tribe; our country against your country"—until we arrive at nation states, NATO, and UN.

Small coalitions tend to be swallowed by larger alliances, and the defeated tend to attain the habits of the victorious. In this way, customs become more and more unified. The final stage of this development is globalization. For many of us and in many aspects, life becomes easier this way: McDonalds, internet, Big Brother, football, and plastic bottles have become ubiquitous parts of everyday life, useful or entertaining. The flipside is that our world is also becoming poorer, as variety is disappearing.

The new developments in cultural zoology establish a yet closer affinity to other organisms than even evolutionarily minded scientists had hitherto realized. This affinity comes with good news: Our common heritage with other animals extends to such a degree that we can perceive them as cultural beings, which produce a mind-boggling variety of lifestyles. But there is equality in a sad way, too: We now know that the disappearance of other organisms is not only loss of genetic diversity—it can also mean a loss of cultural diversity.

Diversity is aesthetically pleasing. But diversity, including cultural diversity, could only be had for the price of change in the first place—as change is the heart-beat of evolution. As individuals, organisms will not normally live through even a single interval of these drumbeats. But in the current case, change is occurring more rapidly. We also know more clearly that there will be consequences.

Stephen Crane, at the turn of the twentieth century, gave this fatality a poetic touch: "Sir, I exist!"/"However," replied the universe,/"The fact has not created in me/A sense of obligation" (Crane [1899] 1998). Yes, change is the heartbeat of evolution, steady and unstoppable and not caring in the least. Our cultural capacities—equally a blessing and a curse—may just speed up the process.

So, could it be that what turns out to be unique about us may be our capacity to know beforehand the dimensions of demise? This thought may well spawn another paradigm: that of animal thanatology. While as yet unexplored, one would be willing to hazard a guess: Animals such as apes and elephants understand that individual life is finite—if the fact is anything to go by that, day after day, they may attend to the corpse of a dead conspecific. But perhaps they are not burdened with the abstract concept of "extinction"? Surely, one would not wish it upon them.

References

Ackerman D (2007) *The Zookeeper's Wife*. Norton, New York, NY

Antweiler C (2009) *Was ist den Menschen gemeinsam? Über Kultur und Kulturen*, 2nd ed. Wissenschaftliche Buchgesellschaft, Darmstadt

Baird RW (2000) The killer whale: foraging specializations and group hunting. In: Mann J, Connor RC, Tyack PL, Whitehead H (eds) *Cetacean Societies: Field Studies of Dolphins and Whales*. University of Chicago Press, Chicago, IL, pp 127–153

Boesch C, Hohmann G, Marchant LF (2002) *Behavioral Diversity in Chimpanzees and Bonobos*. Cambridge University Press, Cambridge, MA

Boyd R, Silk B (2003) *How Humans Evolved*, 3rd ed. Norton, New York, NY

Bshary R, Wickler W, Fricke H (2002) Fish cognition: a primate's eye view. Animal Cognition 5:1–13

Byrne RW (1995) *The Thinking Ape. Evolutionary Origins of Intelligence.* Oxford University Press, Oxford

Caldecott J, Miles L (ed) (2005) *World Atlas of Great Apes and Their Conservation.* University of California Press, Los Angeles, CA

Campbell BG, Loy JD (2000) *Humankind Emerging*, 8th ed. Allyn and Bacon, Needham Heights, MA

Crane S [1899] (1998) *War Is Kind and Other Poems.* Dover Publications, Mineola, NY

de Waal FBM (1997) Are we in anthropodenial? Discover 18:50–53

de Waal FBM (1999) Cultural primatology comes of age. Nature 399:364–365

Emery NJ, Clayton NS (2004) The mentality of crows: convergent evolution of intelligence in corvids and apes. Science 306:1903–1907

Fowler A, Sommer V (2007) Subsistence technology in Nigerian chimpanzees. International Journal of Primatology 28:997–1023

Fragaszy D, Perry S (eds) (2003) *The Biology of Traditions.* Cambridge University Press, Cambridge, MA

Galef B, Heyes C (1996) *Social Learning in Animals: The Roots of Culture.* Academic Press, New York, NY

Goodall J (1971) *In the Shadow of Man.* Collins, London

Goodall J (1994) Conservation and the future of chimpanzee and bonobo research in Africa. In: Wrangham RW, McGrew WC, de Waal FBM, Heltne PG (eds) *Chimpanzee Cultures.* Harvard University Press, Cambridge, MA, pp 397–404

Hall KRL, Schaller GB (1964) Tool-using behavior of the California sea otter. Journal of Mammalogy 45:287–298

Hardus ME, Lameira AR, Van Schaik CP, Wich SA (2009) Tool use in wild orang-utans modifies sound production: a functionally deceptive innovation? Proceedings of the Royal Society London, Series B, published online 5 August 2009

Hohmann G, Fruth B (2003) Culture in bonobos? Between-species and within-species variation in behavior. Current Anthropology 44:563–571

Huffman MA (1996) Acquisition of innovative cultural behaviors in nonhuman primates: a case study of stone handling, a socially transmitted behavior in Japanese macaques. In: Galef BG, Heyes CM (eds) *Social Learning in Animals: The Roots of Culture.* Academic, Orlando, FL, pp 267–289

Hurley S, Nudds M (eds) (2006) *Rational Animals?* Oxford University Press, Oxford

Ingram CJ, Mulcare CA, Itan Y, Thomas MG, Swallow DM (2009) Lactose digestion and the evolutionary genetics of lactase persistence. Human Genetics 124:579–591

Jurmain R, Kilgore L, Trevathan W, Nelson H (2003) *Introduction to Physical Anthropology*, 9th ed. Wadsworth/Thomson Learning, Belmont, CA

Jurmain R, Nelson H, Kilgore L, Trevathan W (2001) *Essentials of Physical Anthropology.* Wadsworth/Thomson Learning, Belmont, CA

Kottak CP (1982) *Anthropology: The Exploration of Human Diversity*, 3rd ed. Random House, New York, NY

Krützen M, Mann J, Heithaus HR, Connor RC, Bejder L, Sherwin WB (2005) Cultural transmission of tool use in bottlenose dolphins. Proceedings of the National Academy of Sciences of the USA 102:8939–8943

Larsen CS (2008) *Our Origins: Discovering Physical Anthropology.* Norton, New York, NY

Lassiter LE (2006) *Invitation to Anthropology.* AltaMira Press, Lanham, MD

McGrew WC (1992) *Chimpanzee Material Culture. Implications for Human Evolution.* Cambridge University Press, Cambridge, MA

McGrew WC (2004) *The Cultured Chimpanzee: Reflections on Cultural Primatology.* Cambridge University Press, Cambridge, MA

McGrew WC, Marchant LF, Nishida T (eds) (1996) *Great Ape Societies.* Cambridge University Press, Cambridge, MA

Miller BD (2005) *Cultural Anthropology.* Pearson, Boston, MA

Morgan BJ, Abwe EE (2006) Chimpanzees use stone hammers in Cameroon. Current Biology 16:R632–R633

Nelson H, Jurmain R (1991) *Introduction to Physical Anthropology*, 5th ed. West Publishing, St. Paul, MN

Oota H, Settheetham-Ishida W, Tiwawech D, Ishida T, Stoneking M (2001) Human mtDNA and Y-chromosome variation is correlated with matrilocal vs. patrilocal residence. Nature Genetics 29:20–21

Parish AR, de Waal FBM (2000) The other "closest living relative": How bonobos (Pan paniscus) challenge traditional assumptions about females, dominance, intra- and inter-sexual interactions, and hominid evolution. In: LeCroy D, Moller P (eds) *Evolutionary Perspectives on Human Reproductive Behavior*. Annals of the New York Academy of Science, vol 907. pp 97–113

Parish AR, Voland E (2001) Cost/benefit considerations in decisions to share in hunter-gatherer societies. In: Haft F, Hof H, Wesche S (eds) *Bausteine zu einer Verhaltenstheorie des Rechts*. Nomos, Baden-Baden

Park MA (2002) *Biological Anthropology*, 3rd ed. McGraw-Hill, New York, NY

Perry S, Manson JH (2008) *Manipulative Monkeys: The Capuchins of Lomas Barbudal*. Harvard University Press, Cambridge, MA

Peterson D (2006) Jane Goodall: The woman who redefined man. Houghton-Mifflin, Boston, MA

Peterson D, Ammann K (2003) *Eating Apes*. University of California Press, Berkeley, CA

Relethford JH (2005) *The Human Species, an Introduction to Biological Anthropology*. McGraw-Hill, New York, NY

Rendell L, Whitehead H (2001) Culture in whales and dolphins. Behavioral and Brain Sciences 24:309–382

Rice PC, Maloney N (2005) *Biological Anthropology and Prehistory: Exploring Our Human Ancestry*. Allyn and Bacon, Needham Heights, MA

Savage-Rumbaugh S, Lewin R (1994) *Kanzi: The Ape at the Brink of the Human Mind*. Wiley, Hoboken, NJ

Schick KD, Toth N, Garufi G, Savage-Rumbaugh ES, Rumbaugh D, Sevcik R (1999) Continuing investigations into the stone tool-making and tool-using capabilities of a bonobo (*Pan paniscus*). Journal of Archaeological Sciences 26:821–832

Sommer V (1996) *Heilige Egoisten. Die Soziobiologie indischer Tempelaffen*. CH Beck, Munich

Sommer V (2008) *Schimpansenland. Wildes Leben in Afrika*. CH Beck, Munich

Sommer V, Ammann K (1998) *Die Großen Menschenaffen: Orang-Utan, Gorilla, Schimpanse, Bonobo*. BLV, Munich

Stanford C, Allen JS, Anton SC (2006) *Biological Anthropology: the Natural History of Humankind*. Pearson/Prentice Hall, Upper Saddle River, NJ

Stein PL, Rowe BM (2003) *Physical Anthropology*, 8th ed. McGraw-Hill, New York, NY

Tebbich S, Bshary R (2004) Cognitive abilities related to tool use in the woodpecker finch, *Cactospiza pallida*. Animal Behaviour 67:689–697

Tomasello M, Call J (1997) *Primate Cognition*. Oxford University Press, Oxford

Vasey P (2006) The pursuit of pleasure: an evolutionary history of homosexual behaviour in Japanese macaques. In: Sommer V, Vasey P (eds) *Homosexual Behaviour in Animals: An Evolutionary Perspective*. Cambridge University Press, Cambridge, MA, pp 191–219

Whiten A, Goodall J, McGrew WC, Nishida T, Reynolds V, Sugiyama Y, Tutin CEG, Wrangham RW, Boesch C (1999) Cultures in chimpanzees. Nature 399:682–685

Whiten A, Goodall J, McGrew WC, Nishida T, Reynolds V, Sugiyama Y, Tutin CEG, Wrangham RW, Boesch C (2001) Charting cultural variation in chimpanzees. Behavior 138:1481–1516

Whiten A, Horner V, Marshall-Pescini S (2003) Cultural panthropology. Evolutionary Anthropology 12:92–105

Wrangham RW, McGrew WC, de Waal FBM, Heltne PG (eds) (1994) *Chimpanzee Cultures*. Harvard University Press, Cambridge, MA

van Schaik CP, Ancrenaz M, Borgen G, Galdikas B, Knott CD, Singleton I, Suzuki A, Utami SS, Merrill M (2003) Orangutan cultures and the evolution of material culture. Science 299: 102–105

Chapter 4
Nothing to Talk About

On the Linguistic Abilities of Nonhuman Primates (And Some Other Animal Species)

Julia Fischer

Abstract Is language a species-specific feature that distinguishes humans from other animals? While monkey and ape calls carry rich information that is potentially available to listeners, callers have little voluntary control over the structure of their calls and are hence unable to use these calls in a symbolic fashion. Likewise, combinations of calls do occur, but again, these do not appear to be driven by a set of rules applied by the sender. In contrast, listeners are able to attribute meaning to single calls as well as to specific call combinations. Although nonhuman primate communication may function very effectively with regard to social and ecological affordances, it is fundamentally distinct from human speech, where members of a linguistic community agree on the referential content of utterances by convention, and where syntactical rules provide a means to generate infinite meaning. Hence, the idea that speech and language are species-specific human traits is probably not an illusion. Future studies should address the selective pressures that shape communicative behavior, while genetic studies are needed to uncover the constraints that apparently play a role in animal communication.

4.1 A Very Old Question

In late 1892, the self-taught scientist Richard Garner set sail for West Africa to study the vocal behavior of wild chimpanzees and gorillas. His quest was to tear down the language barrier—the assertion that only man had language. Garner believed that language had no origin, but existed to a lesser or greater degree in all forms of life (Garner 1900). Whether or not language is shared with other creatures is indeed a very old question, and intimately tied to the origins issue. Indeed, in 1789, the *Berliner Akademie der Wissenschaften* posed the famous prize question, asking whether humans—solely based on their natural abilities—would be able to invent

J. Fischer (✉)
Cognitive Ethology Laboratory, German Primate Center, 37077 Goettingen, Germany
e-mail: fischer@cog-ethol.de

U.J. Frey et al. (eds.), *Homo Novus – A Human Without Illusions*,
The Frontiers Collection, DOI 10.1007/978-3-642-12142-5_4,
© Springer-Verlag Berlin Heidelberg 2010

language. The prize was later awarded to Johann Gottfried Herder for his treatise on the origin of language. Herder proposed that words arose from imitations of animal sounds, and that the imitation was linked to recognition and categorization of objects in the world (Herder 1772). The idea that imitation of sounds played a crucial role in the evolution of language is one of the classic themes in the origin debate; others have stressed the importance of gestural communication, or viewed language as originating in the expression of emotion. With the advent of evolutionary theory, the idea of a common biological substrate of human and (other) animal communication began to take hold (Darwin 1872). Yet, a number of cultural anthropologists, such as Franz Boas, rejected evolutionary accounts of the origin of language (see Radick 2008) and continued to fuel the debate about the origin of language. In contrast to the growing diversity of speculations and conjectures revolving around the origin of language, the data unfortunately remained scarce, until scholars seriously began to study the putative linguistic abilities of other species.

After the Second World War, the fledgling science of ethology provided a niche within which comparative studies began to contribute substantially to the understanding of the communicative abilities of different species, including our closest living relatives, the monkeys and apes. Peter Marler, who began his studies in Britain but soon moved to the United States, initiated a vigorous research program investigating the foundations of song learning in birds (Marler 1957), as well as the principles of nonhuman primate communication (Marler 1955, 1961). In contrast, North American evolutionary biologists, such as Steven J. Gould brushed this issue aside and maintained that language was probably just a byproduct of our large and complex brains, while Noam Chomsky argued that speech and language were simply too complex to have evolved (but see Hauser et al. 2002 for a more recent perspective). In the wake of a seminal article by Pinker and Bloom (1990), scholars from different disciplines finally joined to revisit the question of the origin of human language from a multitude of perspectives. Although the debate is alive and well, and partly sustained by a biannual conference on the evolution of language, the issue is still not settled—and probably never will be. In the meantime, however, we continue to learn a great deal about the principles governing animal communication, including the genetic foundations of communicative and social behavior.

4.2 Design Features of Speech

Language in general is characterized by a set of features (instead of one particular trait) that distinguish it from other means of communication. One fundamental aspect is its symbolic nature, another the existence of a set of rules (syntax) that gives rise to novel meanings by systematic composition of the units that make up the language. Both symbolism and syntax are based on conventionalization, and hence learning plays a major role. Spoken language, in addition, is characterized by a linear sequence (in contrast to sign languages, for instance, which operate in space and time) as well as by its use of the vocal–auditory channel (Hockett 1960). This list is by no means exhaustive. For instance, categorical perception of phonemes was long believed to be a hallmark of speech perception, but more recently, a number of studies have shown that nonlinear phenomena in auditory categorization is a widespread phenomenon across diverse taxa (see Fischer 2007 for a review). Below, I will first address the question of the semantic content in nonhuman primate vocalizations. I will include the cognitive mechanisms underlying vocal production of sounds in nonhuman primates, because these help us to understand the constraints in vocal flexibility. Despite the largely innate nature of the vocalizations, there is increasing evidence for a limited degree of plasticity. Next, I will review a number of studies that have dealt with the comprehension of calls and the attribution of meaning. I will then turn to the question of the syntactic abilities of monkeys and—from a comparative perspective—starlings. Because a number of scholars have recently resurrected the idea that spoken language evolved via gestural communication, I will very briefly discuss the evidence for this conjecture, before concluding with an outlook on future research perspectives.

4.3 Semantics in Nonhuman Primate Communication

Do monkeys or apes talk about things? Are there any indications that they refer to objects and events in the external world? Does their communication have a symbolic quality? These questions were brought to the forefront when Tom Struhsaker reported that vervet monkeys, *Chlorocebus aethiops*, living in the Amboseli National Park in Kenya produce three acoustically distinct alarm calls and three different adaptive escape strategies in response to their three main predators, leopards, eagles, and snakes (Struhsaker 1967). This led to the question of whether the vervets denote the predator type when they give a predator specific alarm call. Playback experiments revealed that upon hearing an alarm call, the monkeys responded appropriately, even when there was no predator around. The authors concluded that the vervets' calls in fact denoted the predator type, or else the required response (Seyfarth et al. 1980). Either way, the alarm calls were deemed to be referential, because they apparently referred to either objects or escape strategies external to the signaler. In due course, the question of semantic communication became a hot topic, and numerous studies were initiated that addressed the question of the meaning of animal signals.

John Smith was one of the first to stress the importance of distinguishing between the role of the sender who generates the message and the recipient who interprets the signal (Smith 1977). In fact, the problem with the initial conclusion that callers may have denoted the predator type was that the information provided in the signal was inferred from listeners' responses. The early studies on semantic or referential communication failed to differentiate between these two roles, and traces of this confusion can still be found today. Subsequent analyses, however, pointed out that for such a seemingly referential communication system to function signals simply have to be sufficiently specific (Macedonia and Evans 1993). If this is the case, recipients can use these signals to predict subsequent events and choose their responses accordingly (Macedonia and Evans 1993; Seyfarth and Cheney 2003). Two sets of questions follow from this insight: firstly, how much control do primates have over their vocal production to produce specific sounds in specific situations; and second, how do primates (or other animals) attach meaning to sounds?

4.4 Call Production

Studies of the ontogeny of vocal production as well as the neurobiological foundations of vocal control in nonhuman primates suggest that the structure of primate vocalizations is largely innate (Hammerschmidt and Fischer 2008). Unlike in most song-birds, exposure to species-specific calls is not a prerequisite for the proper development of the vocal repertoire. Nevertheless, developmental modifications occur. However, most of these changes can be attributed to growth. Ey and colleagues, for instance, analyzed the changes in the structure of contact barks in a cross-sectional study of 58 free-ranging Chacma baboons, *Papio hamadryas ursinus*. Baboon clear calls are tonal and harmonically rich. With increasing age, animals uttered longer calls with a lower fundamental frequency. These two variables also showed a significant interaction between age and sex, indicating that the profiles of age-related variations differed between the sexes (Ey et al. 2007a). In sum, there is ample evidence that primate vocalizations experience only minor structural changes during ontogeny, and that most of the changes can be attributed to growth (Hammerschmidt et al. 2000; Pfefferle and Fischer 2006) and some also to hormonal changes during puberty, which may affect the usage and structure of certain signals (Fischer et al. 2004b).

4.5 Vocal Plasticity

Several studies have suggested the existence of group-specific call characteristics in nonhuman primates that apparently could not be attributed to ecological or genetic variation. One of the earliest such reports came from Green (1975) who observed that three different populations of Japanese macaques produced acoustically different calls. Pygmy marmosets, *Cebuella pygmaea*, modified their trill vocalizations when placed with a new partner in such a way that the calls of both partners became more similar to each other than before pairing (Snowdon and Elowson 1999). The alarm calls given by members of two populations of Barbary macaques, *Macaca sylvanus*, revealed significant variation between sites. Playback experiments in which calls from their own or the other population were broadcast suggested that this observed variation was perceptually salient (Fischer et al. 1998). Chimpanzees from different groups also reveal group-specific differences in the acoustic structure of their pant-hoots (Mitani et al. 1992; Crockford et al. 2004).

Since learning is not a prerequisite to develop the species-specific repertoire, it is not quite clear how group-specific calls come about. Some experimental studies suggest an influence of auditory feedback on vocal output. For instance, Japanese macaques' coo calls produced in response to a playback coo were more similar to the playback coo than to spontaneous coos (Sugiura 1998). Common marmosets, *Callithrix jacchus*, increase their sound level as well as their call duration in response to increased levels of white noise in their environment (Brumm et al. 2004). Cotton-top tamarins, *Saguinus oedipus*, uttered shorter combinations of long calls with fewer pulses when they were exposed to white noise. In contrast, when they were exposed to modified real-time versions of their own vocal output, their calls became louder and had longer inter-pulse intervals (Egnor et al. 2006). Interestingly, olive baboons produce significantly longer grunts when they range in a habitat with poor visibility compared to open habitats (Ey et al. 2009), suggesting that they adapt their calls to environmental conditions.

There are at least two sources that appear to contribute to the adjustment of vocal output, one is socially mediated and probably effective over longer periods, leading to group specific calls, and the other is an immediate adjustment to perturbations of auditory feedback. Whatever the details of small scale adjustments of vocal output, neither suggests that learning is a prerequisite for the development of the species-specific vocal repertoire. Further support for this notion comes from the fact that there is no evidence that nonhuman primates incorporate novel sounds into their repertoire. In contrast, marine mammals such as bottlenose dolphins, *Tursiops truncatus*, were found to be able to imitate the whistles of other conspecifics (Janik 2000), parrots, *Psittacus erithacus*, can be taught to imitate human speech sounds (Pepperberg 2000), mockingbirds mimic all sorts of sounds in their environment; and there is even an odd anecdote about Hoover, the harbor seal who was reported to be talking, albeit with a heavy Bostonian accent (Deacon 1997). That is, flexible vocal production as a result of experience has evolved several times independently, but we do not appear to share it with our closest living relatives (Fischer 2003; Egnor and Hauser 2004).

4.6 Call Comprehension

In contrast to vocal production, experience appears to play a major role in the development of the comprehension of and correct responses to calls. Playback experiments revealed that vervet infants gradually develop the appropriate responses to alarm calls, and at an age of about 6 months, they behaved like adults (Seyfarth and Cheney 1980). Further evidence for a gradual development of responses comes from a study of infant baboons. Subjects developed the ability to discriminate between calls that fall along a graded acoustic continuum, as evidenced by the time spent looking towards a concealed speaker broadcasting the calls. At 2.5 months of age, infants did not respond at all to the playback of alarm or contact barks. At 4 months of age, they sometimes responded, but irrespective of the call type presented. By 6 months of age, infants reliably discriminate between typical variants of alarm and contact barks (Fischer et al. 2000). Further experiments showed that infants of 6 months and older exhibit a graded series of responses to intermediate call variants. They responded most strongly to typical alarm barks, less strongly to intermediate alarm calls, less strongly still to intermediate contact barks, and hardly at all to typical contact barks (Fischer et al. 2000). A study in Verreaux's sifakas, *Propithecus verreauxi verreauxi*, also reported that from an age of about 6 months, young responded in an adult-like fashion to the playback of alarm calls (Fichtel 2007). There appears to be some flexibility, because vervet monkey infants who were exposed to specific alarm calls frequently developed the appropriate response earlier than infants who were rarely exposed to it (Hauser 1988). Furthermore, a study of the development of maternal recognition showed that, from as early as 10 weeks of age, Barbary macaque infants respond significantly more strongly to play-backs of their mothers' calls than to playbacks of unrelated females from the same social group (Fischer 2004), strongly suggesting that infants are able to recognize their mothers by voice from this early age onwards. Taken together, these findings corroborate the argument that the structure of the vocalizations is largely innate, whereas call comprehension is based on learning (Seyfarth and Cheney 1997).

How do infants learn to attach the correct meaning to sounds in their environment? To date, there is only indirect and partly contradictory evidence of the degree to which infant responses to calls are influenced by adult behavior. For instance, Seyfarth and Cheney reported that vervet infants were more likely to respond correctly to the different alarm calls when they had first looked at an adult (Seyfarth and Cheney 1986). In contrast, infant baboons responded to intermediate alarm and contact barks that were typically ignored by adults, suggesting that in these situations, infants did not simply copy adult behavior (Fischer et al. 2000). In Barbary macaques, infant responses—looking toward the speaker—were not influenced by the behavior of their caretakers, that is, whether or not he or she looked towards the speaker (Fischer 2004). Similarly, after playback of conspecific alarm calls, infant and juvenile Barbary macaques more frequently ran away or climbed into trees than did adults (Fischer et al. 1995; Fischer and Hammerschmidt 2001). Apparently, social learning is not a prerequisite for the development of appropriate responses.

4.7 Fast Mapping in a Domestic Dog

The asymmetry between production and perception is shared with most other terrestrial mammals, and taken to an extreme in the example of Rico, a domestic dog. Rico became famous through the German TV Show *Wetten, dass?*, in which he demonstrated his knowledge of the names of (at that time) 70 toys by correctly retrieving them on command. We then contacted the owner and asked for permission to study his word learning and other cognitive abilities (Kaminski et al. 2008) in more detail. When we did the first experiments with him, his "vocabulary" had already grown to 200 names of toys, the limiting factor being the size of the owner's apartment. While it quickly became clear that Rico instantly learned to attach a new word to a new object, we were interested in whether he would also be able to map a word onto its referent indirectly, a feature known as "fast mapping". Fast mapping is another trait that was long believed to be restricted to humans, and a special adaptation required for the rapid growth of the vocabulary of toddlers (Carey and Bartlett 1978). The initial studies in 3–4 year old children consisted of a situation in which the child was asked to bring one of two objects. The color name of one of the objects was known (e.g., red), while the other color was for instance a dark green. The child was then requested to bring, for example, the "chromium" object, on the assumption that the children did not know this name yet. Most of the children of about 3.5 years and older were correctly able to pick out the object that was not red. Moreover, later on, they were able to tell the name of the color when confronted with other objects of the same color, or point to the correct objects when asked which one was the "chromium" one (Carey and Bartlett 1978).

Since we could not ask Rico to produce the name of an object, we had to modify the experimental paradigm accordingly. An array of toys that he already knew was scattered in one of the rooms, and one novel toy was added. From an adjacent room, the owner asked Rico to retrieve first one or two of the familiar objects. Next, she would ask him to fetch, say, the "Lessing", on the assumption that this was a new word for Rico. In most of the cases, Rico did indeed fetch the new object, demonstrating that he was able to learn by exclusion. This form of exclusion learning is not that surprising though, and has been demonstrated in a number of operant conditioning paradigms (Kastak and Schusterman 2006). More interesting to us was whether he would actually be able to remember that the new word referred to the new object. To test this, the "Lessing" and the other items were hidden for 4 weeks, until the next round of experiments commenced. In these trials, an array of familiar and novel items were scattered in the room, including the particular to which the name "Lessing" had previously been inferred by Rico. In half of the (few) test cases, Rico delivered the correct toy. When he was tested on the same day, his performance was even better. Overall, his performance matched that of children 3.5 years of age (Fischer et al. 2004a; Kaminski et al. 2004). Notably, both children and Rico outshine adult humans, who rapidly forget what they have learnt 4 weeks ago in equivalent experimental situations.

4.8 Recursive Syntactic Patterns

So far, we have seen that the vocal production of nonhuman primates and a range of other terrestrial mammals is confined to a set of relatively fixed vocal patterns, the use of which can be modified to some degree. One evolutionary strategy to overcome the constraints of a limited set of communicative units is to make use of signal combinations. For instance, male Campbell's monkeys, *Cercopithecus campbelli*, which belong to the guenons, give predator-specific alarm calls in response to leopards and eagles. Members of another guenon species, the Diana monkey, *C. diana*, attend to these hetero-specific alarm calls by responding with own species-specific alarm calls. The interesting observation is that male Campbell's monkeys respond to "less dangerous" situations such as falling trees with low-frequency resonating "boom" calls. If these booms precede alarm calls, listener responses are much weaker (Zuberbühler 2002). According to Zuberbühler, this constitutes an example of a syntactic rule in monkey communication. To fully assess the power of this system, however, solid observational and not just experimental data are needed, to see whether the boom regularly occurs in other contexts as a form of modifier. Otherwise, one would need to concede that this system provides little generative power.

More recently, Zuberbühler and Arnold published a series of papers in which they examine call combinations in male putty-nosed monkeys, *C. nictitans martini*. These monkeys combine two loud alarm calls, "hacks" and "pyows" into different call series depending on external events. To elucidate the alarm call patterns, the monkeys were presented with either acoustic stimuli (playbacks of leopard growls and eagle shrieks (Arnold and Zuberbühler 2006)) or stuffed models (Arnold et al. 2008). Eagle trials mainly elicited series of hacks, sometimes followed by pyows, and only one series of pyows only. Leopard trials elicited mainly pyow series and a few series of hacks followed by pyows. There were also pure pyow as well as mixed series produced in "unknown contexts". That is, the single pyow or hack calls revealed little context specificity while the series as a whole provided more reliable information. In a further study, Arnold and Zuberbühler showed that playbacks of such mixed sequences elicited group travel, while pure pyow series evoked cryptic behavior in female monkeys (Arnold and Zuberbühler 2008). The authors concluded that meaningful combinatorial signals have evolved in primate communication.

The question that remains is whether the combinatorial call system of the forest monkey constitutes an analogy to human syntax; or put more broadly, what we can learn about the evolution of human syntax by studying such a system. One necessary precondition, again, is to distinguish first between the caller and listener, i.e., between the production side and the perception/comprehension side. The question thus breaks up into two: firstly, what are the necessary (cognitive or motivational) preconditions to generate such sequences, and secondly, how much does it take to attribute differential meaning to the different types of sequences? Presently, it appears conceivable that the call sequences are in fact related to the motivational state of the animal in a probabilistic way. Again, the "smart part" is taken on by the listeners, who are able to integrate the acoustic information over longer periods of time.

A number of experimental laboratory studies have addressed the question of whether animals can distinguish between sequences of calls (or syllables) that are assembled according to some simple syntactical rules. In the first such study, Fitch and Hauser (2004) tested cotton-top tamarins. They generated two categories of "grammars", first a so-called finite state grammar (FSG), represented by the formula $(AB)^n$, which yields sequences such as ABAB or ABABAB; and secondly a context-free grammar, represented by A^nB^n, which yields sequences with the pattern AABB or AAABBB. The units were drawn from a pool of syllables and marked as belonging to either the A or B category by speaker gender. To assess whether the monkeys discriminated between these two "grammars", Fitch and Hauser adopted the habituation/discrimination paradigm, in which the monkeys were first familiarized with one of the grammars and then presented with a "test" grammar (Fitch and Hauser 2004). The results showed that the monkeys discriminated context-free grammars from finite state grammars, but not the other way around. A relatively simple explanation for the observed results might be that the monkeys processed the differential transitional probabilities for A and B in the two grammars. In the FSG grammar, A must be followed by B and vice versa, i.e., there has to be a change in category from item to item. Testing with CFG violates this rule: A can be followed by either A or B. In contrast, training with PSG establishes a rule that A can be followed by either A or B, thus testing with FSG is consistent with this rule.

A follow-up study in starlings adopted the same general setup (Gentner et al. 2006), but this time, the starling were trained in an operant paradigm. One of the first striking results is that it took those animals that learned to discriminate the two grammars more than 25,000 trials to do so. Humans tested in a similar setting need about 35 min (Bahlmann et al. 2008). While at least some of the starlings apparently mastered the task, a more critical reading of the results suggests that they presumably combined a strategy of attending to the first two syllables in a sequence (same: context-free grammar; different: finite state grammar), plus the total number of syllables in one string (Corballis 2007; Számadó et al. 2009). Taken together, these results corroborate the view that animals can attend to sequential features of sounds, but it remains unclear whether the paradigms used in these studies truly tap into the processing of recursive information (Perruchet and Rey 2005). Furthermore, it is important to keep in mind that human syntax is based on words that carry symbolic meaning, while it remains unclear whether human syllables have any meaning to tamarins, or whether single syllables in starling song refer to anything in the world. In sum, animal call sequences may be combinatorial but they are not semantically compositional in the sense that the elements that make up the utterance carry specific meaning (Számadó et al. 2009).

4.9 Gestural Communication

Possibly as a consequence of the frustration with the lack of flexibility in the vocal production of nonhuman primates, researchers have become more interested in the gestural communication of apes, and recently also monkeys. The reasoning is that while primates lack voluntary control over their vocal output, they do have excellent voluntary control over their hands. Therefore, it has been hypothesized that human speech possibly evolved via a proto-sign language (Corballis 2002; Arbib 2005). In addition, the brain area in monkeys (area F5) that controls manual movements is proposed to be homologous to Broca's area, which is involved in speech production, further fueling the interest in this modality of communication.

In this context, it is important to distinguish between enculturated subjects raised by humans, captive, and wild animals. Some enculturated apes, such as the chimpanzee Lana, are taught to use certain gestures, e.g., "open" and "more"; Lana used these in a range of different contexts (Kellogg 1968). Detailed studies of the signaling behavior of chimpanzees, bonobos, *Pan paniscus*, gorillas, *Gorilla gorilla*, and siamangs, *Symphalangus syndactylus* in captivity, in contrast, did not provide any evidence that gestures referred to certain objects or events in the subjects' environment (Call and Tomasello 2006). The majority of flexibly used gestures occurred in the play context—a context that by definition is characterized by highly variable behaviors. The evidence for gestural communication among wild subjects is still somewhat sketchy, and no study to date has reported that the animals signaled about objects or events in their surroundings. This finding suggests that a lack of voluntary control over the vocal apparatus is only one of the constraints that prevent a more elaborate communication among our closest living relatives.

4.10 Outlook

There is now convergent evidence to support the notion of a strong asymmetry between animals' abilities to intentionally transmit vs. to acquire information. Despite the fact that the structure of primate vocalizations appears fixed from birth, primate calls nevertheless provide rich information about sender attributes such as size, age, sex (reviewed in Ey et al. 2007b), hormonal levels (Pfefferle et al. 2008), and also motivational state (Fichtel and Hammerschmidt 2003). Specific relations between motivational state and certain contextual situations allow listeners to use signals as predictors of upcoming events. There is, however, no conclusive evidence that signalers have the intention to provide information to others. While nonvocal signals such as gestures and postures show a higher degree of variability (particularly in apes), there is no convincing evidence that gestures are used to intentionally communicate about objects or events, either (Fischer 2008).

To elucidate the constraints operating on the vocal communication of other animals, the genetic foundations of communication have attracted increasing attention. One approach is to identify humans with language impairments and then determine any underlying genetic anomalies. The most famous case is found in a British family in which a specific language impairment appeared to be inherited in an autosomal dominant fashion. Initially, linkage was found to a region of chromosome 7. Subsequently a case of chromosome translocation was found in another patient with a similar phenotype and this allowed the geneticists to eventually identify a point mutation in the FOXP2 gene (Fisher and Marcus 2006). This gene had not previously been thought to be implicated in brain function, but was subsequently shown to play a role in development of regions of the frontal lobe, subcortical structures, and cerebellum. It is not, however, a language gene; rather it is a transcription factor that affects the function of many other genes, and is involved in the development of the lung, heart, and other organs (Fisher and Marcus 2006). Strikingly, the FOXP2 gene appears to be highly conserved (Enard et al. 2002), with only three amino acid changes in the corresponding protein during the last 70 million years or so. Intriguingly, two of these changes occurred after the split between humans and chimpanzees, tentatively suggesting that they might have played a crucial role in the development of speech. Therefore, a large consortium of researchers led by Wolfgang Enard and Svante Pääbo from the Max Planck Institute for Evolutionary Anthropology in Leipzig set out to study the effects of this "humanized" version of the FOXP2 gene in a mouse model (Enard et al. 2009). We contributed to the acoustic analyses of the ultrasonic vocalizations of these animals. Somewhat disappointingly, but maybe not so surprising, the genetic modification had only minor influence on vocal output. The most striking difference was an increase in neural plasticity in the basal ganglia (Enard et al. 2009). While there remains a long way to go before we can track the pathway from single genes to specific features of communicative behaviors, this approach appears increasingly promising. For instance, mice that lack a specific gene that has been implicated in the development of autism spectrum disorders show little interest in conspecifics and show highly reduced vocal communication (Jamain et al. 2008). Taken together, careful behavioral observations of the communicative behavior of wild animals provides insights into what the animals really do, and what they don't—and studies involving genetically modified animal models contribute to our understanding of why the animals don't do certain things that we may have expected to find. Both avenues, as well as experimental studies of the cognitive abilities of monkeys and apes, both in the wild and in laboratory settings, are needed to fully comprehend the complexity but also the limitations of the communicative abilities of our closest living relatives.

Acknowledgments I thank Eckart Voland for his willingness to consider views of the world that sometimes differ radically from his own, the organizers of this volume for insisting on my contribution, Tabitha Price for comments on the manuscript, and Kurt Hammerschmidt for the many discussions we shared.

References

Arbib M (2005) From monkey-like action recognition to human language: an evolutionary framework for neurolinguistics. Behavioral and Brain Sciences 28:105–167

Arnold K, Pohlner Y, Zuberbühler K (2008) A forest monkey's alarm call series to predator models. Behavioral Ecology and Sociobiology 62:549–559

Arnold K, Zuberbühler K (2006) The alarm-calling system of adult male putty-nosed monkeys, *Cercopithecus nictitans martini*. Animal Behaviour 72:643–653

Arnold K, Zuberbühler K (2008) Meaningful call combinations in a non-human primate. Current Biology 18:R202–203

Bahlmann J, Schubotz RI, Friederici AD (2008) Hierarchical artificial grammar processing engages Broca's area. NeuroImage 42:525–534

Brumm H, Voss K, Koellmer I, Todt D (2004) Acoustic communication in noise: regulation of call characteristics in a New World monkey. Journal of Experimental Biology 207:443–448

Call J, Tomasello M (2006) *The Gestural Communication of Apes and Monkeys*. Lawrence Earlbaum Associates, Mahwah, NJ

Carey S, Bartlett E (1978) Acquiring a single new word. Child Language Development 15:29

Corballis MC (2002) *From Hand to Mouth*. Princeton University Press, Princeton, NJ

Corballis MC (2007) Recursion, language, and starlings. Cognitive Science 31:697–704

Crockford C, Herbinger I, Vigilant L, Boesch C (2004) Wild chimpanzees produce group-specific calls: a case for vocal learning? Ethology 110:221–243

Darwin C (1872) *The Expression of the Emotions in Man and Animals.* Murray, London

Deacon T (1997) *The Symbolic Species.* Penguin, London

Egnor SER, Hauser MD (2004) A paradox in the evolution of primate vocal learning. Trends in Neurosciences 27:649–654

Egnor SER, Iguina CG, Hauser MD (2006) Perturbation of auditory feedback causes systematic perturbation in vocal structure in adult cotton-top tamarins. Journal of Experimental Biology 209:3652–3663

Enard W, Gehre S, Hammerschmidt K, Holter SM, Blass T, Somel M, Bruckner MK, Schreiweis C, Winter C, Sohr R, Becker L, Wiebe V, Nickel B, Giger T, Muller U, Groszer M, Adler T, Aguilar A, Bolle I, Calzada-Wack J, Dalke C, Ehrhardt N, Favor J, Fuchs H, Gailus-Durner V, Hans W, Holzlwimmer G, Javaheri A, Kalaydjiev S, Kallnik M, Kling E, Kunder S, Mossbrugger I, Naton B, Racz I, Rathkolb B, Rozman J, Schrewe A, Busch DH, Graw J, Ivandic B, Klingenspor M, Klopstock T, Ollert M, Quintanilla-Martinez L, Schulz H, Wolf E, Wurst W, Zimmer A, Fisher SE, Morgenstern R, Arendt T, de Angelis MH, Fischer J, Schwarz J, Paabo S (2009) A humanized version of Foxp2 affects cortico-basal ganglia circuits in mice. Cell 137:961–971

Enard W, Przeworski M, Fisher SE, Lai CSL, Wiebe V, Kitano T, Monaco AP, Paabo S (2002) Molecular evolution of FOXP2, a gene involved in speech and language. Nature 418:869–872

Ey E, Hammerschmidt K, Seyfarth RM, Fischer J (2007a) Age- and sex-related variations in clear calls of Chacma baboons (*Papio hamadryas ursinus*). International Journal of Primatology 28:947–960

Ey E, Pfefferle D, Fischer J (2007b) Do age- and sex-related variations reliably reflect body size in non-human primate vocalizations—a review. Primates 48:253–267

Ey E, Rahn C, Hammerschmidt K, Fischer J (2009) Wild female olive baboons adapt their grunt vocalisations to environmental conditions. Ethology 115:493–503

Fichtel C (2007) Ontogeny of conspecific and heterospecific alarm call recognition in wild Verreaux's sifakas (*Propithecus verreauxi verreauxi*). American Journal of Primatology 70:127–135

Fichtel C, Hammerschmidt K (2003) Responses of squirrel monkeys to their experimentally modified mobbing calls. Journal of the Acoustical Society of America 113:2927–2932

Fischer J (2003) *Vokale Kommunikation bei nichtmenschlichen Primaten: Einsichten in die Ursprünge der menschlichen Sprache?* Habilitationsschrift. Universität Leipzig, Leipzig

Fischer J (2004) Emergence of individual recognition in young macaques. Animal Behaviour 67:655–661

Fischer J (2007) Categorical perception. In: Brown K (ed) *Encyclopedia of Language & Linguistics,* 2nd ed. Elsevier, Oxford

Fischer J (2008) Transmission of acquired information in nonhuman primates. In: Byrne JH (ed) *Learning and Memory: A Comprehensive Reference.* Elsevier, Oxford, pp 299–313

Fischer J, Call J, Kaminski J (2004a) A pluralistic account of word learning. Trends in Cognitive Sciences 8:481–481

Fischer J, Cheney DL, Seyfarth RM (2000) Development of infant baboons' responses to graded bark variants. Proceedings of the Royal Society of London, Series B 267:2317–2321

Fischer J, Hammerschmidt K (2001) Functional referents and acoustic similarity revisited: the case of Barbary macaque alarm calls. Animal Cognition 4:29–35

Fischer J, Hammerschmidt K, Todt D (1995) Factors affecting acoustic variation in Barbary macaque (*Macaca sylvanus*) disturbance calls. Ethology 101:51–66

Fischer J, Hammerschmidt K, Todt D (1998) Local variation in Barbary macaque shrill barks. Animal Behaviour 56:623–629

Fischer J, Kitchen DM, Seyfarth RM, Cheney DL (2004b) Baboon loud calls advertise male quality: acoustic features and their relation to rank, age, and exhaustion. Behavioral Ecology and Sociobiology 56:140–148

Fisher SE, Marcus GF (2006) The eloquent ape: genes, brains and the evolution of language. Nature Reviews Genetics 7:9–20

Fitch WT, Hauser MD (2004) Computational constraints on syntactic processing in a nonhuman primate. Science 303:377–380

Garner RL (1900) *Apes and Monkeys. Their Life and Language.* Ginn, Boston, MA

Gentner TQ, Fenn KM, Margoliash D, Nusbaum HC (2006) Recursive syntactic pattern learning by songbirds. Nature 440:1204–1207

Green S (1975) Variation of vocal pattern with social situation in the Japanese monkey (*Macaca fuscata*): a field study. In: Rosenblum LA (ed) *Primate Behavior*, vol 4. Academic, New York, NY, pp 1–102

Hammerschmidt K, Fischer J (2008) Constraints in primate vocal production. In: Griebel U, Oller K (eds) *The Evolution of Communicative Creativity: From Fixed Signals to Contextual Flexibility.* MIT Press, Cambridge. MA, pp 93–119

Hammerschmidt K, Newman JD, Champoux M, Suomi SJ (2000) Changes in rhesus macaque 'coo' vocalizations during early development. Ethology 106:873–886

Hauser MD (1988) How infant vervet monkeys learn to recognize starling alarm calls: the role of experience. Behaviour 105:187–201

Hauser MD, Chomsky N, Fitch WT (2002) The faculty of language: what is it, who has it, and how did it evolve? Science 298:1569–1579

Herder JG (1772) *Abhandlung über den Ursprung der Sprache.* Reclam, Leipzig

Hockett CF (1960) The origin of speech. Scientific American 203:88–96

Jamain S, Radyushkin K, Hammerschmidt K, Granon S, Boretius S, Varoqueaux F, Ramanantsoa N, Gallego J, Ronnenberg A, Winter D, Frahm J, Fischer J, Bourgeron T, Ehrenreich H, Brose N (2008) Reduced social interaction and ultrasonic communication in a mouse model of monogenic heritable autism. Proceedings of the National Academy of Sciences of the USA 105:1710–1715

Janik VM (2000) Whistle matching in wild bottlenose dolphins (*Tursiops truncates*). Science 289:1355–1357

Kaminski J, Call J, Fischer J (2004) Word learning in a domestic dog: evidence for 'fast mapping'. Science 304:1682–1683

Kaminski J, Fischer J, Call J (2008) Prospective object search in dogs: mixed evidence for knowledge of what and where. Animal Cognition 11:367–371

Kastak CR, Schusterman RJ (2006) Sea lions and equivalence: expanding classes by exclusion. Journal of the Experimental Analysis of Behavior 78:449–465

Kellogg WN (1968) Communication and language in the home-raised chimpanzee—gestures, words, and behavioral signals of home-raised apes are critically examined. Science 162: 423–427

Macedonia JM, Evans CS (1993) Variation among mammalian alarm call systems and the problem of meaning in animal signals. Ethology 93:177–197

Marler P (1955) Characteristics of some animal calls. Nature 176:6–8

Marler P (1957) Specific distinctiveness in the communication signals of birds. Behaviour 11: 13–39

Marler P (1961) Logical analysis of animal communication. Journal of Theoretical Biology 1: 295–317

Mitani JC, Hasegawa T, Gros-Louis J, Marler P, Byrne RW (1992) Dialects in wild chimpanzees? American Journal of Primatology 27:233–243

Pepperberg IM (2000) *The Alex Studies: Cognitive and Communicative Abilities of Grey Parrots.* Harvard University Press, Cambridge, MA

Perruchet P, Rey A (2005) Does the mastery of center-embedded linguistic structures distinguish humans from nonhuman primates? Psychonomic Bulletin and Review 12:307–313

Pfefferle D, Brauch K, Heistermann M, Hodges JK, Fischer J (2008) Female Barbary macaque (*Macaca sylvanus*) copulation calls do not reveal the fertile phase but influence mating outcome. Proceedings of the Royal Society of London, Series B 275:571–578

Pfefferle D, Fischer J (2006) Sounds and size—identification of variables that reflect body size in Hamadryas baboons. Animal Behaviour 72:43–51

Pinker S, Bloom P (1990) Natural-language and natural-selection. Behavioral and Brain Sciences 13:707–726

Radick G (2008) *The Simian Tongue*. University of Chicago Press, Chicago, IL

Seyfarth RM, Cheney DL (1980) The ontogeny of vervet monkey alarm calling behavior: a preliminary report. Zeitschrift für Tierpsychologie 54:37–56

Seyfarth RM, Cheney DL (1986) Vocal development in vervet monkeys. Animal Behaviour 34:1640–1658

Seyfarth RM, Cheney DL (1997) Some features of vocal development in nonhuman primates. In: Snowdon CT, Hausberger M (eds) *Social Influences on Vocal Development*. Cambridge University Press, Cambridge, MA, pp 249–273

Seyfarth RM, Cheney DL (2003) Signalers and receivers in animal communication. Annual Review of Psychology 54:145–173

Seyfarth RM, Cheney DL, Marler P (1980) Monkey responses to three different alarm calls: evidence of predator classification and semantic communication. Science 210:801–803

Smith WJ (1977) The Behavior of Communicating: An Ethological Approach. Cambridge: Harvard University Press

Snowdon CT, Elowson AM (1999) Pygmy marmosets modify call structure when paired. Ethology 105:893–908

Struhsaker TT (1967) Auditory communication among vervet monkeys (*Cercopithecus aethiops*). In: Altmann SA (ed) *Social Communication Among Primates*. University of Chicago Press, Chicago, IL, pp 281–324

Sugiura H (1998) Matching of acoustic features during the vocal exchange of coo calls by Japanese macaques. Animal Behaviour 55:673–687

Számadó S, Bishop D, Deacon T, d'Errico F, Fischer J, Hurford JR, Okanaya K, Szathmáry E, White S, Bickerton D (2009) *What Are the Possible Biological and Genetic Foundations for Syntactic Phenomena? Biological Foundations and Origin of Syntax*. MIT Press, Cambridge, MA, pp 207–236

Zuberbühler K (2002) A syntactic rule in forest monkey communication. Animal Behaviour 63:293–299

Illusion Number 2
We Are Independent of Our Sociobiological Roots

Chapter 5
The True Egoist Is Cooperative

Ethical Problems Seen Against a Background of Evolutionary Biology, Behavioral Research, and Sociobiology

Christian Vogel

Abstract Biological facts and moral norms are closely related to each other, although the one does not follow the other as night follows day. This always produces problems when an attempt is made to introduce newly discovered biological facts into moral and ethical arguments. Examples include social Darwinism, classical behaviorism, and sociobiology. Obviously, our ethics must reach beyond our natural inclinations—even be ready to come into conflict with them. Equally obviously, our "oughts" and our "ables" must remain bound up with one another. It is in this field of tension between facts and norms, "can" and "ought", that the following speculations lie.

5.1 Introduction

Insofar as they involve mankind, anthropology and behavioral research may be jeopardized by ideological influence in the fields of morality and ethics, since the social behavior of man has an unmistakable moral dimension. The closeness of the subject of the investigation to its object—man—makes that investigation virtually unavoidable. Personal self-awareness is inextricably mixed up with the work, at the same time inspiring and endangering it. The distinction between scientific motivation, formulation, degree of approximation, concepts, construction of hypotheses, and attempts at interpretation on the one hand, and nonscientific motives, illusion, valuation, conventions, desires, and ideologies on the other is less rigid than in most disciplines. This applies as much to cultural anthropology as to human ethology.

Christian Vogel is deceased

U.J. Frey et al. (eds.), *Homo Novus – A Human Without Illusions*,
The Frontiers Collection, DOI 10.1007/978-3-642-12142-5_5,
© Springer-Verlag Berlin Heidelberg 2010

5.2 "Normative Biologism" as a Hazard

The ideologically induced hazards emerge not only from within the research itself, they can also be provoked and introduced from outside. In an otherwise already disorientated spiritual world, belief in science seduces people again and again to the pious hope that "right" principles and moral norms for human society can be established or legitimized by scientific means. Such a belief forces evolutionary biologists, ethologists, and anthropologists to embrace the "naturalistic fallacy". They encourage sociopolitical ideologies, the application of factual knowledge to human conduct, and the wish to derive moral values and corresponding maxims from descriptions of nature. This has been given the name "normative biologism" by the American psychologist Donald T. Campbell. It constitutes a tempting delusion that is always ready to reappear, and is a constant ideological threat to our political security.

5.3 Social Darwinism

The classical example is the so-called *social Darwinism*, which arose immediately after Darwin's theory of natural selection in 1859. The temptation to apply the principle of *natural selection* to the social and political relationships of human beings has been present from the very beginning. There is no doubt that Darwin himself encouraged this development in his book *The Descent of Man* (Darwin 1982), although he was certainly aware of the ethical dangers of simply transferring it wholesale to the social life of men and women.

Put as briefly as possible, Darwin's descriptive concept runs something like this: all organisms have the ability and the tendency to increase exponentially and, if outside circumstances do not prevent them, this is what they do. If, however, the resources become short—which inevitably will happen with increasing population density—the participants in this competitive situation who have achieved a greater chance of survival and a higher reproductive rate become better adapted to prevailing conditions. If characteristics determining different degrees of adaptation are inheritable, natural selection will of course lead to greater adaptability in the next generation. In short, the directing power of evolution is natural selection, which manifests itself in the suitability or fitness associated with the different reproductive rates of different individuals.

With a few modifications, Darwin's theory of natural selection is accepted today by evolutionary biologists throughout the world. From the biological point of view it is a purely descriptive, not a normative theory. However, Darwin's theory quickly became normative, ideological, and therefore explosive in nature. Darwin himself borrowed the slogan "survival of the fittest in the struggle for existence" from Herbert Spencer, who saw this competition for limited resources as a kind of bloody battlefield on which the "more suitable", the "more efficient", the "better", and even the "more valuable" must survive. These are the original natural conditions that have produced the unique qualities of the animal man. The opposite to this is an "artificial" situation brought into being by culture and civilization, by which improved hygiene, advanced medical treatment, and the accepted moral necessity of cherishing the sick, the weak, and the poor have set aside the power of natural selection and thereby produced a degenerate population. This is an idea which to some extent touches upon Rousseau's *Discourse sur l'inégalité*, published in 1755. Starting from the proposition that natural selection brings about progress—a point of view not altogether endorsed by the modern evolutionary biologist!—one conclusion seems obvious: that every hindrance to natural selection must result in backsliding and degeneration. The cure for this must be an artificial return to natural selection, a doctrine which soon became the catchword of the eugenicist.

I should like at this point to emphasize that Darwin himself protested against such conclusions on moral grounds. Civilized people, he pointed out in 1871,[1] use their medical skills to keep the sick and the incapacitated alive, look after the poor, and enable the weak, bodily, and mentally or morally less gifted members of society to reproduce themselves.

Nobody, he insisted, who had any experience of breeding pet animals can doubt that this is extremely bad for the species itself. It is astonishing how soon insufficient or excessive care leads to the degeneration of a domestic breed, and nobody, except in the case of man himself, would be so crazy as to allow his least satisfactory animals to breed.

The fact that we do this, and must do this, with other human beings is a result of our deep-seated "sympathetic instinct". To ignore this instinct is to violate our noblest nature and sense of values. We have to live with the admittedly disadvantageous consequences of the preservation and reproduction of the weak. It is just this barrier which separates the follower of Social Darwinism from the eugenicist and the racial hygienist, who, like animal breeders, do not regard individual welfare as supremely important, but rather "the common good" and the "health of the corporate body", instead of "individual health" and "social hygiene", in order to preserve and possibly improve the "quality of the race".

In the early days of the politically ideological treatment of Darwin's theory a whole spectrum of hangers-on and prophets arose, showing all colors from the extreme left to the ultra-conservative. It is well-known how readily Karl Marx and Friedrich Engels took on the theory. For politically conservative Rudolf Virchow, Darwin's theory of descent was almost a socialist doctrine. Against this, the zoologist and social Darwinian biologist Ernst Haeckel claimed that Darwinism is anything but a socialist doctrine. If one wanted to attribute a particular political color to this English theory—which certainly would be possible—then it would be aristocratic. It is in no way democratic, and least of all socialist. The theory of natural selection teaches that in the life of man, as in the life of animals and plants everywhere and at all times, there is only a small privileged minority which is able to exist and flourish, whereas the vast majority starves and comes to a miserable and more or less untimely end (Haeckel 1902, cited by Altner 1984, p. 103f).

---

[1] Translator's note: All citations of English authors are paraphrases of the German rendering, and not direct transcriptions of the original text. Passages taken from German authors are intended only to convey the original meaning. No attempt has been made to produce a literal translation.

Furthermore, he continued, we wish to take this opportunity of mentioning how dangerous it is to carry such scientific theories over directly into the realm of practical politics (Haeckel 1902, cited in Altner 1981, p. 103f).

We observe how quickly and imperceptibly Darwin's theory of natural selection became incorporated into political ideology. The normative appeal to natural worth belongs here to the basic armamentarium of sociopolitical discourse. The principles of biological evolution provided social Darwinism and eugenics with a model for both the organization of society and for laying down rules of morality.

It is now evident that social Darwinism of all political flavors has violated three basic principles of Darwin's scientific theory of selection.

First and foremost, sociopolitical ideology has transferred direct competition from the level of the individual to that of tribes, peoples, nations, races, or social classes. Natural selection is now supposed to improve directly behavior that serves the needs of the people, race, or class, apart from any advantages or disadvantages that it may award to the individual concerned. From the point of view of modern evolutionary biology, that does not make sense at all. Darwin himself had certainly been unsure about it.

Secondly, the ideology of Darwin's once teleology-free concept of natural selection has become teleological. Survival and, particularly, the greater chances of reproduction enjoyed by the "more efficient" are no longer the inevitable result of natural selection, but a goal towards which organisms strive.

Thirdly, this "goal of selection" is interpreted morally and made into a value judgment. It becomes a sought-after and wished-for objective of political thinking and negotiation, following the aphorism "whoever acts in accordance with the goals of natural selection does well from both the biological and moral standpoints"—a classical example of the naturalistic fallacy, a leap from "is" to "ought"!

This sort of social-Darwinian thinking led to the eugenic idea of healing the "social body", which was supposed to replace the vanished, civilizing path of natural selection artificially. The eugenicist and geneticist Fritz Lenz, who had great influence both before and during the era of National Socialism, actually wished to abolish individualism as a *Weltanschauung*.

I would now like to consider two contemporary problems in behavioral research: ethical discussion in classical German ethology and in Anglo-Saxon social biology.

5.4 Classical German Ethology

In 1954, Konrad Lorenz developed the concept of "morally analogous" behavior in animals. This signifies an instinctive repertoire of behavioral patterns that, because of its survival value, arose by natural selection as a matter of survival expediency. He also meant that much of what we would happily attribute to our "rationally responsible morality" was, in fact, anchored deeply in our biogenetically developed "repertory of instincts". There is in this concept an important and revolutionary ideological significance that we would do well to examine more closely.

Self-sacrifice for the community is considered to be the highest virtue, selflessness being among the exemplary personal characteristics. "Service before self!", a principle which is written on all flags of "national" and state-oriented ideologies. The classical models for these exemplary aspirations have always been and will always be taken from nature. Since antiquity, the most obvious have always been the communities of bees and ants, with their apparently unselfish citizens. The total behavior of these individual animals completely and entirely corresponds to the norm of selflessness towards the state, and therefore towards all other representatives of the people (Hassenstein 1983, p. 77). Without selfless altruism towards the community there can be none of the well-lubricated functioning of social life upon which so many living creatures depend for their existence.

The classical ethology of the German tradition has always reasoned this way, the concepts of its evolutionary biology obviously providing a serviceable scientific basis. Classical ethology observed the species, and often, by extension, the population, a social organization, a state, or a "social body" as a "superorganism" standing above the individual and comparable to a complex living creature. Individuals more or less correspond to its cells, the families, guilds, castes, professional bodies, etc. to its organs. Their goal, objective, and purpose are the preservation and well-being of the body or the community—a point of view very similar to the subordination of the individual to the "social body" put forward by classical eugenics.

In 1955, Lorenz wrote that the individual, like the species, is a system made up of different parts or members, and in which each of these members exercises a two-way influence on all the others. This reciprocal relationship is regulatory since it maintains itself and, after being disturbed, strives to restore the original equilibrium again (Lorenz 1955, p. 105).

Therefore, the species is also a homeostatic, self-maintaining higher-order organism. As Konrad Lorenz pointed out, it is almost a biological truism to speak of a "survival expedient" that controls the events of evolution. This means that anything which benefits or serves survival is adaptive, whereas anything which does not is maladaptive. When necessary, the same argument can be applied to subspecies or races, and especially to populations and social units such as families, peoples, states, classes and the like. From this point of view natural selection must reward the sort of behavior that supports the survival of species and societies, but will inevitably work against anything which in principle damages the self-interest of these supraordinate units.

According to the *consensus gentium*, our morality must be orientated towards the common good. The precepts of classical ethology therefore teach us that our ethical system is aimed only in that direction which is rewarded by the natural selection of those social organisms that are evolutionarily rewarded by natural selection. Public welfare before self-interest! We desire survival, and therefore need to preserve the power of evolutionarily orientated ethics, wrote Eibl-Eibesfeldt. As members of a group, men must always keep the general welfare of their group in mind (Eibl-Eibesfeldt 1984). The moral "ought", apparently following the example of natural selection, favors the interests of supraordinate groups! Here also is the "social body" given priority over its individual members, as German eugenicists had observed as far back as the turn of the 20th century.

5.5 Morally Analogous Behavior

The "morally analogous behavior" of social animals plays a central role in the ethical discussions of Konrad Lorenz. In his famous work "Moral-analoges Verhalten geselliger Tiere" (Lorenz 1954), he emphasized the importance of the analogous functioning of many instinctive drives and inhibitions of animals to the "rationally responsible morality" of man.

He wrote in 1955 that the functional similarity between drives and inhibitions and the exercise of rational morality often made it difficult to decide whether the compulsion to a specific activity is to be attributed to the deepest pre-human layers of our personality, or whether it results from the cogitations of our highest reason. Since it is drummed into us from early youth on that the latter are of great value and the former virtually worthless, we tend to believe that reason is the basis of our actions that may be, in fact, merely due to healthy instincts (Lorenz 1955, p. 140).

This kind of "healthy" instincts may be the result of natural selection—genetically determined evolutionary adaptations—and such biological adaptations may be the essential background to our moral behavior. The semantic confusion between the words "adapted", "adaptive", and "healthy" is easy to see and carries the risk of becoming identified with "good", "right", and "worthwhile".

This danger is considerably increased by the fact that classical German ethology associates adaptive conditions with the typological concept of normality. Behavior that departs from this "normality" is then regarded as "deviant" or even pathological. For each group there is only one kind of generally accepted normal behavior, and no alternative strategy or mixture of strategies is considered. According to Lorenz, behavior is "normal" if it has been selected for the survival of the species in this and no other form.

The same author states, however, that we are not to understand the arithmetic average of all individual cases, but rather a "type" reconstructed from specific variations, which, for obvious reasons, is in fact seldom or never observed. People who possess the gift of perceiving form can immediately recognize ideal structure or behavior (Lorenz 1963, p. 277).

The "ideal type" becomes the "ideal norm" and achieves normative function. We recognize the ideal norm, judged as conformism, well enough, wrote Eibl-Eibesfeldt (Eibl-Eibesfeldt 1984, p. 867). In order to find the norm, it is important to think in terms of survival value (Eibl-Eibesfeldt 1984, p. 883).

According to Lorenz, "morally analogous behavior" among animals shows itself in various forms of "loyalty" towards other members of the group; in the readiness of a socially superior individual to serve (and even on occasion at his own risk to champion) an inferior member, as well as in the rules that govern ritualized token fights (attack only begins when the other party is ready, the most dangerous weapons are not used, and the "submissive posture" is accepted as ending the fight), recognition of the "possessions"—food or sexual partner—of a "chum" and, last but not least, the natural reluctance to kill other members of the same species (*natürliche Tötungshemmung*), which was discovered by Konrad Lorenz, and which only man, considered as a "civilized" being, has largely forfeited. All these behavioral patterns have been developed to ensure survival of the species (Lorenz 1954).

If, under the conditions of natural selection, behavior that *serves* the species is fundamentally "adaptive", "normal", "healthy", and therefore "right" and "good", then all behavior that *damages* the species must naturally be interpreted as "maladaptive"—as "abnormal", "deviant", "pathological" and therefore also "wrong" and "bad". Members of a group who behave in such *abnormal* fashion can hardly be tolerated any longer as members.

Eibl-Eibesfeldt, for instance, asserts that those individuals in the animal kingdom breaking "tournament" rules within their own species, and which thereby resist in a harmful way the restraining nature of those rules, are treated as outsiders by the group (Eibl-Eibesfeldt 1984, p. 876).

It is alleged that, here within the animal kingdom, rules are put into effect that serve the best interests of the group and/or species, and take on a normative function. Indeed, this type of behavior has virtually become acclaimed among human beings as exemplary. Service beneficial to the community is biologically adaptive and thus "good", selfish behavior is maladaptive and "morally inferior". This was also the point of view of German eugenicists, of racial hygiene, and National Socialist ideology. It was in connection with the "antisocial" that Konrad Lorenz made his fatal statement:

> From the far-reaching biological analogy of the relationship between a neoplastic growth and the body on the one hand, and between the member of a community who has "dropped out" to become antisocial and the community itself on the other, it is possible to derive the significant parallels in the necessary precautions . . . every attempt to rehabilitate those fallen away from complete acceptance being therefore hopeless. Fortunately, their elimination is easier to accomplish for a "healer of the people"—and the effect upon the social body less dangerous—than a surgical operation on a single individual. (Lorenz 1940, p. 69)

We know only too well to what kind of ideology and social politics this false understanding of nature led: "You are nothing! Your community is everything!" Racially or nationally "inferior" individuals are isolated as "foreign parasites" and eliminated.

5.6 Normative Consequences

Two further points of view associated with Lorenz must be considered.

If natural selection automatically produces altruistic behavior to support survival, then our rational morality should be orientated towards this example. This becomes ever more necessary the further civilized man distances himself from natural man; and because of this, there will be decay in natural morality, against which our rational morality is bound to react.

If it is some genetically programmed species-specific mechanism which in this way brings about "morally analogous" behavior in support of survival, then it must apply universally and equally to all members of the species. It cannot be variably distributed within one species. All members of the species obey the same genetic program: they are treated all equally, the principle of equality being thus widely guaranteed. In fact, many ethologists believe that man is the first to have violated this principle.

In the course of man's early cultural development one inborn prohibition prevailed, equally applicable to all members of the species, "thou shalt not kill any member of the species". Out of this another moral rule was developed: "thou shalt not kill any member of the group, although killing strangers is perfectly permissible, at least under certain special conditions: we are true men, the others do not seem to be!" (Leyhausen 1974, p. 228). That is how Paul Leyhausen formulated the custom, practiced since human thought began. He saw it as a purely cultural product after natural conditions had been lost. In accordance with this, he hopes that the natural, biologically adaptive, and survival-promoting condition of *Tötungshemmung* with regard to other members of the species could, in some not too distant future, become artificially reconstituted. Here also, the putative natural conditions of "morally analogous" behavior serving survival are seen as something that has been forfeited, so that the "degenerative" deficiency must be made up for intellectually. Natural morality is a profoundly selfless morality that serves the needs of the community.

People in general and even some scientists cling to the belief that human behavior that serves the community rather than the individual must be the result of responsible conscious reasoning. This view is demonstrably false (Lorenz 1963, p. 228).

Natural selection and the human concept of moral responsibility are directed towards the same goal: the optimization of altruistic behavior in the service of the community!

5.7 First Objections

Modern evolutionary biology has completely lifted the "moral ideology" of classical German ethology off its hinges. The belief that biological species, populations, or societies are autonomous, self-regulating, and therefore homeostatic systems, steered by the reproduction of their single elements—the individuals—in the interests of survival or the preservation of society, has been disproved. The sole principle that is the basis for reproductive competition is that of competition between individuals. How, then, is it possible for an individual to affect such a population genetically, when all reproductive advantages must be surrendered in favor of service to the common good? All those others who selfishly achieve advantage will outstrip their "altruistic" fellows. So far as biogenetic reproduction is concerned, individual welfare takes precedence in the natural evolutionary process over community welfare. The deciding factor in evolution is individual selection, and in the end it is not a matter of species, populations, or societies, but of the successful and above average ability to hand on genes to the next generation. Relinquish the selection principle that looks towards the survival of the species and towards the common good and the concept of "morally analogous" behavior among animals must itself be discarded.

These reflections and the empirical evidence for them lead to the conclusion that natural selection is unable to bring about any altruistic behavior in the genetic sense. It must be accepted that the biological evolutionary process knows nothing of morality. The events of evolution are absolutely indifferent to moral considerations.

5.8 Sociobiology

Sociobiology is regarded in lay circles as another form of normative biologism. Many scientists therefore avoid this term and prefer to speak of "the evolutionary biology of social behavior", their scientific beliefs being based entirely upon Darwin's theory of natural selection.

The directing power of evolution according to this theory is natural selection, which depends upon the success of the organism in competition against a background of limited resources. This implies that those individuals are evolutionarily successful who can introduce a greater number of genes into the next generation of their immediate descendants than would be achieved by chance. Evolutionary fitness, as Darwin called it, depends upon the characteristics (and particularly the behavior) of each individual. In other words, natural selection "values" the individual (or in more precise genetic terminology, the individual phenotype) and selects according to its individual qualities. The social biologist Richard Dawkins, slightly tongue in cheek, described the individual as the "vehicle of the gene" or as its "retail distributor" (*Ausbreitungsmaschine*), by which the spreading of the gene naturally depends upon the spreading of its power of replication and the efforts of its "vehicle" (Dawkins 1978). For evolution, however, it is not the fate of this "vehicle"—the individual—that matters, but the suprapersonal survival of following generations. Furthermore, what survives throughout the generations is not a collection of individuals but the gene itself, and this means the replication and further replication of its structural and behavioral programs. It follows without further discussion that these replications or genetic programs have a better chance of spreading throughout future generations the more their "vehicles" can help succeeding generations to grow up and survive, and for this reason it is the behavioral programs that are decisive.

The following paradoxical question had already presented itself to Charles Darwin: how is it possible for cooperative or "altruistic" behavior—behavior that supports others at a cost to the active individual—to arise on the evolutionary basis of ever-present competition, even allowing for the fact that the existence of social animals shows that this frequently takes place?

The sociobiologists have been the first to offer a convincing answer to this question!

5.9 Inclusive Fitness

If genetic programs propagate successfully, this need not be the result of each individual reproduction, which would imply a highly personal or Darwinian "fitness". It is more in this way: the successful distribution of the programs also serves the behavior of their individual bearers, since the reproductive efforts of close relatives are also supported. The inherited programs become fixed in other individuals because of their common ancestry. The probability of this is greater for parents, children, and siblings than for grandparents, grandchildren, uncles, aunts, first cousins, and the like. Natural selection does not therefore take into account only those behavioral genetic programs that help single individuals to multiply, but also those that bring about a higher rate of reproduction in their close relatives. It can under certain circumstances—still in the interest of the allelic genes, of course—even be phylogenetically advantageous to sacrifice temporarily or permanently the individual's chance of survival in order to contribute to that of a close relative. This is "altruistic" behavior in the sociobiological sense: surrendering individual reproductive advantages in favor of another's—in this case a close relative. Evolutionary events obviously affect not only the personal or Darwinian fitness mentioned above, but also the increase of *inclusive fitness*, which is orientated towards the genes, and which is concerned not only with personal reproductive success but also with that of blood relatives in order of their proximity. This is the deciding factor that causes animal societies to be built up almost entirely on the principle that "blood is thicker than water". It is therefore from the point of view of biological evolution virtually predictable that such "nepotistic" behavior will be found in human societies, and that the pattern of graded kinship plays a central role in the form and intensity of social relations.

I must emphasize that this supporting of close kin is to be regarded as "genetically selfish", and that it does indeed facilitate the distribution of its own inherited programs.

5.10 Reciprocal Altruism

There is a second way in which cooperative and "altruistic" behavior could have arisen and become established through the strict mechanism of natural selection, and one which is entirely independent of genetic kinship. It was described by Robert Trivers as *reciprocal altruism.*

The prerequisite for this evolutionary principle is that the (reproductive) loss due to an altruistic action does not exceed its (reproductive) gain. The likely way of achieving this is by the short-term beneficiary of the altruistic action repaying it by a similar action when the opportunity arises. This type of relationship is known to occur in many mammalian groups. The efficiency of such a direct or indirect reciprocal relationship increases the longer the social partners live together in conditions of mutual trust; and in just those circumstances that are typical for the higher primates, including man. One can perceive that this is also, in the final analysis, a principle of "genetic selfishness", rather after the style of the aphorism "the true egoist is cooperative", with which we are indeed familiar from many theoretical models (Hofstadter 1983, p. 10).

The outcome of these considerations and the manifold empirical examples prove that natural selection is *ipso facto* incapable of bringing about (1) any kind of unselfish "altruism" in the genetic sense, or (2) any purely egalitarian or selected "altruistic" behavior, or reciprocal dependability, which is not associated with either the degree of kinship or a mutual interdependence.

This is, of course, diametrically opposed to the doctrines of classical German ethology. The matter has been summarized by Richard Alexander: up to now, no sign of a true altruism reaching out to include the whole species or population has been found in any living creature (Alexander 1983, p. 166). This is not the result of a biologically degenerating human civilization, it is an ancient component of our nature!

5.11 What Does This Mean for Us?

What do the conclusions of the sociobiological perspective mean for our ethical systems?

Start with the fact that our nature simply does not contain—any more than morally analogous behavior—the seeds of responsible ethics reaching out to include the whole of mankind and the accompanying environment. With this, the basis for a naturally double morality is laid down, a moral double standard that distinguishes between near kinship and membership of one of the sexes. In many cultures, "male" and "female" signify two utterly different worlds, each with its widely differing customs and rules. The corresponding double morality (which in our society also has not been disposed of, in spite of all our efforts at emancipation!) is culturally superimposed upon biological reproductive strategies specific to the two sexes. This conflict—this "war of the sexes"—is historically as old as the necessarily cooperative compromise required for sexual reproduction.

The reciprocal dependability and finely graded preferential treatment of the next of kin is already reflected in Henry Sidgwick's book *The Methods of Ethics*, published in 1874.

> We are all of the opinion that everyone has a duty to be kind and helpful to his parents, wife and children, and also to other relations (albeit to a lesser degree), to those he has accepted into his close circle and to neighbors and fellow countrymen, more than to other people, and to his own more than to other races; in general, indeed, to all people, the more so the closer they are. (Sidgwick 1874, cited in Patzig 1984, 31:680)

The other side of the coin is represented by the increasing aversion, refusal, exclusion and aggression which appear at increasing genetic distances. "I against my brother; my brother and I against our cousins; my brother, my cousins and I

against those who are not related to us; my cousins, my friends and I against our enemies in the village; all of us together with the whole of the village against the next
village" is an ancient Arabic proverb. One finds this gradation all over the world;
it is obviously not a result of civilized decadence! Peter Kropotkin denounced this
double morality in 1902:

> Therefore life in the raw is divided into two lines of action and appears in two different
> ethical forms—the relationships within a family and relationships outside the family; and
> intertribal justice (like our popular justice) is very different from general "Justice". If it
> comes to a fight, it is the most excessive cruelties that call forth the highest admiration of
> the family. This double morality penetrates the whole of mankind and is still extant today.
> (Kropotkin 1975, p. 115)

Against this, the proclamation of Human Rights, binding all people together,
including all peoples and races equally as one brotherhood, seems rather far from
nature. It is this aspect that classical German ethology interpreted very differently.

Sociobiology also lies under the threat of normative biologism, and particularly
when the empirical "is" is recast into the normative "ought". I am struck by a drastic
example from the political right in West Germany:

> What is required is the development of a human morality derived from that which has
> proved itself to be successful for organic life during the hundreds of thousands of years
> of its existence; i.e. "ingroups" and "outsiders" with altruism inside one's own group, and
> defence or indifference towards foreign groups of the same species. The human race has
> evolved by means of group selection; its death will be brought about by neither atomic
> catastrophes nor population explosions, but by a pre-scientific morality which conceives of
> all men as brothers, regardless of which part of the world they come from. This is anti-
> evolutionary, even though it corresponds (in earlier centuries admittedly more in precept
> than in practice) to occidental ethics. In short, if our attitude here remains unchanged, the
> matter will, in the evolutionary sense of the word, be disposed of. (Rieger 1987)

There is one thing I wish to emphasize beyond all possible doubt. Morality and
ethics do not require to be justified by biological evolution; and if they did, it would
certainly not be possible, since biogenetic evolution has no moral dimension. To this
extent there is no chance of giving human moral law a scientific basis.

Obviously our ethics must go beyond our natural inclinations. There is no legitimate way of getting from the indicative "is" condition of our nature to the imperative
"ought". On the other hand, it cannot be overlooked that "ought" and "can" must be
bound together if we are not to be content with an ethic that is merely postulated.
The philosopher Günther Patzig put it like this:

> Moral principles can be culled from within pure reason; moral rules for the behavior of real
> men and women and institutions of justice must fit in with human nature. (Patzig 1984,
> p. 684)

We are not able to remove from our genome the behavioral tendencies that we
have acquired though natural selection. We have, however, enough freedom of judgment and thinking capacity to redirect these tendencies again and again into new
beginnings.

Acknowledgements English translation by Joyce and Francis Steel (revised by Ulrich Frey).

References

Alexander RD (1983) Biologie und moralische Paradoxa. In: Gruter, M, Rehbinder, M (eds) *Der Beitrag der Biologie zu Fragen von Recht und Ethik*. Duncker & Humblot, Berlin

Darwin CR (1982) *Die Abstammung des Menschen [The Descent of Man* 1871], 4th ed. Kroner, Stuttgart

Dawkins R (1978) *Das egoistische Gen*. Springer, Berlin [Translation of: Dawkins R (1976) *The Selfish Gene*. Oxford University Press, Oxford]

Eibl-Eibesfeldt I (1984) *Die Biologie des menschlichen Verhaltens. Grundriss der Human-ethologie*. Piper, Munich

Haeckel E (1902) Deszendenztheorie und Sozialdemokratie. In: *Gemeinverständliche Vorträge und Abhandlungen aus dem Gebiet der Entwicklungslehre*, 2nd ed."Gesammelten populären Vorträge", 1st & 2nd issues, Emil Strauss, Berlin [Cited in Altner G (1981) *Der Darwinismus: Die Geschichte einer Theorie*. Wissenschaftliche Buchgesellschaft, Darmstadt]

Hassenstein B (1983) Evolution und Werte. In: Riedl RJ, Kreuzer F (eds) *Evolution und Menschenbild*. Hoffmann und Campe, Hamburg

Hofstadter DR (1983) Metamagikum. Kann sich in einer Welt voller Egoisten kooperatives Verhalten entwickeln? Spektrum der Wissenschaft 8:8–14

Kropotkin P (1975) *Gegenseitige Hilfe in der Tier- und Menschenwelt*. Ullstein, Frankfurt

Leyhausen P (1974) The biological basis of ethics and morality. Science, Medicine and Man 1:228

Lorenz K (1940) Durch Domestikation verursachte Störungen des arteigenen Verhaltens. Zeitschrift für angewandte Psychologie und Charakterkunde 59:2–81

Lorenz K (1954) Moralanaloges Verhalten geselliger Tiere. Forschung und Wirtschaft 4:1–23

Lorenz K (1955) Über das Toten von Artgenossen. In: *Jahrbuch der Max-Planck-Gesellschaft*. Max-Planck-Gesellschaft, Göttingen

Lorenz K (1963) *Das sogenannte Böse*. Borotha Schoeler, Vienna

Patzig G (1984) Verhaltungsforschung und Ethik. Neue deutsche Hefte 31:675–686

Rieger J (1987) Book review. Neue Anthropologie 15(3):74

Sidgwick H (1874) *The Methods of Ethics*. Macmillan, London [Cited by Patzig G (1984) Verhaltungsforschung und Ethik. Neue deutsche Hefte 31:675–686]

Chapter 6
The Social Brain and Its Implications

Robin Dunbar

Abstract We sometimes imagine that the human social world is unlimited, a product of our species' unique capacity for culture. In fact, it turns out to be a great deal more restricted in size than we suppose. In this respect, we humans are simply part of a continuum of social complexity that runs through the primates. Our social world is (a little) bigger than that of other primates only because we have a (proportionately) larger brain. Even so, it is still a very small-scale world—our natural social world is no more than about 150 people, even when we live in modern cities. There are nonetheless some aspects of the human condition that are unique. These mainly have to do with the fact that ensuring the cohesion of groups as large as those we find in modern humans has required new mechanisms. And it is here that culture has played its biological role.

6.1 Introduction

We humans are as much the product of evolution as any other species. As a result, what we are and what we do is as much a product of our evolutionary history as is the behavior and anatomy of the earthworms on which Darwin lavished so much time, energy, and literary composition. Although there has been a longstanding fashion for seeing humans as set apart from other animals by virtue of their capacity for culture, we should not be misled into assuming that, however different culture might seem from the more lackluster behavior of other animals, human culture somehow divorces us from our biological and evolutionary roots. Far from it, for our capacity for culture is itself just as much the product of the processes of Darwinian selection acting in our past as the behavior of the humblest earthworm or barnacle.

That said, we must also beware the converse temptation of seeing humans as "nothing but" ordinary animals. I am happy to accept that much of the motivational

R. Dunbar (✉)
Institute of Cognitive & Evolutionary Anthropology, University of Oxford,
OX2 6PN Oxford, UK
e-mail: robin.dunbar@anthro.ox.ac.uk

U.J. Frey et al. (eds.), *Homo Novus – A Human Without Illusions*,
The Frontiers Collection, DOI 10.1007/978-3-642-12142-5_6,
© Springer-Verlag Berlin Heidelberg 2010

and cognitive underpinnings of what humans do is not really so different from what we find in many other advanced vertebrates, and especially other anthropoid primates. I am, however, much less impressed by the claims that other species share a meaningful capacity for culture with humans in the way that some have argued (see for example McGrew 1992; Whiten et al. 1999; Rendell and Whitehead 2001; van Schaik et al. 2003; Whiten 2005). I don't doubt for a minute that most of those species on behalf of which such claims have been made (mainly great apes and cetaceans) have the capacity for cultural transmission, perhaps even full blown imitation, but that is manifestly not the same thing as attributing to them full human culture.

Culture, ultimately, is about the *meaning* that is attached to elements of behavior or ritual, and that is almost certainly about the kinds of cognition you can bring to bear on the matter and not about the way you learn things. The capacity to associate meaning with behavior is something that almost certainly requires higher orders of social cognition (in particular) than any animal other than humans possesses. The mode of transmission is surely interesting in a generic sense (and it is comforting, from a biologist's perspective, that these capacities do have some kind of phylogenetic continuity), but it is not what culture is about in humans or why culture makes a difference to humans—and, by extension, why it enabled humans to become so successful, assuming that we can attribute our success as a species to the products of culture.

My point is, rather, that humans have created levels of social embeddedness that do not, as a matter of evident fact, exist elsewhere, and indeed cannot exist elsewhere because they are crucially dependent on aspects of cognition that are unique to humans. That is something we have to explain. Those traits we share in common with other species require no explanation, but those by which we seem to differ do need some kind of explanation as to why humans, and only humans, possess them. And that explanation must nonetheless form part of the panoply of evolutionary processes that make up the Darwinian universe. We cannot plead special creation: we cannot imagine that humans were magically touched by the wand of special circumstances. However special human capacities might seem, they are necessarily the product of everyday evolutionary processes, and we must be able to show why that is so. I shall suggest that the core to the explanation we seek lies in the social brain hypothesis, a unifying theme in the evolution of primate cognition and primate sociality.

So let me first begin by sketching out the background that is common to both humans and other animals in terms of brain evolution. Having outlined the social brain hypothesis, I will then try to show why humans are unique.

6.2 The Social Brain Hypothesis

The social brain hypothesis provides a generic explanation for the evolution of large brains in higher vertebrates (birds and mammals). In its current incarnation (Shultz and Dunbar 2007; see also Pérez-Barbería et al. 2007), its central tenet is that the cognitive demands of pair bonds selected for increased brain size (and, specifically, neocortex volume) across the birds and mammals (as evidenced by the fact that species that mate monogamously have significantly larger relative brain/neocortex volumes than species that mate promiscuously or polygamously). In a very small number of mammalian taxa (but mainly primates), the cognitive mechanisms involved in managing pair-bonded relationships seem to have been generalized to other members of the group, creating what are in effect "friendships" (Silk 2002; Shultz and Dunbar 2007). As a result, these taxa exhibit a different pattern of brain size distribution: neocortex volume correlates with social group size (Dunbar and Shultz 2007). Although it is still possible to detect a residual influence due to monogamy in primates (Dunbar 2010), this monogamy signature is overridden by the very strong effect due to group size per se and is not obvious on superficial examination of the data. We assume that the same kinds of cognitive mechanisms underpin both pair-bonded (i.e., reproductive) relationships and (nonreproductive) close friendships, although this has yet to be established.

The fact that group size correlates with relative brain (or neocortex) size in primates implies, under the social brain hypothesis (formerly the Machiavellian Intelligence Hypothesis, as originally conceived by Byrne and Whiten 1988) that there is some kind of cognitive constraint on group size. Strictly speaking, we do not know what this constraint might be (i.e., whether it is a constraint of memory or of the manipulation of information about relationships) or, indeed, where the constraint bites.

On the first point, we can be fairly sure that memory per se is not the issue (though obviously there must be some kind of memory component since past interactions with specific individuals must form a core element in all relationships). More importantly, it is clear, at least from work on humans, that memory for faces and names far exceeds the limits on the size of the personal social network (i.e., the number of individuals with whom one has regular personalized social relationships). But whether the constraint is purely one of time invested in a relationship or whether it has a cognitive component in terms of how individuals build relationships in their minds remains unclear.

On the question of where the constraint acts, there are at least three possibilities. One is that the limit is set by the total number of relationships an individual can handle and keep track of as these change dynamically over time (i.e., as individuals fall in and out of "friendship" with each other). A second possibility is that it is not the total number of relationships (in the sense of total group size) that sets the limit, but rather a subset of those relationships. Lindenfors et al. have shown that, across primates as a whole, relative neocortex volume is better predicted by the number of adult females in the group than by the number of adult males or by total group size, thereby providing some prima facie evidence in support of the suggestion that it is only *some* relationships in the group that individuals worry about (Lindenfors et al. 2007). Since females are the core to most primate groups, then it is natural that it should be female relationships that set the limit, and the group as a whole is simply an emergent property of this subset. This is not unreasonable, since each female will have a species-typical number of offspring associated with her and there is good evidence to suggest that the number of males in a group is a simple linear function of the number of females (Dunbar 2000). The third possibility is that the cognitive constraint is imposed by some even smaller subset of the group. In this case, the group as a whole is an emergent property of the effectiveness with which this subset is bonded. Whereas the second suggestion is driven by the conventional theoretical supposition that females form the core of primate social groups, and males and infants are just hangers-on, this third proposal arises from an explicit hypothesis about the internal dynamics of primate social groups and follows from the finding by Kudo and Dunbar that, in primates, grooming clique size correlates both with total group size and with relative neocortex size (Kudo and Dunbar 2001). The reason why this might hold is more complex and has to do with the particular nature of primate sociality (for details, see Dunbar 2010), so I will summarize the argument.

In a nutshell, primates form groups to solve the ecological problems that they face (foraging, predation risk, etc.), and the size of group that is optimal for each species reflects the balance of selection pressures that they are under. However, forming groups incurs consequential costs, both in terms of ecological competition and in terms of female reproductive biology (female fertility declines with increasing group size because increasing levels of harassment result in unstable female menstrual cycles). This, in effect, gives rise to a trade-off between the costs and benefits of large groups that the animals have to resolve if they are to be able live in larger groups. Primates solve this problem by forming coalitions, which buffer them against these costs (Dunbar 2010). The key issue at this point is the suggestion that how well the coalitions function determines how well the overall group can be held together as a coherent entity, with its corollary that the size of the clique then determines overall group size, as shown by Kudo and Dunbar (Kudo and Dunbar 2001; see also Lehmann et al. 2010).

I suspect that this third explanation is in fact the right one, not least because our work on human social networks suggests likewise that most people's social world is actually focused on a small subset of individuals (typically about 15) (Roberts et al. 2009). More importantly, the data on the size of grooming cliques within primate species reveals that they typically have an inverted-U shaped distribution when plotted against total group size (Dunbar 1984; Kudo and Dunbar 2001; Dunbar 2003). Figure 6.1 illustrates this with data from macaques (genus *Macaca*) and gelada (*Theropithecus gelada*). In both cases, clique size increases as group size increases, but then reaches an asymptotic value after which it declines. Despite this, the amount of time devoted to grooming by individual animals increases linearly with group size, both within and between species (Lehmann et al. 2007), so the decline in clique size in larger groups is not due to animals spending less time grooming. Rather, it is because they are concentrating an increasingly higher proportion of their grooming time on fewer individuals. Similarly, Lehmann et al. (2010) found that, among the cercopithecine primates, social groups become more highly structured and less interconnected as their size increases, apparently reflecting the fact that coalitions become increasingly more important as the stresses imposed by group size become

Fig. 6.1 Mean grooming clique size (defined as the number of other females that account for more than 10% of an individual's grooming bouts) for (**a**) individual groups of *Macaca* and (**b**) individual reproductive units (harems) of *Theropithecus gelada*. Source: Kudo and Dunbar 2001

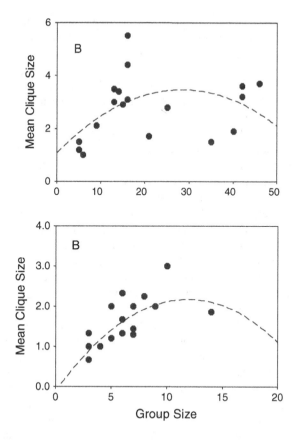

more intrusive. This point is nicely illustrated by what happens when a female's time budget comes under pressure from the demands of lactation and she is forced to sacrifice social time in the interests of increased feeding time demand. Among gelada baboons, for example, lactating females withdraw from casual grooming interactions and concentrate what time they have available on their core grooming partners when the infant's demand for milk starts to force them to withdraw time from social interaction (Dunbar and Dunbar 1988).

I interpret this as reflecting the fact that coalitions (i.e., grooming cliques) have to function much more effectively and reliably in very large groups because of the pressure that individuals (especially females) are under when they are surrounded by more competitors, and this means that they have to invest more time in their core relationships. This is a simple consequence of the limits on an individual's time, and the amount of time required to create a relationship of a particular quality and stability (see Seyfarth 1977). This inevitably means that individuals' interactions are less widely distributed and the bonds that hold the group as a whole together are necessarily less intimate. This has two important implications. First, it must inevitably place an upper limit on the size of group that can be held together as a coherent social entity. Second, some mechanism is needed to overcome this constraint if group size is to be able to increase beyond this limit.

At present, we have very little idea how this constraint works (and hence what the natural limits on group size might be) or how different species have managed to solve the problem. However, there seems to be a natural phase shift among the monkeys between those constrained to living in small social groups (most South American monkeys and the less social Old World monkeys such as the colobines) and those species where large social groups are more typical (mostly the baboon/macaque/guenon taxon). Significantly, this latter group is also the one characterized by a conspicuously intense form of matrilinearity in which females are not only philopatric but an individual's grooming relationships are confined almost entirely to a small subset of close female relatives. Presumably, these species have managed to find a way through this particular glass ceiling that allows them to support a higher level grouping. We may assume that glass ceilings of this kind reoccur at successive intervals of increasing group size. Indeed, it may be precisely this problem that creates the regular layering pattern that we find in both multilevel social systems among humans as well as other mammals (see Zhou et al. 2005; Hill et al. 2008).

If so, this might explain the fact that primate societies typically have a multi-level structure, although the number of layers may vary between species. This is obvious in species that have the kind of fission–fusion sociality found in species such as hamadryas and gelada baboons. Here, small family-like harems, which themselves consist of subsets of grooming-based coalitions, gather together in higher level units (clans), several of which in turn form an upper tier of social organization (the band). Minimally, these species have a four-tier structure. A more general analysis of grooming coalitions suggested that, among primates in general, a three-tier structure was common: Kudo and Dunbar found that grooming cliques were scaled to total group size by a factor of 9 (Kudo and Dunbar 2001), suggesting that there was an intermediate hidden layer that would yield the standard scaling ratio of 3 between successive layers identified by Hill et al. (2008).

There is some suggestion that this multilayering might in fact be part of a very general phenomenon, at least within the primates. If we plot a cumulative distribution for primate group sizes (see Fig. 6.2), then the overall rather nice Zipfian power law distribution has a number of quite distinct phase shifts that indicate a change in intercept. The data suggest that these phase shifts occur at group sizes of about 15, 30, 50, and 150—values that correspond very closely with those identified by Zhou et al. as forming a natural sequence of grouping levels in human social networks (Zhou et al. 2005). The only exception is the 30 grouping level, although this does seem to feature in both the distribution of ethnographic group sizes and ego-centric networks (Hill and Dunbar 2003) even though it remains hidden within the fractal analysis of Zhou et al. (2005) (probably because fractal analysis will only pick up signatures for naturally scaled patterns, and 30 is not a natural multiple of the base unit of 5).

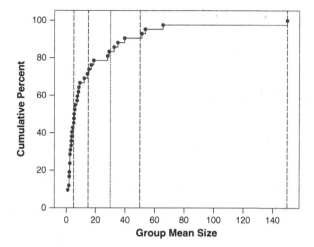

Fig. 6.2 A cumulative plot of primate group sizes, showing discrete phase shifts in the regression line that appear to correspond to known grouping layers in human groups. The *right-hand* data point is for humans. *Dashed lines* indicate the successive grouping layers at 5, 15, 50, and 150 identified in natural human networks by Zhou et al. (2005). The *dotted line* marks a group size of 30, since there appears to be an additional natural break at this point. The step changes at the phase shifts are marked by the horizontal segments in the cumulative distribution. Source: group size data from Dunbar (1992a), except for revised group sizes for semi-solitary prosimians based on nest group sizes (from Bearder 1987 and Kappeler and Heymann 1996); group size for humans is taken to be 150, following Dunbar (1992b)

There are two interesting implications of these results. One is that there is a natural scaling pattern in primate group sizes with a series of glass ceilings. In other words, you can increase group size as ecological requirements dictate, but only up to a limit. This limit is, presumably, set by the animals' capacities to bond with each other, thereby creating coherent, socially stable groups. In order to break through that limit, you need to solve the bonding problem. Since, in primates at least, bonding appears to be a function of two core components—a time element (generally reflected in grooming and the endorphin surge generated by grooming) and a cognitive element (reflected in the social brain relationship)—it is not entirely obvious which mechanism imposes the constraint or why it does so. The second implication is that social complexity apparently arises by adding successive layers to the network. It is not so much that there are species differences in social group size as such, but rather that there are species differences in the number of layers you can have in the network. Because the scaling ratio is the same (approximately 3) across both humans and mammals, Hill et al. assumed that the differences in total group size (the outermost layer) must be due to the fact that each species had a different baseline group size for its innermost layer, such that the universal scaling ratio then produced different final group sizes after the same appropriate number of layers (Hill et al. 2008). In fact, the data in Fig. 6.2 suggest that the differences between species might actually lie in the number of layers they can sustain as coherent units, with each of the layers being of pretty much the same size across all species. In other words, all primates have a typical standard inner-core grouping layer of about 5 individuals, more or less irrespective of mating system and social type.

6.3 The Bonding Problem in Human Communities

It turns out that humans fit the general distribution of the primate social brain graph very neatly. The predicted group size for humans (based on the ape grade equation from the brain-size/group-size graph) is approximately 150, and 150 turns out to be an extremely common value in human social groupings (Zhou et al. 2005; Hamilton et al. 2007; Dunbar 2008). It is also the typical size of human ego-centric (or personal) social networks, defined as the number of people with whom you have a personalized relationship (Hill and Dunbar 2003; Roberts et al. 2009). Human social networks, like those of primates, are highly structured, containing a series of layers that increase in size but decline in relationship quality (Hill and Dunbar 2003; Zhou et al. 2005; Roberts et al. 2009). Detailed analyses of a large sample of such networks suggest that the inner two layers of the network (roughly 5 and 15 individuals in size) behave very differently from the outer two layers (Roberts et al. 2009): relationships with inner-layer members are intrinsically more intimate and based on regular interactions over time. Outer-layer relationships seem to be less intensely personalized, and are based more on indirect relationships of obligation, often mediated by language (e.g., kinship ties or friends of friends).

Humans face a particular problem in this respect because of the exceptionally large size of their communities (150, compared to a maximum of ca. 50 among the nonhuman primates). Since larger groups incur more stresses that threaten to drive members apart, the stability of very large social groupings will ultimately depend on how tightly bonded they are and on the kinds of relationships that are needed to maintain coherence in the outer layers of the group (i.e., those beyond which regular personal day-to-day interactions are possible).

The problem that human communities explicitly face is their sheer size. The observed size of human communities (approximately 150 individuals: Dunbar 1992b, 2008) far exceeds the size of group that can easily be bonded by social grooming among primates. Indeed, at both the within-species and between-species levels, primate groups that exceed this natural limit become increasingly fragmented (see, for example, Dunbar 1992c; see also Lehmann et al. 2007). Using data from a large sample of primate species, Lehmann et al. obtained the following equation for the relationship between grooming (or social) time and mean social group size (Lehmann et al. 2007):

$$\text{Log}_{10}(\%G) = 0.05 + 0.56 \ \log_{10}(N) - 0.06 \ SR + 0.24 \ D \qquad (6.1)$$

where $\%G$ is the percentage of day time devoted to grooming, N is group size, SR is the adult sex ratio (indexed as the ratio of adult females:males), and D is female dispersal indexed as a binary factor ($0 = $ females disperse, $1 = $ females are philopatric). Taking the data for primates as a whole, there seems to be an upper limit on the amount of time that primates devote to grooming at 20% of total day time (Dunbar 1992b). If we set grooming time in Eq. (6.1) at this value and take the sex ratio to be 50:50, the maximum size of group that can be maintained as a coherent unit by grooming alone is 219 for a female-dispersal species and 82 if females are philopatric.

Although it has often been assumed that humans are a female dispersal species, in fact this view seems increasingly untenable and probably reflects an over-emphasis on agricultural societies where males are able to monopolize resources (in the form of land), thus making it advantageous for females to be willing to move between groups. Because most contemporary human populations are agricultural, it creates the impression that humans are naturally patrilineal and female-dispersal. But when these particular economic constraints are removed, humans seem universally to revert either to a form of bilateral dispersal or to a form of matrilineal social structure based on small groups of very closely related females who live closer to the wife's parental home than to the husband's. The suggestion that humans are naturally female-philopatric is reinforced by the social networks data, which indicate that social networks tend to be strongly gender-biased (males prefer males, and females prefer females) and that females tend to include more kin in their networks than males do (Dunbar and Spoors 1995; Roberts et al. 2008). Thus, it is probably the female philopatry form of Eq. (6.1) that we should apply to humans, and this suggests an upper limit of approximately 80 to the size of group that can be bonded via social grooming alone. Groups of this size were probably characteristic of *Homo ergaster/erectus* (Aiello and Dunbar 1993; Dunbar 2009b) and represent a temporally very early phase of (pre-)human evolution.

In order to increase community size above this limit, humans would have needed to find some other way of bonding groups. I have argued for a three-step process involving first laughter, then music, and finally the rituals of religion (Dunbar 2009a). All three of these activities are very effective at triggering the release of endorphins in the brain (the same process that appears to allow grooming to act as an efficient bonding mechanism in monkeys and apes) (Keverne et al. 1989). By being able to trigger the same underlying mechanism, these activities function as a kind of extension of the grooming processes that underpin primate social bonding. They seem to work by creating a sense of communality, of belonging to the group, through the endorphin effect, just as grooming does. But they have the added advantage that they do not require physical contact between interactants, and thus can effectively allow the circle of interaction to be increased from the one-to-one characteristic of primate grooming to the one-to-many exchange that is more characteristic of humans. Human conversation groups have an effective size (i.e., broadcast size) that is three times the size of a grooming interaction (Dunbar et al. 1995).

What is not clear as yet is whether there is, in turn, a limit on the size of group that can be bonded using these activities. Laughter works well as a bonding device, but the circumstances under which it occurs might well involve quite small groups (conversation groups of 4–5 individuals). This might suggest that its scope, while greater than that for grooming, is nonetheless limited, permitting community size to be increased beyond the limit that grooming imposes at ca. 80 individuals, but perhaps imposing a subsequent limit at, say, 110–120. Music (notably singing and dancing) and religion seem to have a greater potential for involving more individuals.

However, humans also seem to exploit a cognitive mechanism that completely sidesteps the endorphin mechanism, and that is kinship. Our analyses of contemporary human social networks suggest that kin occupy a privileged position in our networks, and especially in the outer layers (Dunbar and Spoors 1995; Roberts et al. 2009). If you are a member of a large kinship group (extended family), you tend to have fewer unrelated friends in your social network than people who come from smaller extended families. While close friends and close kin may share a great deal in common in terms of the psychological bases of the relationship (it mostly seems to reflect time spent in each other's company), this is not true of distant friends and distant kin. In this case, absence of contact has a differential effect on the quality of the relationship in that distant friends who are contacted rarely are deemed less intimate than distant friends contacted more frequently, whereas distant kin are treated as equally intimate irrespective of how often they are contacted (Roberts et al. 2009). Somehow, kin are given special status, and this status can only come through language-based identification, since it seems to have little to do with actual rates of contact.

What makes this intriguing is the fact that kinship relationships rely on a purely linguistic identifier ("This is your second-cousin Mary") being enough to establish a relationship that carries weight in terms of trust, reciprocity, and obligation. More importantly in functional terms, it does so in a way that makes the outer layers of the network especially cohesive—and maybe resistant to facture. Verbal criteria of this kind have another interesting property that may be important: they massively reduce the cognitive demands of maintaining knowledge about relationships. In other words, you need to know only one thing about an individual to know how to behave towards them and what to expect from them in turn, and that is simply how they fit into your pedigree. In contrast, a relationship with friends requires the investment of time and effort; as a result, lack of investment results in the relationship degrading. Thus, humans seem to have capitalized on a natural structural feature of human communities to manage the information demands involved more efficiently. I say "natural" here, because in traditional hunter–gatherer societies, everyone in a community is related, directly or indirectly as biological kin or as affinal kin, to everyone else. Indeed, Dunbar showed that the number of living descendents (i.e., members of the three living generations: offspring, parents, grandparents) in a natural-fertility population practicing exogamy would amount to almost exactly 150 if they were descended from an apical pair in the great-grandparental generation (Dunbar 1995). What makes this pedigree structure interesting in the present context is the fact that this is about as far back as any living member can remember from personal knowledge (grandmother's grandmother), something that may be essential for being able to specify degrees of relatedness.

6.4 Cognitive Afterword

The role that kinship classification might play in the coherence of human's unusually large social groups raises questions about the cognitive demands involved. We have shown (1) that, in humans, individuals' competences on multilevel false belief tasks correlate with the size of their social network (Stiller and Dunbar 2007) and (2) that, across primate species, these competences seem to correlate with frontal lobe volume (Dunbar 2009b). The scale of these competences is such that while monkeys can seemingly aspire only to first-order intentionality, and great apes can probably just about manage second order (O'Connell and Dunbar 2003; Cartmill and Byrne 2007), humans can cope with fifth order (Kinderman et al. 1998; Stiller and Dunbar 2007). This massive difference in cognitive processing seems to be what underpins humans' capacity for culture, in that it seems to make a radical difference to the kinds of stories (fictional or otherwise) that can be composed and to the kinds of religious statements that can be constructed (Dunbar 2008). The difference between being able to do this with fifth-order intentionality even compared to fourth-order intentionality is substantial: in terms of story-telling, it makes the difference between a mere narrative and something that really challenges the audience's intellect and emotions.

In the context of language and its grammatical structure, fifth order is interesting because looks suspiciously similar to the widely recognized limit on number of embedded clauses with which normal adult humans are able to cope (Dunbar 2009b). Similarly, five layers correspond both to the number of generations in typical pedigrees from the ethnographic literature and to the number of generations needed to link all members of a community in a natural-fertility society (Dunbar 1995). I suspect that this is no coincidence and reflects the fact that all these elements (cultural activities such as story-telling and religion, and kinship naming) are exploiting modern humans' full cognitive capacity in this respect, and that, in essence, the modern human mind/brain evolved explicitly for this purpose.

And so, finally, it is precisely because of the massive differences in computational power represented by the contrast between second- versus fifth-order intentionality that humans have culture in the proper sense of the term and apes and monkeys don't (Dunbar 2008). And, in turn, it is because of these combined capacities that humans are able to maintain such large social communities as coherent units.

References

Aiello LC, Dunbar RIM (1993) Neocortex size, group size and the evolution of language. Current Anthropology 34:184–193

Bearder S (1987) Lorises, bushbabies and tarsiers: diverse societies in solitary foragers. In: Smuts B, Cheney D, Seyfarth RR, Wrangham R, Struhsaker T (eds) Primate Societies. Chicago University Press, Chicago, IL

Byrne R, Whiten A (eds) (1988) Machiavellian Intelligence. Oxford University Press, Oxford

Cartmill EA, Byrne RB (2007) Orangutans modify their gestural signaling according to their audience's comprehension. Current Biology 17:1345–1348

Dunbar RIM (1984) *Reproductive Decisions: An Economic Analysis of Gelada Baboon Social Strategies*. Princeton University Press, Princeton, NJ

Dunbar RIM (1992a) Neocortex size as a constraint on group size in primates. Journal of Human Evolution 22:469–493

Dunbar RIM (1992b) Coevolution of neocortex size, group size and language in humans. Behavioral and Brain Sciences 16:681–735

Dunbar RIM (1992c) Time: a hidden constraint on the behavioural ecology of baboons. Behavioral Ecology and Sociobiology 31:35–49

Dunbar RIM (1995) On the evolution of language and kinship. In: Steele J, Shennan S (eds) *The Archaeology of Human Ancestry: Power, Sex and Tradition*. Routledge, London

Dunbar RIM (2000) Male mating strategies: a modelling approach. In: Kappeler P (ed) *Primate Males*. Cambridge University Press, Cambridge, MA

Dunbar RIM (2003) Evolution of the social brain. Science 302:1160–1161

Dunbar RIM (2008) Mind the gap: or why humans aren't just great apes. Proceedings of the British Academy 154:403–423

Dunbar RIM (2009a) Mind the bonding gap: constraints on the evolution of hominin societies. In: Shennan S (ed) *Pattern and Process in Cultural Evolution*. University of California Press, Berkeley, CA

Dunbar RIM (2009b) Why only humans have language. In: Botha R, Knight C (eds) *The Prehistory of Language*. Oxford University Press, Oxford

Dunbar RIM (2009c) The social brain and its implications for social evolution. Annals of Human Biology 36:562–572

Dunbar RIM (2010) Brain and behaviour in primate evolution. In: Kappeler PH, Silk J (eds) *Mind the Gap: Tracing the Origins of Human Universals*, pp. 315–330. Freeman, San Francisco, CA

Dunbar RIM, Dunbar P (1988) Maternal time budgets of gelada baboons. Animal Behaviour 36:970–980

Dunbar RIM, Duncan N, Nettle D (1995) Size and structure of freely forming conversational groups. Human Nature 6:67–78

Dunbar RIM, Shultz S (2007) Understanding primate brain evolution. Philosophical Transactions of the Royal Society, London 362B:649–658

Dunbar RIM, Spoors M (1995) Social networks, support cliques and kinship. Human Nature 6:273–290

Hamilton MJ, Milne BT, Walker RS, Burger O, Brown J (2007) The complex structure of hunter-gatherer social networks. Proceedings of the Royal Society, London, Series B 271:2195–2202

Hill RA, Bentley A, Dunbar RIM (2008) Network scaling reveals consistent fractal pattern in hierarchical mammalian societies. Biology Letters 4:748–751

Hill RA, Dunbar RIM (2003) Social network size in humans. Human Nature 14:53–72

Kappeler PM, Heymann EW (1996) Nonconvergence in the evolution of primate life history and socio-ecology. Biological Journal of the Linnean Society 58:297–326

Keverne EB, Martensz N, Tuite B (1989) Beta-endorphin concentrations in cerebrospinal fluid of monkeys are influenced by grooming relationships. Psychoneuroendocrinology 14:155–161

Kinderman P, Dunbar RIM, Bentall RP (1998) Theory-of-mind deficits and causal attributions. British Journal of Psychology 89:191–204

Kudo H, Dunbar RIM (2001) Neocortex size and social network size in primates. Animal Behaviour 62:711–722

Lehmann J, Andrews K, Dunbar RIM (2010) Social networks and social complexity in female-bonded primates. In: Dunbar RIM, Gamble C, Gowlett JAJ (eds) *Social Brain, Distributed Mind*, pp. 57–83. Oxford University

Lehmann J, Korstjens A, Dunbar RIM (2007) Group size, grooming and social cohesion in primates. Animal Behaviour 74:1617–1629

Lindenfors P, Nunn CL, Barton RA (2007) Primate brain architecture and selection in relation to sex. BMC Biology 5:20

McGrew WC (1992) *Chimpanzee Material Culture*. Cambridge University Press, Cambridge, MA

O'Connell S, Dunbar RIM (2003) A test for comprehension of false belief in chimpanzees. Evolution and Cognition 9:131–139

Pérez-Barbería J, Shultz S, Dunbar R (2007) Evidence for intense coevolution of sociality and brain size in three orders of mammals. Evolution 61:2811–2821

Rendell L, Whitehead H (2001) Culture in whales and dolphins. Behavioral and Brain Sciences 24:309–382

Roberts S, Dunbar RIM, Pollet T, Kuppens T (2009) Exploring variations in active network size: constraints and ego characteristics. Social Networks 31:138–146

Roberts S, Wilson R, Fedurek P, Dunbar RIM (2008) Individual differences and personal social network size and structure. Personality and Individual Differences 44:954–964

Seyfarth RM (1977) A model of social grooming among adult female monkeys. Journal of Theoretical Biology 65:671–698

Shultz S, Dunbar R (2007) The evolution of the social brain: anthropoid primates contrast with other vertebrates. Proceedings of the Royal Society, London, Series B 274:2429–2436

Silk JB (2002) The 'F'-word in primatology. Behaviour 139:421–446

Stiller J, Dunbar RIM (2007) Perspective-taking and memory capacity predict social network size. Social Networks 29(1):93–104

van Schaik CP, Ancrenaz M, Borgen G, Galdikas B, Knott CD, Singleton I, Suzuki A, Utami SS, Merrill M (2003) Orang utan cultures and the evolution of material cultures. Proceedings of the National Academy of Sciences of the USA 299:102–105

Whiten A (2005) The second inheritance system of chimpanzees and humans. Nature 437:52–55

Whiten A, Goodall J, McGrew WC, Nishida T, Reynolds V, Sugiyama Y, Tutin CEG, Wrangham RW, Boesch C (1999) Culture in chimpanzees. Nature 399:682–685

Zhou W-X, Sornette D, Hill RA, Dunbar RIM (2005) Discrete hierarchical organization of social group sizes. Proceedings of the Royal Society, London, Series B 272:439–444

Chapter 7
Why Most Theories Get It Wrong

Altruistic Intentions as an Explanation of the Evolution of Genuine Altruism

Julia Pradel and Detlef Fetchenhauer

Abstract The existence of human altruism is a paradox, as evolutionary theory teaches us that natural selection should never have evolved it. Indeed, sociobiology has shown that a lot of altruism in humans emerges as ultimate self-interest. In this chapter we argue, however, that genuine altruism does exist and that it can be a result of natural selection. On the basis of experimental results we show that individuals do behave genuinely altruistically in one-shot interactions when no one is watching. We argue that previous evolutionary theories fail to explain this finding, because their analysis focuses on the biological concept of altruism only (i.e., altruism defined in terms of *behavioral consequences*), and leaves the altruistic *intentions* of actors aside. Referring to the commitment model of Frank we argue that individuals have evolved dispositions for altruistic intentions. We hypothesize that (1) these stable altruistic intentions are recognizable, and (2) that altruists cooperate electively with each other on the basis of their recognition abilities. Providing empirical evidence for our hypotheses, we argue that genuine altruism could have evolved through assortative processes between cooperating altruists whose synergetic benefits outcompeted the selfish advantage of egoists.

7.1 Why Most Theories Get It Wrong: Altruistic Intentions as an Explanation of the Evolution of Genuine Altruism

With the stated aim of enhancing healthcare and reducing poverty, the Bill & Melinda Gates Foundation, which is the largest privately operated foundation in the world, has granted commitments amounting to $20.1 billion since its inception in 1999 (Bill & Melinda Gates Foundation 2009). Besides, its noble founders, the IT mogul Bill Gates, who is said to be the richest man in the world (Forbes 2009), and his wife, Melinda, have announced their intention to donate 90–95% of their

J. Pradel (✉)
Institut für Wirtschafts- und Sozialpsychologie, Universität zu Köln, 50931 Köln, Germany
e-mail: julia.pradel@uni-koeln.de

U.J. Frey et al. (eds.), *Homo Novus – A Human Without Illusions*,
The Frontiers Collection, DOI 10.1007/978-3-642-12142-5_7,
© Springer-Verlag Berlin Heidelberg 2010

entire possessions before they die. When news like this makes the rounds through the media, critical voices are frequently raised, stating that such self-publicizing charity can never be truly good-natured; rather, people like Bill Gates, who perform self-promotion by helping others, are assumed to be driven by the selfish motive to improve their reputation and social status and—en passant—their company's revenues.

Indeed, the view that all human behavior is ultimately selfishly motivated has a long tradition in many scientific disciplines such as politics, economics, and biology. In political philosophy, Machiavelli (1498/2009) acted on the assumption that men are cleverly deceitful and unscrupulous creatures, and Hobbes (1651/2009) not only visualized humans to be egoistic and greedy for power, but balefully driven by rampant passions. Hobbes markedly expressed this notion in his often cited metaphor "homo homini lupus est"—"man is a wolf to man". Even though economic theory excludes passions as the driving force of human behavior and rather refers to reason, the economic idea of man, the "homo oeconomicus", is close to the idea of the philosophers Machiavelli and Hobbes in assuming the existence of thoroughly self-interested actors—implicitly rejecting the existence of genuine altruists.

The biological view is based on the principle of natural selection, which assumes the "survival of the fittest", i.e., those individuals who are best adapted to the local environment are the ones that are going to survive and to reproduce. As egoists are better adapted than altruists in that they acquire more resources, egoists should have more offspring than altruists. Consequently, on Darwinian grounds, individuals with an altruistic trait should have become extinct through the process of natural selection. To emphasize, the evolutionary perspective thus holds that genuine altruism cannot exist, and that any form of altruistic behavior finally has to emerge as self-interest. The phrase "scratch an altruist and watch a hypocrite bleed" (Ghiselin 1974, p. 247) vividly depicts the consensus of sociobiologists that the assumption of altruism is a gross illusion.

However, even Darwin (1871) himself, referring to daily observations of human altruism, had the intuition that humans are altruistic indeed, but Darwin was assailed by doubts how he could explain such a biologically impossible phenomenon conclusively.

7.2 Solving the Paradox of Altruism

7.2.1 The Initial Perspective: In the Very End, Every Altruistic Act Comes Down to Evolutionary Selfishness

A considerable contribution to ending the debate on whether altruism exists or not was provided by Sober and Wilson (1998), who drew attention to the fact that "altruism" has different meanings in the three fields of everyday life, sociobiology, and psychology. In everyday life the concept of altruism seems to require an element of both action and motivation: individuals who never help others are rarely considered altruists, and people similar to Bill Gates who do help others will only be awarded the title of an altruist if their behavior is the result of truly good intentions.

In contrast to the common conception of altruism, *psychological* altruism is primarily defined on the basis of motives, and relates only derivatively to the behaviors those motives may generate. That is, an act is altruistic if the actor has the *intention* to benefit someone at his own costs (even if this intention at times goes awry).

However, as intentions—no matter whether they are altruistic or egoistic—are only proximate causes (i.e., triggers) of behavior, they do not explain the adaptive value of altruistic acts. According to Sober and Wilson (1998), an evolutionary analysis of altruistic behavior thus has to leave intentions aside and has to focus on ultimate causes of altruistic behavior only. This means that *evolutionary* altruism is solely defined on the basis of consequences, i.e., an act is altruistic if the actor benefits someone at his own costs. The challenge for sociobiologists is to show how such self-sacrificial behavior can evolve, not considering how or even whether the individual thinks or feels as he or she performs the behavior.

All previous theories that have been central to the explanation of the evolution of altruism follow the approach of basing the analysis of altruistic behavior on the behavior's consequences.

For example, Hamilton (1964), in his theory of *kin altruism*, explains that people behave altruistically towards those who are kin because such behavior serves to carry their own genes into the next generation, not by the production of their own descendants but by aiding the reproduction of non-descendant relatives. To give a concrete illustration, if someone saves the lives of his children, he is not behaving altruistically in evolutionary terms, as he increases his *inclusive fitness* by rescuing his kids.

Reciprocal altruism (Trivers 1971) describes situations in which a donor carries out an altruistic act that is reciprocated by the beneficiary at a later time. As the donor does not suffer a net loss out of the interaction, Singer (1981) notes that "reciprocal altruism is not really altruism at all; it could more accurately be described as enlightened self-interest" (p. 42), which, again, is evolutionary adaptive.

In a similar vein, enlightened self-interest applies to altruistic behavior in situations where *indirect reciprocity* may arise (Alexander 1987). Namely, the concept of indirect reciprocity assumes that altruism, which advertises an actor's tendency to cooperate, is advantageous in situations with onlookers, because it attracts cooperation from third parties in the future (Nowak and Sigmund 2005).

Finally, recent experiments on *altruistic punishment* showed that humans are willing to punish the selfishness of others even if the selfish behavior is not directed at themselves but at an unrelated person instead. Keeping these results in mind, altruistic acts often represent social norm compliance, which is a certain way to forego detrimental sanctions by social members (Fehr and Gächter 2002, 2003).

Based on what we have said so far, "moral actors in ultimate analysis are self-interested profit-maximizers" (translated from Voland 2007, p. 118). But is it really true that every altruistic human act comes down to evolutionary selfishness? As this chapter is devoted to proving that genuine altruism does exist, we have to perform the task of giving empirical evidence that people sometimes do behave altruistically in an evolutionary sense, i.e., that they give up resources for which they are never compensated.

7.2.2 The New Perspective: Genuine Altruism Does Exist

Life is full of opportunities to behave selfishly. However, people often do not seize their chance to realize the maximum payoff. For illustration, let us consider the choice confronting a person who wants to dispose of some toxic pesticide (Frank et al. 1993). The explicit regulations of communities and public health require that such toxins are delivered to an official disposal facility. Yet such facilities are often inconveniently situated and there is practically no way someone could be penalized for choosing the easier option of pouring the pesticide down the drain. But do all people take this inconsiderate course? Obviously not, as can be seen from the utilization of official disposal facilities.

To give a second example, consider people in dyadic relationships who want to cooperate in a joint mission, e.g., in starting a company or a family. Both projects call for considerable investment, which each partner fears could be undercut if the other partner decides to defect at a later point of time when he or she faces a "golden opportunity" (Frank 1988, p. 73) to behave self-interestedly without negative consequences, e.g., by embezzling money or having an affair. Without reasonable assurance that this will not happen, neither is willing to invest to make the most out of the joint project. However, people do start businesses and families. The assurance seems to exist. Probably, life teaches people that their counterparts do not behave strictly selfishly.

However, both examples are not watertight; skeptics might develop interpretations showing that cooperation in such circumstances comes down to enlightened self-interest. As this is the case with most examples from everyday life, economists, and more and more frequently also psychologists and biologists, have used a specially designed experimental paradigm, the so-called dictator game (for a summary of research see Camerer 2003), to investigate truly altruistic behavior. In this paradigm, two persons interact with each other only *once* and under conditions of total anonymity. The dictator is given a certain amount of money and has to divide this money between him- or herself and an unknown recipient. The recipient has no means to influence or to veto the decision of the unknown dictator.

Analyzing this situation, what result would the above stated theories predict as the decision of the dictator? According to the theory of kin altruism (Hamilton 1964), altruistic behavior should not be observed, as it can be presumed that the recipient in the dictator game is genetically unrelated to the dictator. Likewise, the theory of reciprocity (Trivers 1971) would not predict any cooperation, as the interaction is non-iterated and does not allow for future compensation. Because the interaction is anonymous, costs in terms of losses of reputation (Milinski et al. 2001) or punishment (Fehr and Gächter 2002, 2003) are irrelevant. The dictator game thus constitutes a golden opportunity to act self-interestedly without any negative consequences. Conclusively, the dictator should keep all the money for himself.

However, a vast number of studies, both in industrialized societies (Camerer 2003) and in hunter–gatherer societies (Henrich et al. 2004), show that the average amount of money given to the recipient is substantially above zero, even when large amounts of money are at stake (Camerer 2003). These results contradict the thesis "Homo homini lupus est".

One possible explanation for the observation of genuine altruism in dictator games could be that unconditional altruism is indeed a product of natural selection, but that it was adaptive only long ago in environments that were different from the one that we live in today. To exemplify, if in ancestral environments individuals lived in small groups, where the interactants were either kin or were likely to meet again, and the interactions were usually under social surveillance, selfishness would have been maladaptive and a general heuristic about unconditional altruism could have developed (Hagen and Hammerstein 2006).

This argument, however, is not satisfactory for two reasons. First, even if ancestral interactions were mainly amongst kin, it is not clear that indiscriminate altruism would have been the evolutionary best strategy: individuals should have learned to adapt their behavior in accordance with the different degrees of relatedness of their interaction partners, i.e., humans should have helped closer family more (Sesardic 1995). Second, self-interested behavior is advantageous in any situation not observed by others, and even if these situations were rare, the loss of resources due to gratuitous altruistic behavior would have been too large to survive the pressures of natural selection. Instead, evolution should have favored the capability of selectively adjusting the performance of self-interested and altruistic behavior according to the risk of being detected and punished by others. As empirical evidence shows, humans are indeed highly sensitive to cues of detection and strategically adjust their behavior (Gintis et al. 2003). Thus, the mismatch hypothesis turns out to be an insufficient explanation of the existence of genuine altruism.

Another explanation for the existence of genuine altruism is group selection, an idea that was already broached by Darwin (1871) himself, proposed again by Wynne-Edwards (1962), and reintroduced by Wilson and Sober (1994). The central idea is that groups consisting of altruists outcompete groups of egoists and are thus favored by selection despite the fact that altruistic behavior, at the individual level, continues to decrease the fitness of any individual. The topic of group selection is a matter of eminent controversy in biology. The central difficulty is that, although it is certainly in the interest of any individual to be part of a group of altruists, the individual himself always fares best by behaving egoistically. Therefore, the current view of most biologists is that group selection is a theoretically possible evolutionary process, but that it is unlikely in reality because it would work only under very special and rarely fulfilled conditions (e.g., if groups are small, if their extinction rate is high, if there is little intergroup migration, etc.) (Johnson et al. 2008; Richerson and Boyd 2001, 2005; Sesardic 1995). Owing to this practical fragility, group selection seems to be an unsatisfactory candidate to explain the evolution of genuine altruism.

Whenever we come to our wit's end with explaining phenomena on the basis of existing theories, it seems worthwhile to call the basic assumptions of these theories into question. Until now, standard evolutionary explanations of altruism have left psychological phenomena, i.e., altruistic intentions and moral emotions, completely aside. Ignoring psychology was surely an important primary step in, first, obtaining a clear-cut picture of an exclusively biological problem, and second, finding solutions to the so-modeled problem. However, classical evolutionary theory would not predict that people behave altruistically when they are unobserved. But real-life shows that they do. To model human behavior in a way that fits the data, it might be about time to integrate a psychological determinant into the analysis—peoples' altruistic intentions and moral emotions—for they might help to explain the existence of genuine altruism.

In his commitment model, the economist Frank (1987, 1988, 2005, 2008) assumes that some humans have evolved stable dispositions for altruistic intentions that lead them to behave altruistically even in situations such as the dictator game. More specifically, this model predicts that evolution has selected for psychologically altruistic dispositions, i.e. stable intentions to behave altruistically, although the resultant altruistic behavior that individuals with these dispositions display is evolutionary disadvantageous and thus *genuinely* altruistic.

The pivotal point within the commitment model consists in the recognizability of altruistic intentions. For illustration, let us consider a population consisting of genuine altruists, who cooperate unconditionally, and cheaters, who opportunistically behave in whatever way maximizes their payoffs. In the first scenario, altruistic intentions are unrecognizable so that altruists and cheaters pair at random. Whenever the same types interact with each other (i.e., if altruists interact with altruists and cheaters interact with cheaters), the interacting partners share the resulting payoff equally. However, if altruists interact with defectors, altruists go away empty-handed and defectors pocket the full return. How will the population evolve over time? As cheaters realize higher average returns than altruists and the two types reproduce according to their average payoffs, cheaters will be selected for and genuine altruists will become extinct.

Now, let us assume the same scenario except that altruists and cheaters are perfectly distinguishable such that their altruistic or self-interested intentions are recognizable. In this case altruists would selectively interact with other altruists only, a process that is called *assortative interaction*, while cheaters have no chance but to stay among themselves. In this case genuine altruism would stand the pressures of natural selection. In fact, if we make the reasonable additional assumption that cooperation between altruists leads to a synergetic extra-payoff that cheaters do not attain, altruists would even make up an ever-growing share of the population.

Assuming further that the recognition of altruists is costly, the commitment model could explain the continuing variation in altruistic tendencies among humans. On the one hand, the higher the percentage of altruists in a given population, the less monitoring of the altruism of others will take place, and the more adaptive it is to act nonaltruistically. On the other hand, the lower the percentage of altruists in a given population, the more monitoring will take place, and the more adaptive it is to act altruistically. This would lead to frequency-dependent selection and equilibrium of altruists and nonaltruists existing side by side.

But how can an altruistic psychological disposition be recognized? Frank assumes that stable altruistic intentions are linked to emotional commitment, and that there is an observable symptom present in people who experience emotional commitment, something like a "sympathetic manner" (Frank et al. 1993, p. 249). Thus, individuals who have this characteristic can be identified as genuine altruists while those who lack the characteristic will be identified as cheaters.

There is one obvious criticism of this argument. When signaling altruism is profitable, self-interested persons should be interested in faking the relevant signal in order to reap the benefits of an altruist without paying the cost of altruistic behavior (Fehr and Fischbacher 2005). In evolutionary terms, this means that natural selection should create a deceptive copy of the altruistic signal. Although this argument is powerful, it can be reversed, as the existence of a deceptive copy should lead to the modification of the original signal in order to prevent plagiarism (Frank 2005). An arms race between the true signal and its fraudulent copy should arise. The question of whether the original signal is in the vanguard of this race at the present moment of our evolutionary history calls for empirical examination, to which we shall return later.

There are certain similarities between the commitment model of Frank and the theory of Geoffrey Miller (2000, 2007), who explains the evolution of human altruism via sexual selection. He argues that altruism is an important attribute of one's mating partner and that therefore humans prefer sexual partners who signal a high degree of reliability and willingness to take care. However, unlike Frank, Miller does not deal very much with the issue of how to identify true altruism.

One important difference between Frank's theory and other theories of the evolution of altruism outlined above is the importance of moral emotions. It is these moral emotions that drive altruistic behavior, even if there is no *rational* reason to do so. To fully recognize this difference it is helpful to take a closer look at how "rationality" is defined by economists, on the one hand, and biologists, on the other.

According to economists, humans are able to anticipate the probable outcomes of their behavior and thus choose the alternative with the highest expected payoff. In doing so, they totally ignore any externalities of their own behavior (i.e., they are totally uninterested in doing harm or good to others if this is not in their strategic self-interest). Even though the decision-making processes of economic actors might proceed unconsciously, they act "as if" they anticipated the future outcomes of their decision. Thus, in economic terms a rational decision is driven by the anticipation of the highest payoff.

In contrast, biologists would not assume that organisms are able to engage in such complex reasoning, be it consciously or unconsciously. For them, rational (i.e., adaptive) behavior is driven by the past success or failure of a certain strategy. Via natural selection those organisms survive and reproduce whose strategy is the most adaptive. Therefore, strategies of bacteria might sometimes be more rational (i.e., leading to optimal results under the given constraints) than those of humans as through the rapid process of natural selection a "strategy" emerges that is optimally adapted to the specific environment.

Interestingly, other theories of the evolution of altruism (i.e., kin altruism, reciprocal altruism, indirect reciprocity, etc.) are purely rational from both a biological and an economic perspective. They get along without assuming the existence of moral emotions. Consider, for example, the concept of reciprocity.

A biologist would assume that individuals, in the context of repeated interactions, behave cooperatively not because they are persistently calculating potential future interactions but because in the course of our evolutionary history a behavioral module for reciprocal altruism proved to be adaptive.

An economist, on the other hand, would not base his argument on the evolutionary adaptiveness of reciprocal altruism, but rather refer to the actor's capability to rationally anticipate the advantage of this strategy. This means that even if actors had no genetic predispositions to act in a certain manner, but would analyze the situation with game theoretical prudence, they would engage in altruistic behavior for their own long-term self-interest in all the situations discussed above, i.e., when they are interacting with kin, as well as when they are under social surveillance and face repeated interactions with non-kin. However, they would be constantly tempted to engage in selfish behavior when nobody is watching (i.e., in situations such as the dictator game).

This is very different in Frank's commitment model, which assumes that (at least some) humans act altruistically even if prudent and strategic reasoning would tell them to act selfishly.

That is why moral emotions are so important. They cause people to behave altruistically even if this not in their material self-interest. As humans are able to identify such true altruists; these altruists are ultimately rewarded with having more and more reliable interaction partners. This is why such altruism has managed to evolve. If it had decreased the ultimate fitness of its bearer it would have gone extinct long ago. But, to repeat, only because altruists do *not* attempt to reach such ultimate profits are they able to gain them.

To summarize, the commitment model shows what sociobiology for a long time assumed to be unfeasible: that genuine altruism, which we now define as a prosocial act that is carried out without the intention to profit from it, can actually be the result of natural selection. The key to the solution of the paradox seems to be that genuine altruism is the disadvantageous manifestation of adaptive moral emotions.

However, to prove the relevance of this theory for the evolution of genuine altruism, we have to show that two conditions are fulfilled: (1) Individuals have to be able to distinguish altruists from egoists, and (2) altruists have to elect like-minded individuals for mutual cooperation, i.e., individuals have to assort themselves along the dimension of altruism, in order to reap synergetic extra benefits. In the following we review empirical findings on these two issues.

7.3 Evidence for the Solution of the Paradox of Altruism

7.3.1 Altruists and Egoists Are Distinguishable

As to the first condition, there is indeed evidence that humans are equipped with a psychological mechanism to detect cheaters (Cosmides and Tooby 1992) and altruists in social interactions (Brown and Moore 2000).

A few studies examined whether humans are also capable of predicting altruistic intentions reliably. In a laboratory experiment, Frank et al. (1993) offered groups of participants the opportunity to get acquainted with each other before playing a prisoner's dilemma game. In the one-shot prisoner's dilemma each of two players has two strategies, cooperate or defect. Each player gets a higher payoff by defecting, no matter what the other player does. Yet when both defect, each gets less than when both cooperate. Subjects were able to discuss the paradigm for 30 min and make nonbinding declarations to their interaction partners about their game decision. Subsequently, the subjects indicated their actual response secretly and were asked to guess the estimates of their counterparts. Estimates were better than chance. However, Ockenfels and Selten (2000) criticized these results, pointing out that the accuracy of the predictions might have been due to the explicit declarations of defectors. As they would have had little interest in deviating from their pronounced decision, estimating their behavior was easy. Brosig (2002), therefore, replicated the experiment under restricted conditions. Pre-communication sessions were filmed so that subjects who explicitly announced their intention to defect were excluded from the analysis. Accuracy of predictions exceeded chance level under these conditions as well.

Going a step further, Price (2006) was able to provide evidence of intention-reading abilities in humans in a real-life collective action setting, i.e., a setting in which a group of people jointly generate a resource to be shared equally among themselves. In this study, the Shuar hunter-horticulturalists were asked to rate the absences of their co-members and it was shown that they were able to distinguish intentional non-cooperators, who could have cooperated but decided not to, from unintentional non-cooperators, who were unable to cooperate for good reasons.

In fact, altruism detection abilities have been evidenced not only in settings where individuals were acquainted with each other. Instead, some studies show that humans are able to rate cooperative intentions in unknown persons spontaneously and on the basis of only minimal visual information. While Yamagishi et al. (2003) asked observers to memorize the faces of unknown target persons who had played prisoner's dilemma games earlier and showed that observers were better able to remember the faces of cheaters than the faces of cooperators, Verplaetse et al. (2007) asked observers to identify targets who had played cooperatively in such a game. Observers succeeded, but only if they responded to event-related pictures that were taken during the decision-making moment rather than when they responded to neutral pictures. This result suggests that individuals can predict the degree of altruistic behavior of strangers by reading signs of emotions evoked in significant social decisions. However, the identification of benevolent emotional states is no guarantee of the existence of permanent altruistic traits, although permanent dispositions are the preferable criterion for selection of good interaction partners.

To test the exact hypothesis of the commitment model (Frank 1987, 1988, 2005, 2008), which is that stable altruistic dispositions are identifiable, Brown et al. (2003) asked subjects to watch videos and make assessments of target persons who had previously indicated their level of altruism in a self-report. The ratings of judges correlated significantly with the self-reported altruistic dispositions of targets.

However, as self-reported altruism might be biased, Fetchenhauer et al. (2010) ran a slightly different experiment. In a first step, target persons were invited to the laboratory. Being asked to present themselves to the camera, target persons were filmed in a neutral setting completely unrelated to altruistic behavior. Afterwards, and unannounced, participants played a dictator game, which functioned as a measure of altruism. In a second step, judges watched the 20-second silent video clips of the unknown target persons and were asked to estimate the behavior of the target persons in the dictator game. Judges could not base their estimates on situational cues related to the dictator game but instead had to draw on stable cues to altruism that were available in the neutral videos. Estimates were significantly better than chance also under these conditions. Trying to find out which cues the judges might have used for their estimations, Fetchenhauer et al. investigated whether the judges simply referred to the sex of the target when estimating his or her level of altruism. Indeed, the sex of the target would have been a good cue as females on average gave more money than males. However, the validity of estimates could not be explained by the identification of the sex difference alone. A regression analysis using judges' average estimates as the dependent variable and both the targets' sex and their actual behavior as independent variables showed that both independent variables were significant predictors, indicating that judges were able to deteremine those variations in the degree of altruism that went beyond differences due to the sex of the target.

In sum, the results indicate that individuals are able to distinguish altruists from egoists. This mechanism raises opportunities for assortment processes such that altruists selectively interact with other altruists. But is this true?

7.3.2 Altruists Elect Like-Minded Individuals for Mutual Cooperation

As to the second condition for the evolution of altruism by assortment, namely, that altruists choose like-minded persons for mutual cooperation, assortation with respect to mating ("assortative mating") is a well-established empirical fact (Mascie-Taylor 1995; Spuhler 1968) that has been shown for a multiplicity of somatic and psychological characteristics. The phenomenon, furthermore, exceeds mate choice, as it has been evidenced for friendships as well (Berscheid 1985). Several studies have investigated assortative partner choice with particular respect to altruism. Sheldon et al. (2000), for example, asked university freshmen to recruit three peers to participate in an N-person prisoner's dilemma game. Subjects with a prosocial value orientation tended to assort with one another. They thus achieved a group-level advantage in the game returns, which counteracted the individual-level advantage of antisocial participants.

Ehrhart and Keser (1999) evidenced assortative processes of cooperative individuals in a public goods game. In this game each participant receives a certain amount of private money and is grouped with other participants under conditions of total anonymity. All participants secretly decide how much of their private money to put into a public pool. They know that the amount of money in the pool will be multiplied by the experimenter and equally distributed among the participants

afterwards. Public good games are played for several rounds. While the best strategy for the group is to put all money into the pot, the best strategy for the individual
is to give nothing but nevertheless receive his or her share from the pot. In the public good game by Ehrhart and Keser subjects were allowed to change groups. More
cooperative subjects continuously fled from the less cooperative ones, while the less
cooperative subjects attempted to get into cooperative groups in order to free-ride
on their contributions.

In another public good game study by Page et al. (2005), subjects were informed
about the other participants' contribution histories. They could then form cooperative groups based on mutual partner preference. The most cooperative subjects chose
to interact with each other and thus received higher returns than less cooperative
co-players did.

Finally, Pradel et al. (2009) investigated the existence of assortation of altruistic
individuals in a real-life setting. Participants were students, aged 10–19 years, from
six secondary school classes. In school classes, relations between students vary in
intensity. A student usually has a few friends among classmates, several he or she
simply likes, and others he or she might dislike. Additionally, the relationship to
a few classmates could be one of indifference. The school setting, therefore, offers
the opportunity to examine, first, whether the validity of judgments of other people's
altruism depends on the social closeness of targets and, second, whether friendships
form along the dimension of altruism. Pradel et al. operationalized altruistic behavior as the decision in a dictator game, in which each student was asked to secretly
divide a sum of money between him- or herself and another anonymous classmate.
Subsequently, and unannounced, each student was asked to estimate how each classmate had divided the money in the dictator game. Estimates were better than chance.
Moreover, sociometry revealed that the accuracy of estimates depended on social
closeness. Friends and disliked classmates were judged more precisely than liked
classmates or those met with indifference. Most importantly, the results confirmed
the existence of assortative processes, as indeed altruists were friends with more
altruistic persons than were egoists.

To sum up, the results show that the second condition for the evolution of altruism
through assortation is supported: altruistic individuals associate with each other.

7.4 Conclusion

The existence of genuine human altruism is a paradox, as evolutionary theory
teaches us that natural selection should have erased this trait. Sociobiology has
shown that a lot of altruism in humans indeed emerges as ultimate self-interest.
The central aim of this work, however, was to show that genuine altruism does exist
and that it definitely can be a result of natural selection.

Our argument is based on results of dictator game studies, which show that individuals do behave genuinely altruistically in one-shot interactions when no one is
watching. In a first instance, we have argued that existing theories on the evolution
of altruism might have failed to explain this finding, because they focus on the
consequences of behavior only, and leave altruistic intentions aside. Referring to

the commitment model of Frank (1987, 1988, 2005, 2008), we have shown subsequently that the whole picture is changed by the integration of the assumption that individuals signal their firm intentions to behave altruistically. For if genuine altruism is recognizable, this gives altruists the chance to cooperate with each other so that the selfish advantage of egoists is outcompeted by extra-payoffs of mutual cooperation between altruists.

We have derived that this selection process is only possible if (1) individuals can distinguish altruists from egoists, and (2) altruists cooperate electively with other altruists. Reviewing a wide range of empirical findings, we have shown that these two conditions are given. Thus, genuine altruism might indeed be a result of assortative processes between altruists who recognize each other's cooperative tendencies.

Although the commitment model successfully explains how genuine altruism could stand the pressures of natural selection, we have to admit at this point that it is questionable whether the commitment model suffices to explain the *initial* evolution of genuine altruism, as there seems to exist a "useless-when-rare" problem: even if 1 day there had been changes in the genetic structure of *one* individual that made this individual become a genuine altruist, this genuine altruist would have had no advantage from being altruistic if he had not been able to find another genuine altruist to cooperate.

This again shows that the explanation of altruism is complex. Thus, in our view, Frank's commitment model might be an especially fruitful explanation in combination with other innovative ideas that have been developed to explain the initial evolution of moral emotions.

As one important example, we mention the theory of the extended phenotype by Voland and Voland (1995), which explains the evolution of conscience. Pointing out that emotional moral conflicts are due to human conscience, Voland and Voland developed the hypothesis that the conscience (which leads to moral emotions) does not suit the selfish-gene interests of the individual possessing such a conscience, but the selfish-gene interests of this individual's parents. To elaborate, the authors assume that conscience evolved due to conflicts that parents and their offspring face with regard to altruistic behavior. Parents and their offspring share 50% of their genes. Therefore, their reproductive interests are quite similar, but not completely identical. For this reason, parents and children may develop different views on how to behave properly; more precisely, parents will demand more altruism from their children than the latter are willing to display. Taking this into account, Voland and Voland assumed the conscience to have evolved as an "extended phenotype" of parental genes, which rules the offspring's behavior in a lifelong way, even when there are no longer any direct opportunities for parental manipulation.

Let us assume that in some individuals an initial kind of conscience indeed evolved, due to the capability of their parents to "extend their phenotype". If this is true, the commitment model would be a perfect candidate to explain why the resultant consciousness-loaded altruists developed even further: because they were able to identify one another and because they started to assort themselves to cooperate.

To conclude, does true altruism exist? If true altruism is defined as a behavior that is harmful to the altruist in both the short run and the long run, the answer is still "no". But if we define altruism as a behavior that is driven by moral emotions and is not motivated by the intention to be profitable, then the answer is "yes". Therefore, Ghiselin and many others might have been too skeptical. Sometimes altruism is more than skin deep, and if you scratch an altruist you see an altruist bleed.

References

Alexander RD (1987) *The Biology of Moral Systems*. Aldine Transaction, Piscataway, NJ

Berscheid E (1985) Interpersonal attraction. In: Lindzey G, Aronson E (eds) *The Handbook of Social Psychology*, 3rd ed. Random House, New York, NY, pp 110–168

Bill & Melinda Gates Foundation (2009) http://www.gatesfoundation.org/about/Pages/foundation-fact-sheet.aspx. Accessed 17 July 2009

Brosig J (2002) Identifying cooperative behaviour: some experimental results in a prisoner's dilemma game. Journal of Economic Behaviour and Organization 47:275–290

Brown WM, Moore CM (2000) Is prospective altruist-detection an evolved solution to the adaptive problem of subtle cheating in cooperative ventures? Supportive evidence using the Wason selction task. Evolution and Human Behavior 21:25–37

Brown WM, Palameta B, Moore C (2003) Are there nonverbal cues to commitment? An exploratory study using the zero-acquaintance video presentation paradigm. Evolutionary Psychology 1:42–69

Camerer C (2003) *Behavioral Game Theory: Experiments on Strategic Interaction*. Princeton University Press, Princeton, NJ

Cosmides L, Tooby J (1992) Cognitive adaptation for social exchange. In: Barkow J, Cosmides L, Tooby J (eds) *The Adapted Mind*. Oxford University Press, New York, NY

Darwin CR (1871) *The Descent of Man, and Selection in Relation to Sex* [Desmond A, Moore J (eds) (2004) The Descent of Man, and Selection in Relation to Sex. Penguin, London]

Ehrhart K, Keser C (1999) Mobility and cooperation: on the run. CIRANO Working Paper no. 99s-24. http://www.cirano.qc.ca/pdf/publication/99s-24.pdf. Accessed 13 November 2009

Fehr E, Fischbacher U (2005) Altruists with green beards. Analyse und Kritik 27:73–84

Fehr E, Gächter S (2002) Altruistic punishment in humans. Nature 415:137–140

Fehr E, Gächter S (2003) Egalitarian motive and altruistic punishment. Nature 425:392–393

Fetchenhauer D, Groothuis T, Pradel J (2010) Not only states but traits—Humans can identify permanent altruistic dispositions in 20 s. Evolution and Human Behavior 31:80–86

Forbes (2009) The world's billionaires. http://www.forbes.com/lists/2009/10/billionaires-2009-richest-people_The-Worlds-Billionaires_Rank.html. Accessed 17 July 2009

Frank RH (1987) If homo oeconomicus could choose his own utility function, would he want one with a conscience? The American Economic Review 77:593–604

Frank RH (1988) *Passions Within Reason*. Norton, New York, NY

Frank RH (2005) Altruists with green beards: still kicking? Analyse und Kritik 27:85–96

Frank RH (2008) On the evolution of moral sentiments. In: Crawford C, Krebs D (eds) *Foundations of Evolutionary Psychology*. Erlbaum, New York, NY

Frank RH, Gilovich T, Regan DT (1993) The evolution of one-shot cooperation: an experiment. Ethology and Sociobiology 14:247–256

Ghiselin MT (1974) *The Economy of Nature and the Evolution of Sex*. University of California Press, Berkeley, CA

Gintis H, Bowles S, Boyd R, Fehr E (2003) Explaining altruistic behaviour in humans. Evolution and Human Behavior 24:153–172

Hagen EH, Hammerstein P (2006) Game theory and human evolution: a critique of some recent interpretations of experimental games. Theoretical Population Biology 69:339–348

Hamilton WD (1964) The genetical evolution of social behavior. Journal of Theoretical Biology 7:1–52

Henrich J, Boyd R, Bowles S, Camerer C, Fehr E, Gintis H, McElreath R (2004) Overview and synthesis. In: Henrich J, Boyd R, Bowles S, Camerer C, Fehr E, Gintis H (eds) *Foundations of Human Sociality: Economic Experiments and Ethno-graphic Evidence from Fifteen Small-Scale Societies*. Oxford University Press, Oxford

Hobbes T (1651/2009) Leviathan. http://en.wikisource.org/wiki/Leviathan. Accessed 27 July 2009

Johnson DDP, Price E, Takezawa M (2008) Renaissance of the individual. Reciprocity, positive assortment, and the puzzle of human cooperation. In: Crawford C, Krebs D (eds) *Foundations of Evolutionary Psychology*. Erlbaum, New York, NY

Machiavelli N (1498/2009) The Historical, Political, and Diplomatic Writings, vol. 3 (Diplomatic Missions 1498–1505). http://oll.libertyfund.org/?option=com_staticxt&staticfile=show.php%3Ftitle=1866. Accessed 27 July 2009

Mascie-Taylor CGN (1995) Human assortative mating: evidence and genetic implication. In: Boyce AJ, Reynolds V (eds) *Human Populations: Diversity and Adaptations*. Oxford University Press, Oxford

Milinski M, Semmann D, Krambeck H-J (2001) Reputation helps solve the 'tragedy of the commons'. Nature 415:424–426

Miller GF (2000) *The Mating Mind*. Vintage, London

Miller GF (2007) Sexual selection for moral virtues. The Quarterly Review of Biology 82:97–125

Nowak MA, Sigmund K (2005) Evolution of indirect reciprocity. Nature 437:1291–1298

Ockenfels A, Selten R (2000) An experiment on the hypothesis of involuntary truth signaling in bargaining. Games and Economic Behavior 33:90–116

Page T, Putterman L, Unel B (2005) Voluntary association in public goods experiments: reciprocity, mimicry and efficiency. The Economic Journal 115:1032–1053

Pradel J, Euler H, Fetchenhauer D (2009) Spotting altruistic dictator game players and mingling with them: The elective assortation of classmates. Evolution and Human Behavior 30:103–113

Price ME (2006) Monitoring, reputation, and 'greenbeard' reciprocity in a Shuar work team. Journal of Organizational Behavior 27:201–219

Richerson PJ, Boyd R (2001) Built for speed, not for comfort: Darwinian theory and human culture. History and Philosophy of the Life Sciences 23:423–463

Richerson PJ, Boyd R (2005) *Not by Genes Alone*. University of Chicago Press, Chicago, IL

Sesardic N (1995) Recent work on human altruism and evolution. Ethics 106:128–157

Sheldon KM, Sheldon MS, Osbaldiston R (2000) Prosocial values and group assortation within an *N*-person prisoner's dilemma. Human Nature 11:387–404

Singer P (1981) *The Expanding Circle: Ethics and Sociobiology*. Oxford University Press, Oxford

Sober E, Wilson DS (1998) *Unto Others: The Evolution and Psychology of Unselfish Behavior*. Harvard University Press, Boston, MA

Spuhler JN (1968) Assortative mating with respect to physical characteristics. Eugenics Quarterly 15:128–140

Trivers RL (1971) The evolution of reciprocal altruism. Quarterly Review of Biology 46:35–57

Verplaetse J, Vanneste S, Braeckman J (2007) You can judge a book by its cover: the sequel. A kernel of truth in predictive cheating detection. Evolution and Human Behavior 28:260–271

Voland E (2007) *Die Natur des Menschen*. CH Beck, Munich

Voland E, Voland R (1995) Parent–offspring conflict, the extended phenotype, and the evolution of conscience. Journal of Social and Evolutionary Systems 18:397–412

Wilson DS, Sober E (1994) Reintroducing group selection to the human behavioral sciences. Behavioral and Brain Sciences 17:585–654

Wynne-Edwards VC (1962) Animal dispersion in relation to social behaviour. Oliver and Boyd, Edinburgh, UK

Yamagishi T, Tanida S, Mashima R, Shimoma E, Kanazawa S (2003) You can judge a book by its cover. Evidence that cheaters may look different from cooperators. Evolution and Human Behavior 24:290–301

Illusion Number 3
The Biological Imperative
Doesn't Matter to Us

Chapter 8
Mortality Crises and Their Consequences for Human Life Histories

Charlotte Störmer and Kai P. Willführ

Abstract Studies on the long-term consequences of crises experienced in early childhood do not draw a uniform picture as to the extent that such crises have an impact on later life. There are studies showing that those who survive a crisis exhibit higher mortality in later life but also that these individuals exhibit lower or unchanged mortality. In this chapter, we develop a hypothesis according to which the impact on later life (or on survival) is caused by the severity of the crisis and that—independent of the nature of the crisis—these effects might be mediated by the immune system.

8.1 Life History Theory and the Illusion of Unconstrained Life Planning

Biological adaptations comprise not only morphological and physiological features, but also the timing of the life phases. Individuals require life courses adapted to their specific living and environmental conditions in order to be able to assert themselves in evolutionary competition. So-called life history theory (LHT) (Stearns 1992; Roff 1992) deals with these temporal aspects of biological adaptivity. A fundamental aspect of this evolutionary approach is the assumption that every organism has only a limited amount of energetic resources at its disposal for the various areas of life. There are allocation conflicts between somatic and reproductive effort, if resources are in scarce supply. Organisms have to adapt their life courses to their respective living environments and have to balance various alternatives. Bogin and Smith sum this up as follows: "For a mammal, it is the strategy when to be born, when to be weaned, how many and what type of pre-reproductive stages of development to pass through, when to reproduce, and when to die." (Bogin and Smith 2000)

C. Störmer (✉)
Zentrum für Philosophie und Grundlagen der Wissenschaft, Universität Gießen, 35394 Gießen, Germany
e-mail: Charlotte.Stoermer@phil.uni-giessen.de

U.J. Frey et al. (eds.), *Homo Novus – A Human Without Illusions*,
The Frontiers Collection, DOI 10.1007/978-3-642-12142-5_8,
© Springer-Verlag Berlin Heidelberg 2010

It is of heuristic value to subdivide the life history effort into three areas: invest-ment into maintenance, growth, and reproduction (Chisholm 1993; McDade 2003). The larger the current effort in one of these areas, the less energy remains for the other areas (e.g., there is a trade-off between current reproductive effort and future survival (see Lummaa 2007); in other words, the more that is being cur-rently invested into reproduction, the more life expectancy will be adversely affected hereby).

In this first of three chapters on LHT, we shall initially examine an aspect of self-preservation comprising one of the most important fields of maintenance, namely the immune system. In particular, we are going to focus on increased immune investment during a crisis and the related consequences for the organism's later life course.

In the next chapter, by Virpi Lummaa, the subject will be the high energetic costs of reproductive effort in women and the significance hereof for senescence processes.

Rebecca Sear argues in the third chapter that investment in growth (and thus the height of an individual in adulthood) has an impact on reproductive success depending on the prevailing environmental conditions.

However, what does this research discipline of evolutionary biology have to do with the illusionary views of human beings about themselves as such? Although not all turning points of human life history can be influenced, much in our life course is under our own personal control—that, at least, is the subjective perception and the common opinion held by many individuals. Accordingly, we do not consider our birth to be capable of being influenced, yet we feel free with regard to our own family planning. LHT teaches us that this is an illusion, because humans cannot set aside their biological conditionality, even after a long phylogenetic history.

Despite all cultural achievements, humans are still not able to free themselves from biological allocation conflicts. Trade-off problems continue to confront us with unchanged severity. Our personal life courses and social parameters are basically shaped by these trade-off problems. Therefore, human life courses are also subject to the biological imperative, even in modern environments.

8.2 Introduction

Demographically speaking, crises are time periods in which a higher than average mortality rate can be measured within populations (Kloke 1998). Crises and times of crisis not only possess a measurable influence on the population level, however, but are also associated with consequences for every individual in the population. Specifically, this means that the individual organism concerned is permanently exposed to harmful physiological and/or psychological stress. And it is precisely this harmful stress which leads to the increased probability of dying for individ-uals, a circumstance which is then reflected on the population level by a directly increased mortality rate. A crisis can, therefore, also be grasped as a time period with an increased risk of death, which can be observed both on the population level and on the individual level.

In addition to direct consequences, crises can also have long-term consequences. Long-term consequences are understood to be effects that have an impact after the crisis for a longer period of time or which only occur some time after the crisis. This chapter will explore the long-term consequences of crises from the perspective of the individual. Theoretically, a crisis can lead to three different scenarios with regard to long-term consequences:

1. The survivors of a crisis are not at all or only marginally affected by long-term consequences. On the population level, *no* changed risk of mortality in later life is detectable between individuals who survived the crisis and individuals who were spared the crisis.
2. The survivors of a crisis are more robust in later life and are characterized by a reduced susceptibility to disease or a reduced overall vulnerability. On the population level, it has been ascertained that individuals who survived the crisis have a *lower* risk of mortality in later life than those individuals who were spared the crisis.
3. The survivors of a crisis are more susceptible to disease or more vulnerable in later life. On the population level, it is ascertained that the individuals who survived the crisis have a *higher* risk of mortality in later life than those individuals who were spared the crisis.

Evidence can be found for all three of the theoretical possibilities listed. For example, it has been observed that the survivors of a crisis are affected by a lasting or temporarily increased mortality in later life (e.g., Catalano and Bruckner 2006), whereas other studies document the opposite, namely that the survivors of a crisis have a lower risk of mortality in later life (e.g., van Poppel and Liefbroer 2005). And the non-occurrence of long-term consequences after the experience of a crisis can also be observed (e.g., Bateson 2001; Willführ 2009). What is the explanation for the fact that crises can vary to such an extent with regard to their long-term consequences? One explanation could be: These are different populations with different living conditions, and therefore they are influenced by different factors. Hence, it would be logical that more differences than similarities are found between the two populations. Our concern, however, is to combine potential long-term consequences of crises into *one* explanatory concept while taking LHT into account. In this chapter, we present a hypothesis according to which the long-term consequences of crises do not vary randomly. It is the severity of the crisis namely that determines whether and in what form long-term consequences will occur.

8.3 Empirical Data

On the population level, an increased risk of mortality can be ascertained, especially during famines, illnesses, and social crises. This is why it appears to make sense to subdivide the crises into three types:

1. Subsistence crises
   These include all famines and longer periods with malnutrition and/or undernutrition in particular.

2. Epidemiological crises
   This type comprises crises that can be traced back to bacterial or viral infectious diseases or parasites.
3. Social crises
   A social crisis can be triggered by the loss of one parent, by a downward social movement, or by increased competition within one's social group, for example.

However, it is not only on the population level that this breakdown is plausible. A distinction can also be made between the burdens of a famine, an illness, or psychological stress (for example, during mourning, see Bartrop et al. 1977) with regard to the individual. The types of crises represent different triggers of a crisis for the individual, even if there can be overlap with regard to the triggers (e.g., if a famine follows an illness).

Table 8.1 summarizes the results of various studies that were able to ascertain long-term consequences after a preceding experience of a crisis. In this table, the studies (Aaby et al. 1993; Bateson 2001; Bengtsson and Lindström 2003, Bengtsson and Broström 2009; Catalano and Bruckner 2006; Flinn 2006; Moore et al. 2006; van den Berg et al. 2009; van Poppel and Liefbroer 2005; Störmer (unpublished), and Willführ 2009) are allocated to the aforementioned three types of crises (subsistence crisis, epidemiological crisis, or social crisis).

Studies documenting a link between the experience of a crisis and long-term consequences can be found for all three types of crises (see Table 8.1). The observed long-term consequences extend from increased to reduced mortality in later life. One could gain the impression that only a few studies show that crises do not cause ascertainable long-term consequences. On the one hand, this is probably based on the fact that many studies primarily deal with the direct consequences and not with the long-term consequences of having experienced a crisis. On the other hand, this is due to a publication bias: Positive results are more likely to be published than negative ones, which is why the absence of long-term consequences after the experience of a crisis is underrepresented in this list (and this applies in particular to crises experienced during childhood).

8.4 Hypothesis

The central element of our hypothesis is that the severity of the crisis is the crucial factor for determining whether and how a crisis will lead to long-term consequences. Accordingly, the potential long-term consequences are aggravated when the crisis is more severe. Table 8.2 compares three different hypothetical crisis scenarios with their long-term consequences.

If a moderate crisis has occurred, the costs or the burdens for the individuals affected tend to be minor. This means that the risk of mortality tends to be slightly increased during the crisis, and therefore there is no noteworthy selection. In addition, the survivors are able to compensate for the disadvantages incurred by them

Table 8.1 Summary of various studies that deal with long-term consequences after a crisis

Type of crisis		Authors	Stressor	Timing of stress	Long-term consequences	Notes
Subsistence crises	"Dutch Famine" and "Siege of Leningrad"	Bateson 2001	Famine	In utero	Increased risk of developing diabetes vs. no long-term consequences	Predictive adaptive response (PAR) (match–mismatch scenario)
	The Gambia	Moore et al. 2006	Undernutrition	Infancy	Increased adult mortality	Seasonality
	Danish Twin Registry	van den Berg et al. 2009	Adverse economic conditions	Early life (around birth)	Increased old-age mortality	
Epidemiological crises	Guinea-Bissau	Aaby et al. 1993	Measles	Before 6 months of age	Delayed increase of mortality risk during childhood	Dose of infection is important for long-term consequences
	Scanian Demographic Database	Bengtsson and Lindström 2003	Disease load	First year of life	Increased old-age mortality	
	Scanian Demographic Database	Bengtsson and Broström 2009	Disease load	During birth year	Increased old-age mortality	
	Krummhörn, East Frisia	Störmer (unpublished)	Smallpox	Pre- and postnatal	Decreased male old-age mortality	

Table 8.1 (continued)

	Type of crisis	Authors	Stressor	Timing of stress	Long-term consequences	Notes
Social crises	Historical Sample of the Population of the Netherlands	van Poppel and Liefbroer 2005	Parental loss	First year of life	Decreased adult mortality	Selection scenario
	Island of Dominica	Flinn 2006	"Early family trauma" (divorce, death, or abuse)	Childhood	Increased adult morbidity	
	Three Historical European Datasets	Catalano and Bruckner 2006	Adverse living conditions	Early childhood	Increased adult morbidity and mortality	
	Krummhörn, East Frisia	Willführ 2009	Parental loss	Infancy, early childhood	No persisting effect on later mortality	Long-term consequences caused by external conditions (e.g. step-parents and mating effort)

Table 8.2 Description of the various consequences of a crisis experienced in early childhood in relation to the severity of the crisis

	Relative strength or weakness of the crisis	Direct consequences of the crisis[a] or which individuals survive the crisis?	Does a selection occur?	Consequence for the survivors' immune system	Long-term consequences of the crisis[a]
Trade-off scenario	Very strong; most individuals die	Only the most robust individuals survive	Yes	Lasting damage, which cannot be rectified or only slowly	The survivors are debilitated and more susceptible to disease; increased risk of mortality
Selection scenario	Strong; many individuals die	Especially the strong and the less susceptible individuals have a survival advantage	Yes	Damage can be rectified	The survivors are stronger and less susceptible to disease; lower risk of mortality
Compensation scenario	Moderate; only a few individuals die	Even many weaker individuals survive	No (or only marginally)	Damage can be compensated for directly	No long-term consequences

[a]In comparison with a control group that was not affected by the crisis.

following the crisis, so that no long-term or late effects are able to be observed. Thus the risk of mortality in the survivors' later life does not significantly vary from those who were spared the crisis. We shall refer to this scenario as the *compensation scenario* in the following. If the crisis is more severe and if the stronger or less susceptible individuals tend to survive it, then a selection process invariably occurs. Once such a crisis has ended, the survivors can compensate for their disadvantages and recover the costs after the crisis. The survivors of the crisis should have a lower risk of mortality in later life compared with those individuals who were spared the crisis, as the crisis has selected the stronger or less susceptible individuals. In the following, this scenario will be called the *selection scenario*. If a crisis is so severe that only the strongest or most robust individuals survive, and if they suffer lasting damage or impairments, the survivors should show a higher risk of mortality in later life. The consequences of such a crisis cannot be compensated for and thus permanently weaken the survivors. In other words, there is a trade-off between current and later survival. In the following, we shall refer to this as the *trade-off scenario*.

8.5 Discussion

In the literature, a few hypotheses attempting to explain the emergence of long-term consequences after an individual has lived through a crisis are discussed. In the following, these hypotheses are compared with ours and subsequently, the extent to which they are compatible is shown.

An explanatory approach (A) says that the crisis causes a physiological deficiency, which leads to lasting damage and/or lastingly disturbs important development processes (Bengtsson and Broström 2009). The result of such crises coincides with the trade-off scenario: The damage directly caused by the crisis or the rectification of such damage is so grave and/or lastingly disrupts important development processes to such a degree that this later leads to a survival disadvantage for the survivors of the crisis. The study by Metcalfe and Monaghan (2001) finds indications that compensation for growth retardation caused by malnutrition or undernutrition is associated with negative long-term consequences (Metcalfe and Monaghan 2001). It is important to emphasis that Metcalfe and Monaghan refer only to compensation for growth retardation and not to compensation for the costs of the crisis in general. Another explanatory approach (B), which is to be found in the literature, assumes that (phylogenetic) adaptation strategies to permanently poor living conditions exist. An early crisis situation thus leads to an adaptive modification of the ontogenesis, namely in the manner that the experience of hunger in the early life of an organism, for example, is interpreted as being an indicator for further such experiences. The adaptation to these (presumed) lasting subsistence difficulties can mean a survival advantage for the organism. This is described as a "predictive adaptive response" (PAR) (Gluckman et al. 2005). Whether mortality-reducing or mortality-increasing long-term consequences are able to be observed depends on the future environmental conditions. If the environment also offers poor living conditions in the future, or if further crises occur, the survivors of the crisis will have a survival advantage, as their organisms have permanently adapted to these poor environmental conditions ("match"). As the survivors of a crisis have a higher probability of survival, such long-term consequences are to be compared with the selection scenario, in terms of their results, even though no selection takes place here, because the individuals are not selected on the basis of varying features in the event of PAR, but adapt ontogenetically.

If, however, the living conditions in the future prove to be different from what was "predicted", the ontogenetic changes made by the organism will automatically prove to be a disadvantage ("mismatch"). A body that, for example, has physiologically adapted to a lack of calories, but later lives with a surplus of calories, will automatically do worse in this environment (see Gluckman and Hanson 2004; Bateson 2001). This result would, therefore, tend to be in line with the trade-off scenario. The main difference to our hypothesis is that our hypothesis (initially) refrains from the assumption that there are phylogenetic adaptation strategies within the meaning of a "predictive adaptive response".

What explanatory approaches A and B have in common is that they see a direct correlation between an early crisis and the corresponding long-term consequences. Moreover, there is another explanatory approach (C), which explains long-term consequences through indirect agency. From this perspective, a crisis causes consequential situations, which in turn also lead to negative consequences for the individual. Having step-parents, for example, is associated with increased child mortality in some studies. So that this social crisis can even occur, it must have been preceded by an event that is also critical, namely the loss of a natural parent (Willführ and Gagnon, unpublished). These "chains of risk" lead to accumulated effects for the individuals concerned (Ben-Shlomo and Kuh 2002). The result of such a situation would correspond to the trade-off scenario, but it must be emphasized that here, in contrast to the trade-off scenario, it is not solely the (original) crisis which causes the damage that adversely affects the individual's later life. Such compounded effects are also conceivable without a causal relationship between a crisis and a subsequent crisis; for example, a subsistence crisis may follow a severe illness independent of that illness. This explanatory approach is also discussed with regard to long-term consequences that occur in connection with epidemiological crises. On the basis of their data, however, Bengtsson and Mineau consider these indirect cause-effect chains to be unlikely (Bengtsson and Mineau 2009).

Finally, it must be noted for the time being that the hypotheses described in the literature to explain crises and their long-term impact are capable of being linked with our model. However, our three-stage model (depending on the severity of the crisis) is partly based on a different functional logic. This will be explained in more detail in the following.

A key issue facing our hypothesis is why different types of crises should be measured solely by their severity with regard to their long-term consequences. What is the link between the long-term consequences of a subsistence crisis, an epidemiological crisis, or a social crisis?

For life histories, McDade makes a distinction between the three areas of maintenance, growth, and reproduction, into which an organism must invest in order to successfully maximize its fitness (McDade 2003; see Chisholm 1993 as well). The development and maintenance of the immune system is one of the most cost-intensive areas of maintenance (Lochmiller and Deerenberg 2000), which is why a key role is ascribed to it in the life history. Consequently, a well-functioning immune system is indispensible for longevity. Could it not be, therefore, that especially adverse long-term consequences and an increased risk of mortality in later life are caused by reduced functionality of the immune system? An immune system that is permanently damaged or unable to fully develop will not exactly represent a survival advantage in the future. If this is true, it is plausible to assume that the long-term consequences of all of the aforementioned types of crises are conveyed by the immune system, provided that a direct link can be traced between the crisis and the immune system. Moore et al. were able to show that a scarcity of food in early childhood and malnutrition lead to increased mortality in adulthood (Moore et al. 1999, 2006). These authors also explain this as follows: The early lack of food impairs developmental processes and therefore the immune system is permanently damaged (Moore et al. 1999); thus, one could speak of a type of weakening of the immune defenses following subsistence. However, not only subsistence crises provide empirical evidence that long-term consequences are linked to the immune system. Some diseases early in life increase the risk of mortality in later life, or, at least, increase it after a time lag. Aaby and his colleagues were able to show that an early measles infection increases the mortality of the children concerned for some years even after their recovery (Aaby et al. 1993). In their work, Bengtsson and Broström explain the later increased risk of mortality by the fact that "physiological damage from severe infections at start of life leads to higher mortality at older ages" (Bengtsson and Broström 2009, p. 1583). The difference between the interpretation of Bengtsson and Broström and our hypothesis lies in the fact that we assume that the physiological damage is caused directly by an impaired immune system. The study by Fridlizius already directly established a connection between a baby being infected by smallpox in the first year of life and irreversible damage to the immune system (Fridlizius 1989). Even with physiological damage that is ascertained after an epidemiological crisis, and which initially cannot be directly associated with an impairment of the immune system, it is empirically unproven to date whether this is actually directly linked to the illness or whether this is caused by a suboptimal or disturbed immune system. Evidence of such correlations between a (temporarily) weak immune system and physiologically lasting damage definitely exist. With a so-called asymptomatic course, the Epstein-Barr virus increases, for example, the risk of nasopharyngeal carcinoma (Crawford 2001; Amon and Farrell 2005). An inflammation of the heart muscle with lasting damage to the heart, favored by a high endogenous level of cortisol, occurs as a result of protracted colds, especially in top athletes (Sack 2004).

Once again: In this perspective, the damaged or disturbed immune system is directly responsible for the physiological consequences. Can such circumstances that harm the immune system also be detected when a social crisis is experienced that initially has a psychological impact and not a direct impact on the physiology? In actual fact there is evidence of this. Bartrop et al. showed that the immune system of mourners is significantly disturbed (Bartrop et al. 1977). A direct link between social crises and the immune system can therefore be deemed proven. Furthermore, Flinn was able to show that children who grew up with step-parents have a permanently elevated hormone titer that is relevant for the immune system (Flinn 2006). This is the case even when both the step-parent and the step-child indicate that they have a good relationship with one another. Flinn identified cortisol as the key hormone in reaction to both psychological and physiological stress. To what extent cortisol is linked to the long-term consequences cannot be discussed any further here, especially because Flinn shows, in a current study, that traumatized individuals do not automatically suffer from a disturbed hormone balance in later life (Flinn 2009). It is known from medical studies, however, that the permanent administration of cortisol or a lastingly elevated, endogenous cortisol level (such as when under permanent stress) can lead to serious and lasting damage (Wolf 2007).

Figure 8.1 shows the hypothetical impact of the various types of crises on the immune system. It can be taken for granted that all three types of crises are connected to the immune system and it can be assumed that they utilize the same physiological/biochemical pathways. In turn, this means that the long-term consequences differ only through their triggers, not in their modes of action.

We are of the opinion that in addition to a heuristic classification system, our hypothesis is capable of explaining the variance in the results of the studies with regard to the long-term consequences without solely referring to the different natures of the datasets. The central hypothesis is the assumption that the long-term consequences of a crisis are determined solely by the severity of the crisis. The immune system, or the damage to the immune system, could play a central role here. If a crisis causes costs or damage to the immune system that can be compensated for by the organism after surviving the crisis, a reduced mortality of the survivors of the crisis in later life is observed, provided that the crisis selects the strong and less susceptible individuals. More moderate crises, which do not lead to a selection process, do not entail long-term consequences. Crises leading to irreparable damage or to developmental damage to the immune system that cannot be compensated for mean higher mortality in later life for the survivors of the crisis.

As the severity, not the type, of crisis determines whether and in what form long-term consequences will occur, there is no need to strictly separate the three types of crisis. If our hypothesis is correct, then it does not matter whether one is able to or must distinguish whether a child is now suffering from the loss of his father or from his nutritional status, which has thus become worse, and/or due to the possibly increased susceptibility to parasites and diseases as a result of the worsening nutritional status. The severity of the crisis overall is decisive. A subdivision into individual types of crises is not necessary, as they all create a comparable situation for the immune system (see Fig. 8.1).

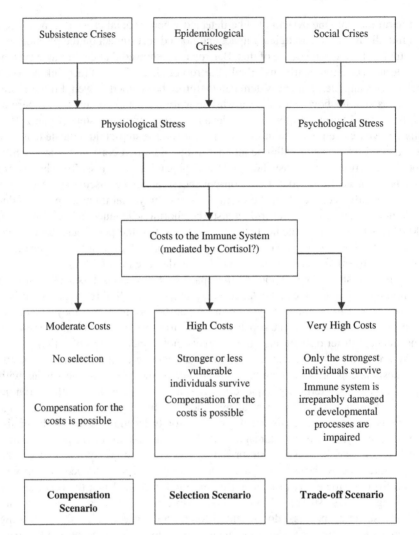

Fig. 8.1 Hypothetical representation of the influence of the various types of crises on the human immune system and the long-term impact resulting herefrom. A key mediation function could be played by the stress hormone cortisol here

8.6 Outlook

If it were possible to develop a uniform measure for the severity of the crisis that allows crises to be classified according to their severity, then our hypothesis would be able to assess the occurrence and the nature of the long-term consequences of a crisis. Such a measure would have to be developed on the basis of the direct consequences of a crisis. On the population level, this could be on the basis of the directly increased risk of mortality (with regard to its temporal dimension). This measure was already proposed in 1979 by Hollingsworth for the demographic determination of the severity of a crisis (Hollingsworth 1979). With the aid of this proxy, crises could then be classified hierarchically by their severity.

However, the aforementioned compounded effects and subsequent crises represent a problem here, in our opinion, because in this case the measurement of the severity of the crisis and its temporal extension would be much more difficult. This is likely to apply especially to social crises. Apart from the long-term consequences, a famine can be considered over once the daily intake of calories permanently lies about the target value; and an infectious disease can be assumed to be over, once the patient has recovered. However, many social crises, such as the early loss of a parent, can only be separated from the crises that might possibly follow with great difficulty. (On the basis of two historical populations, for example, Willführ and Gagnon were able to show not only that the daughters' increased risk of mortality is directly caused by the early loss of their mothers, but also that this mortality is increased in addition by life with a stepmother. Moreover, the later increased mortality of these daughters can mainly be associated with having to live with their stepmothers (Willführ and Gagnon, unpublished).) Such complex facts would have to be dealt with the aid of complex methods, such as multivariate analyses.

Living and growing up under poor living conditions is and was never a rarity in the human phylogenesis. The Darwinian theory of evolution predicts adaptive strategies for recurring events. Such an adaptation could exist in a strategic change in life history. It is conceivable, for example, that there is an increased investment in reproduction at the expense of longevity. Lycett et al. demonstrated that the maternal lifespan falls with rising number of children (Lycett et al. 2000). However, this effect was only measurable if the increased fertility coincided with worse living conditions. Our model is initially able to do without the hypothesis that there is an adaptive trade-off between current and later survival. Nevertheless, trade-offs could exist that reduce longevity and yet are adaptive at the same time. There is no empirical evidence to answer this question yet.

We hope that our model is helpful for a better understanding and for further analyses of crises and their long-term consequences.

References

Aaby P, Andersen M, Knudsen K (1993) Excess mortality after early exposure to measles. International Journal of Epidemiology 22(1):156–162

Amon W, Farrell PJ (2005) Reactivation of Epstein-Barr virus from latency. Reviews in Medical Virology 15:149–156

Bartrop RW, Luckhurst E, Lazarus L, Kiloh LG, Penny R (1977) Depressed lymphocyte function after bereavement. The Lancet 309:834–836

Bateson P (2001) Fetal experience and good adult design. International Journal of Epidemiology 30:928–934

Bengtsson T, Broström G (2009) Do conditions in early life affect old-age mortality directly or indirectly? Evidence from 19th-century rural Sweden. Social Science & Medicine 68: 1583–1590

Bengtsson T, Lindström M (2003) Airborne infectious diseases during infancy and mortality in later life in southern Sweden, 1766–1894. International Journal of Epidemiology 32:286–294

Bengtsson T, Mineau GP (2009) Early-life effects on socio-economic performance and mortality in later life: a full life-course approach using contemporary and historical sources. Social Science & Medicine 68:1561–1564

Ben-Shlomo Y, Kuh D (2002) A life course approach to chronic disease epidemiology: conceptual models, empirical challenges and interdisciplinary perspectives. International Journal of Epidemiology 31:285–293

Bogin B, Smith BH (2000) Evolution of the human life cycle. In: Stinson S, Bogin B, Huss-Ashmore R, O'Rourke D (eds) *Human Biology: An Evolutionary and Biocultural Perspective.* Wiley-Liss, New York, NY

Catalano R, Bruckner T (2006) Child mortality and cohort lifespan: a test of diminished entelechy. International Journal of Epidemiology 35:1264–1269

Chisholm JS (1993) Death, hope, and sex. Life history theory and the development of reproductive strategies. Current Anthropology 34(1):1–24

Crawford DH (2001) Biology and disease associations of Epstein-Barr virus. Philosophical Transactions of the Royal Society London B: Biological Sciences 356:461–473

Flinn MV (2006) Evolution and ontogeny of stress response to social challenges in the human child. Developmental Review 26:138–174

Flinn MV (2009) Are cortisol profiles a stable trait during child development? American Journal of Human Biology 21:769–771

Fridlizius G (1989) The deformation of cohorts: nineteenth century mortality in a generational perspective. Scandinavian Economic History Review 37:3–17

Gluckman PD, Hanson MA (2004) Living with the past: evolution, development, and patterns of disease. Science 305:1733–1736

Gluckman PD, Hanson MA, Spencer HG (2005) Predictive adaptive responses and human evolution. Trends in Ecology and Evolution 20(10):527–533

Hollingsworth TH (1979) A preliminary suggestion for the measurement of mortality crises. In: Charbonneau H, Larose A (eds) *The Great Mortalities: Methodological Studies of Demographic Crises in the Past*, Ordina Editions. Liège, Belgium

Kloke IE (1998) Säuglingssterblichkeit in Deutschland im 18.und 19. Jahrhundert am Beispiel von sechs ländlichen Regionen. http://webdoc.sub.gwdg.de/ebook/diss/2003/fu-berlin/1998/19/indexe.html. Accessed 3 January 2010

Lochmiller RL, Deerenberg C (2000) Trade-offs in evolutionary immunology: just what is the cost of immunity? Oikos 88:87–98

Lummaa V (2007) Life-history theory, reproduction and longevity in humans. In: Dunbar RIM, Barrett L (eds) *Oxford Handbook of Evolutionary Psychology.* Oxford University Press, Oxford

Lycett JE, Dunbar RIM, Voland E (2000) Longevity and the costs of reproduction in a historical human population. Proceedings of the Royal Society London B: Biological Sciences 267:31–35

McDade TW (2003) Life history theory and the immune system: steps toward a human ecological immunology. American Journal of Anthropology 122(S37):100–125

Metcalfe NB, Monaghan P (2001) Compensation for a bad start: grow now, pay later? Trends in Ecology and Evolution 16(5):254–260

Moore SE, Cole TJ, Collinson AC, Poskitt EME, McGregor IA, Prentice AM (1999) Prenatal or early postnatal events predict infectious deaths in young adulthood in rural Africa. International Journal of Epidemiology 28:1088–1095

Moore SE, Collinson AC, N'Gom PT, Aspinall R, Prentice AM (2006) Early immunological development and mortality from infectious disease in later life. Proceedings of the Nutrition Society 65:311–318

Roff DA (1992) *The Evolution of Life Histories—Theory and Analysis.* Routledge, Chapman & Hall, London

Sack S (2004) Der Tod im Sport—ein internistisches Problem? Herz 29(4):414–419

Stearns SC (1992) *The Evolution of Life Histories.* Oxford University Press, Oxford

Störmer C (unpublished) Sex differences in the consequences of early-life exposure to epidemiological stress—A life-history approach

Willführ KP, Gagnon A (unpublished) Surviving early parental loss and the consequences of remarriage

Willführ KP (2009) Short- and long-term consequences of early parental loss in the historical population of the Krummhörn (18th and 19th century). American Journal of Human Biology 21:488–500

Wolf A (2007) Chronischer Stress. Pathophysiologie und neurobiologische Folgen. Journal of Preventive Medicine 3:170–178

van den Berg GJ, Doblhammer G, Christensen K (2009) Exogenous determinants of early-life conditions, and mortality later in life. Social Science & Medicine 68:1591–1598

van Poppel F, Liefbroer AC (2005) Living conditions during childhood and survival in later life—study design and first results. Historical Social Research 30:265–285

Chapter 9
Costs and Consequences of Reproduction

Virpi Lummaa

Abstract The life history of women is characterized by several unusual patterns: women have a relatively late age at maturity compared to other primates, they produce offspring at short inter-birth intervals, and typically have many dependent offspring of varying ages to care for simultaneously. Women then lose their potential to bear children at menopause but can live a few decades afterwards. Such a reproductive strategy involves several trade-offs and costs of reproduction to future success that have to be optimized across the entire lifespan. This chapter summarizes evidence from humans on the costs of reproduction. First, I discuss the short-and long-term effects of investment in reproduction on the survival patterns of individuals. Second, I address how current reproductive investment affects the ability to invest in future reproductive events. Third, I review the evidence for such costs of reproduction and trade-offs changing with the age of the individual and across different environments. Trade-offs are predicted to be most severe among the very young and senescing females, and when resources are limited. Finally, I investigate the heritable genetic basis for individual differences in the consequences of reproduction, and how heritabilities and genetic trade-offs between traits vary with age and across environmental conditions.

9.1 Introduction

The fundamental evolutionary role of reproduction is to ensure genetic contribution to future generations. Ideally, females should start reproduction at maturation and continue increasing their family size with short inter-birth intervals across a long lifespan in order to maximize their lineage persistence across generations. However, such a reproductive strategy is rarely achieved and may not eventually give the highest fitness return because of several constraints and costs involved. First, although

V. Lummaa (✉)

Department of Animal and Plant Sciences, University of Sheffield, Sheffield S10 2TN, UK
e-mail: V.Lummaa@sheffield.ac.uk

U.J. Frey et al. (eds.), *Homo Novus – A Human Without Illusions*,
The Frontiers Collection, DOI 10.1007/978-3-642-12142-5_9,
© Springer-Verlag Berlin Heidelberg 2010

early age at first reproduction is among the most important life-history traits affecting between-individual variation in fitness and has a pivotal effect on family size and reproductive success in humans (Käär et al. 1996; Pettay et al. 2007), it is predicted to evolve to maximize fitness subject to both the benefits and costs of delayed reproduction (Stearns 1992). The costs of delaying reproduction include increased accumulated mortality hazard before reproduction, reduced reproductive span, reduced reproductive output, and longer generation time, whereas the benefits include larger body size, higher initial fecundity and lower offspring mortality brought about by longer growth (Kawecki 1993; Kozlowski 1992; Migliano et al. 2007; Stearns 1992). Second, because of constraints and costs of rapid reproduction to an individual's future reproductive success and survival, large overall family sizes in humans may not necessarily bring the highest fitness returns either (Gillespie et al. 2008).

The costs and consequences of reproduction in human women have interested scientists for over a century (Beeton et al. 1900). Reproduction is predicted to be associated with future health, breeding success, and survival because resources available to individuals in nature are usually limited. Consequently, production of offspring can be "costly" by reducing an individual's ability to invest time and resources in other important body functions, such as growth, immune defense, and body maintenance, and may lead to accelerated reproductive senescence and shorter lifespan (Stearns 1992; Williams 1957). Senescence is manifested as a decline in an individual's physiological and cellular function with age. Evolutionary theories suggest that physiological function decreases with age because, first, genes that have positive effects on fitness early in life will be selected for even if they have negative effects later in life and, second, because a weakening of selection with increasing age leads to an accumulation of mutations (Hamilton 1966; Kirkwood 1977; Medawar 1952; Partridge and Harvey 1985; Williams 1957).

Individuals vary in their consequences and costs of reproduction. In long-lived, iteroparous species that typically reproduce more than once, the relative cost of reproduction may change with age (Charmantier et al. 2006; Réale et al. 1999; Williams 1957), given that social rank, resource access, body condition, and residual reproductive value likely also vary with age. Young growing and old senescent individuals should thus suffer higher costs than prime-age ones, given that they may struggle to meet the energetic demands made by reproduction. The costs of reproduction may also change radically across different environments with differing resource levels (Monaghan et al. 2008), or with different amounts of help available from other individuals with raising the offspring, such as a partner or helpers-in-the-nest, which affect the level of parental investment made and, potentially, ageing rates (Bourke 2007). Rates of senescence might also differ between different life-history traits, or in males versus females, given that females typically invest more in reproduction by producing larger ova and, in mammals, through gestation and lactation (Trivers 1972). Finally, the costs of reproduction may also differ for genetic reasons: individuals may be differently able to bear the consequences of reproduction across different ages, and may also be differently genetically suited to reproduce in a given environment (genotype × environment interactions; reviewed in Wilson et al. 2008).

Humans are a particularly exciting subject with which to study effects of reproduction on senescence. This is because whereas most animals reproduce until they die, in humans, females can survive long after becoming unable to reproduce themselves (Hamilton 1966; Williams 1957), and, although males may remain reproductively capable, they do not often use their potential (Käär et al. 1998). In both historical and traditional hunter-gatherer populations, 30% or more of adult individuals are usually beyond the age of 45, given that most who survive childhood live past their childbearing years (Hawkes 2004). In comparison, in chimpanzee females fertility declines at about the same age as in humans to virtually zero at age 45 (Nishida et al. 2003), but their survival rates follow fertility so that in the wild less than 3% of adults are over 45 (Hill et al. 2001). In humans, aging is slowed down by allocation of more resources to cell maintenance and repair than is done by the nearest primate relatives (Hawkes 2003). Evolutionary life-history theory offers a framework to understand individual differences in the costs and consequences of reproduction and, in humans, such results have implications also for the social sciences, humanities, and public health. This is because although the human life-history is in many ways unusual and modern technology has allowed us to stretch the boundaries of reproduction, ultimately, the same biological principles that underlie life-history evolution in other species have also been documented to apply in humans.

This chapter will summarize evidence from humans on the costs of reproduction. First, I will discuss the short- and long-term effects of investment in reproduction on the survival patterns of individuals. Second, I will address how reproduction affects the ability to invest in future reproductive events. Third, I will review the evidence for such costs of reproduction changing with age and across different environments. Finally, I will investigate the heritable genetic basis for individual differences in the consequences of reproduction, and how these vary with age and across environmental conditions. Most research on the topic is conducted on women, but I will draw comparisons to men wherever such data is available.

9.2 Immediate and Delayed Effects of Reproduction on Survival

Both laboratory experiments on model species and a growing number of studies on natural populations of vertebrates have demonstrated that increases in reproductive effort, such as lower age at first reproduction or higher early-life fecundity, can result in reduced subsequent survival rates (reviewed in Nussey et al. 2008). In humans, the immediate risk of a woman dying from childbirth varies widely: In 2005, on average 9 women per 100,000 births died worldwide in developed regions, whereas the risk in developing regions was 450/100,000 and more than 2,000/100,000 in countries such as Sierra Leone (www.who.int). The probability that a 15-year-old female will die eventually from an (immediate) maternal cause is currently highest in Africa (at 1 in 26), while the developed regions have the smallest lifetime risk at 1 in 7,300. The major causes of maternal death worldwide include severe bleeding/hemorrhage (25%), infections (13%), unsafe abortions (13%), eclampsia (12%), obstructed labor (8%), other direct causes (8%), and indirect causes (20%). Maternal death, in turn, can have downstream effects, not only by cutting short the personal reproductive career, but also by affecting the quality of already produced offspring: several studies show the negative effect that mother loss has on their own infant and child survival (reviewed in Sear and Mace 2008). Thus, particularly in conditions with no modern medical care access, pregnancy and childbirth can reduce a women's longevity by exposing her to early death and have consequences for her fitness.

In addition to the immediate consequences of childbirth for survival, invest-ment in pregnancy and breastfeeding is predicted to carry delayed long-term costs to female longevity by reducing resources available for body maintenance. Costs of pregnancy include resources invested in fetal growth, growth and maintenance of maternal supporting tissues, and maternal fat accumulation, while the costs of lactation include resources invested in milk synthesis and the maintenance of metabolically active mammary glands, and these costs generally outweigh the costs of pregnancy in women who exclusively breastfeed their baby (reviewed in Jasienska 2009). Studies documenting how increases in reproductive effort affect later-life survival have however been equivocal (reviewed in Le Bourg 2007). For example, in historical populations, some studies have been able to establish the expected negative effects of high total reproductive effort on female post-reproductive longevity (e.g. Gagnon et al. 2009; Jasienska et al. 2006a; Westendorp and Kirkwood 1998); but also many studies find no association or a positive correla-tion between total number of children and post-reproductive survival (Korpelainen 2000; Le Bourg et al. 1993; Müller et al. 2002); and sometimes this trade-off is manifested only among the poorest women (Dribe 2004; Lycett et al. 2000). Similar mixed results arise from studies on the association between total family size and longevity in contemporary populations (e.g. Doblhammer 2000; Hurt et al. 2004; Kumle and Lund 2000), or studies on the relationship between age at first reproduction and longevity (reviewed in Helle et al. 2002a, 2005).

Such results are puzzling because if limited resources should promote trade-offs between reproduction and survival, then we would expect to find the strongest evidence for such trade-offs from pre-healthcare populations. Moreover, although short inter-birth intervals are predicted to be particularly detrimental for future sur-vival, studies investigating this association also show mixed results (e.g. Grundy and Tomassini 2005; Menken et al. 2003), and studies on the effects of reproduction on body condition or health in later life are also unequivocal (Tracer 2002). One possibility is that within each population, high-quality individuals have both higher breeding performances across their lifespan and higher probabilities of survival (van de Pol and Verhulst 2006), resulting in positive or "phenotypic" correlations (Daan and Tinbergen 1997) that are difficult to control for in a no-experimental study design. However, this does not explain why most studies that have investigated sim-ilar relationships between family size and longevity in men have generally found no significant relationships at all (e.g. Doblhammer and Oeppen 2003; Helle et al. 2004; but see Penn and Smith 2007) or positive effects of offspring of only one sex (e.g. Jasienska et al. 2006a). Across countries, however, birthrates are indeed related to a sex difference in lifespan: birthrate per female explains 17% of the variation in relative sex differences in lifespan across countries, and low birthrate results in females living relatively longer than males (Maklakov 2008).

An additional complication is that although the cumulative costs of reproduction are related to negative health outcomes such as increased risk of cardiovascular diseases, diabetes, and strokes even in women of good nutritional status, young age at first reproduction and high fertility may also lead to decreased mortality from certain diseases, such as breast and reproductive cancers (reviewed in Jasienska 2009). Depending on the population and overall risk of certain diseases as well as individual differences in the risk of suffering from specific health problems, the costs of reproduction may or may not outweigh the benefits of reproduction to health. Finally, it has also been suggested that extended periods of endogenous estrogen production following continued reproduction and breastfeeding until old age could stimulate biological systems to positively affect survival and health, and could also foster better survival chances through adoption of healthy behaviors or through social support in old age from younger children (Yi and Vaupel 2004).

Nevertheless, when socio-economic status is controlled for in the analyses and high reproductive effort is measured in terms of production of twins over singletons (Helle et al. 2004) or energetically more expensive sons over daughters (Beise et al. 2002; Helle et al. 2002b; Hurt et al. 2004; van de Putte et al. 2004), rather than by production of a large family size per se, high investment in reproduction appears to reduce a women's post-reproductive survival rates, for example due to increased susceptibility to infectious disease (Helle et al. 2004). It is thus possible that in a species such as humans mothers might be able to adaptively adjust their birth intervals or overall family size to match their available resource levels and current body condition to avoid having to pay high costs resulting from too expensive state-dependent reproductive investment, whereas production of twins over singletons or sons over daughters is less under the active control of the mother and more likely to lead to a realization of the costs of reproduction. Resolving the controversy surrounding this topic would benefit from twin designs comparing longevity of genetically similar individuals differing in their reproductive investment patterns, or studies involving large pedigree data sets that allow the associations between reproduction and survival to be investigated not only at the phenotypic, but also at the genetic level (for one example, see Pettay et al. 2005).

Many mechanisms have been identified as contributing to age-related deterioration in function (Nemoto and Finkel 2004). Processes that are widely believed to play an important role in this and could underlie life-history trade-offs involving longevity include the accumulation of oxidative damage to lipids, proteins, and DNA, which then interfere with cell and tissue function, as well as telomere attrition (reviewed in Monaghan et al. 2008). Increases in reproductive effort can impair immune function in the long term (Ardia et al. 2003), and an effective immune system itself may be costly to maintain (Sheldon and Verhulst 1996), and may constrain individual reproductive decisions. Direct evidence that high reproductive effort accelerates immunosenescence or facilitates telomere attrition in humans is however still lacking (but see e.g. Hanna et al. 2009; Helle et al. 2004).

9.3 Costs of Reproduction to Future Breeding Success

In addition to effects on survival, an increase in reproductive effort may manifest as reproductive senescence, i.e. reduced subsequent breeding success. In support of this, female collared flycatchers (*Ficedula albicollis*) that were subjected to experimentally enlarged broods early in life laid smaller clutches in old age (Gustafsson and Part 1990), and female red deer (*Cervus elaphus*) with high early-life fecundity showed stronger subsequent declines in offspring birth weight and delayed calving dates (Nussey et al. 2006). Such relationships between early life reproductive investment and later life survival, maternal performance, and rates of senescence have also been shown to have a genetic basis in natural populations of vertebrates (Charmantier et al. 2006; Nussey et al. 2008; Pettay et al. 2005; Wilson et al. 2008).

Few studies on humans have directly investigated the effects of maternal reproductive investment on her future breeding success, and to my knowledge such studies are entirely absent in men. That such effects are possible at least in women is illustrated by studies showing that increases in current reproductive effort (production of twins or more expensive sons) can reduce the mother's chances of a successful future reproduction (Lummaa 2001) and the quality of the subsequent offspring (Rickard 2008; Rickard et al. 2007, 2009). For example, inter-birth intervals tend to be longer after giving birth to a son than after giving birth to a daughter (Mace and Sear 1997), and mothers who previously produced a son thereafter give birth to offspring that have lower birth weight (Côté et al. 2003; Rickard 2008), smaller size at adulthood (Rickard 2008), and reduced mating and reproductive success (Rickard et al. 2007, 2009) compared to offspring born after a daughter. Moreover, increasing offspring quantity may reduce offspring quality in the short term (Meij et al. 2009) as well as their eventual contribution to maternal fitness (numbers of grand-offspring produced) (Gillespie et al. 2008). This evidence thus predicts that high investment in a current attempt has negative effects on investment in future reproduction, but studies investigating e.g. the effects of offspring birth weight (pregnancy investment) and breastfeeding length (post-natal investment) on future fertility and offspring quality are currently rare.

In women, "reproductive" senescence may also manifest post-reproduction, given that post-reproductive women continue gaining fitness by increasing the survival and/or reproductive capacity of both their own offspring and grand-offspring (reviewed in Sear and Mace 2008). As women age, their ability to affect positively the reproductive success of their adult offspring decreases and young grandmothers are the most helpful ones in aiding survival of their grandchild (Lahdenperä et al. 2004). However, it is not known whether this ability to gain fitness post-reproduction through grandmothering is modified by or traded-off with patterns of reproductive investment prior to menopause. Such effects would have important repercussions for calculating optimal allocation between reproduction and post-reproductive longevity, but are generally ignored in all the current models of menopause evolution (Rogers 1993; Shanley et al. 2007).

9.4 Costs of Reproduction with Age

Both the measures of reproductive performance and costs of reproduction can change with age. First, conception rate and baby birth weight are reduced in older mothers while offspring developmental and genetic problems and neonatal mortality increase (reviewed in Ellison 2001). In pre-industrial women, general offspring quality, as measured by their eventual contribution to the grand-offspring generation, has been documented to decline by as much as 30% from offspring born to mothers aged 17 to offspring born to mothers 40+ years (Gillespie et al. submitted). Such declines in fitness benefits of offspring produced at different ages were caused by both biological and social effects: while maternal age best explained declines in offspring survival to adulthood, increasing birth order and thus competition with siblings (Faurie et al. 2009) became more important in explaining declines in recruitment of adult offspring to reproduction (Gillespie et al. 2008).

Second, also the costs of reproduction to survival may vary with age: effects of birth order of the child on (immediate) maternal mortality risk have been found to follow a J-shaped function, with the risk of dying from childbirth declining after the first birth and then rising again in high-parity women (Knodel 1988). For example, in historical Utah, the risk of mortality after childbirth increased with age, and this risk was greater for women than men (Penn and Smith 2007). Changes in the costs of reproduction with age are predicted, given that not only social rank and thus external resource access commonly vary with age in all human societies, but also changes in body condition and individual residual reproductive value are likely to vary and lead to differences in the personal costs of reproduction. However, little is currently known of such processes, or the causes creating variation between individuals in their reproductive success with age and how previous reproductive effort modifies this. Without assessing the impact of early-life reproductive effort on survival and reproductive outcomes across all reproductive ages, the fitness implications of the long-term effects of early reproductive effort cannot be determined (Nussey et al. 2008).

9.5 Costs of Reproduction Across Environments

Ecological, social, and demographic conditions are likely to interact with the costs of reproduction and influence the rate at which women senesce. Environments with plentiful resources are predicted to be associated with earlier age at maturity and higher reproductive success and survival, whereas resource scarcity and limited energy available for reproduction is predicted to lead to constraints for the expression of optimal combination of these traits (Roff 2002; Stearns 1992).

Limited availability of resources should also promote trade-offs between fitness-related traits, and thus the optimal within-individual allocation of resources is likely to change across resource regimes (Noorwijk and de Jong 1986). In line with this, evidence from both wild animals and humans show that resource availability during reproduction not only influences the costs of reproduction for the breeder, but also the reproductive potential of offspring (Lummaa and Clutton-Brock 2002).

Costs of reproduction can be interpreted in a meaningful way only when they are analyzed in relation to the overall energy budget of the woman: high costs of reproduction will not have the same effects on women who have good diets and low levels of physical activity as on women in poor energetic condition (Jasienska 2009). Such physiological consequences of reproduction for women on different energy budgets are well-documented in humans (reviewed in Jasienska 2009). However, less is currently known of how resource variation affects the strength and direction of selection on human life history, despite this being one of the basic premises of life-history theory. One example is provided by a study on preindustrial Finns, demonstrating that the costs of reproduction were greater in inland areas, where winters are harsh and food was unpredictable, than in milder coastal areas where fish supplemented the diet: selection favored heritable dizygotic twinning in populations enjoying predictable food supply, whereas such an increase in reproductive effort was selected against in populations suffering from frequent famines. The differing selection pressure on multiple births likely led to the observed significant differences in twinning rates between populations with differing access to resources (Lummaa et al. 1998).

Moreover, in agreement with life-history theory, the opportunity for total selection, the strength of natural selection on life-history traits, and trait means differed in the same Finnish populations between women belonging to different wealth classes and thus with differing access to resources. Women from the poorest social class were more likely to have a reduced lifespan due to increased risk of dying from infectious diseases following increased reproductive effort, and experienced in general more profound trade-offs between life-history traits (Pettay et al. 2007). Similarly, even in modern developed countries, individuals with low socio-economic status have, on average, lower life-expectancy (e.g. Martikainen 1995). In historical times, opportunity for selection was higher and selection on earlier age at first reproduction stronger among the poorest mothers compared to wealthier mothers. This is in line with the prediction that selection should favor early reproductive effort in conditions where mortality is high. Further evidence that resource availability may affect selection on life-history traits in humans comes from studies on historical Germans and Swedes. In these populations, a negative relationship between parity and post-menopausal lifespan existed among poor landless women only, whereas in wealthier women, the relationship between parity and post-menopausal lifespan was positive (Dribe 2004; Lycett et al. 2000). A negative relationship between fertility and longevity may, therefore, be expected in women who, due to multiple pregnancies and breastfeeding, pay high costs of reproduction that cannot easily be compensated by increases in dietary intake and reduction in physical activity (Jasienska 2009).

Comparable differences in the costs of reproduction resources arising from social class differences between individuals could also be created, for example, by different amounts of help available from other individuals with raising the offspring, such as partners, grandparents, or other helpers-in-the-nest, that affect the level of investment made by the mother. Unfortunately, there are currently no studies investigating how such differences in the environmental conditions of breeders affect their life-history trade-offs, reproductive costs, and senescence in humans (but see Bourke 2007).

Environmental conditions during early life can have similar long-term effects, but little is known of whether the costs of reproduction and rates of senescence might be modified by the prevailing ecological conditions particularly at critical early stages of life. Poor early-life conditions such as maternal nutrition during gestation and lactation could lead to unfavorable developmental conditions for the offspring in utero (reviewed in Lummaa and Clutton-Brock 2002). This could represent an insult to the developmental process, ultimately reducing overall adult physiological condition. It is thus predicted that the life-history trade-offs will be most severe and the rates of senescence greatest among those women living in harsh ecological conditions, because the importance of contributing heavily to care will be greater in such circumstances and hence reproductive investment will be higher. That such effects are likely is suggesting by the growing evidence across species documenting density-dependent and density-independent aspects of the early environment accounting for large portions of variance in important life-history traits and, consequently, fitness (reviewed in Lindstrom 1999). For example, in humans, varying early environmental conditions, such as month or season of birth, predict longevity and reproductive performance (reviewed in Lummaa 2003). Furthermore, dietary intake prior to birth is associated with subsequent risk of adverse health (Hamdoun and Epel 2007), age at sexual maturation (Walker et al. 2006), ovarian function (Jasienska et al. 2006b), and lifespan (Moore et al. 1999), suggesting that early-life conditions influence development and that this has adverse effects later in life.

However, whether and how such early condition differences vary across ages and relate to senescence patterns is generally unknown. Evidence from red deer shows that rates of reproductive senescence are modified by both early environmental conditions (Nussey et al. 2007) and reproductive investment early in life (Nussey et al. 2006), with individuals born in poor conditions or investing heavily early in life showing greater rates of reproductive decline later in life. In many human societies, individuals reproduce across a long time-span and a range of environmental conditions. Given the ongoing demographic and nutritional transitions world-wide to higher energy diets and advancing maternal ages, such relationships are of potential importance in humans too to researchers from social, political, biological, and medical sciences and warrant further exploration.

9.6 The Genetic Basis for Costs of Reproduction Across Ages and Environments

What constrains the evolution of both high reproductive effort and long lifespan? To determine the potential evolutionary response to selection on traits such as reproductive output at different ages and across different environments, information concerning the genetic structure of the traits in question is necessary. This is because phenotypic correlations between reproductive traits and survival are particularly interesting only if they have a genetic basis, given that natural selection can only lead to an evolutionary response when it acts on a heritable character. For example, studies on wild populations of long-lived mammals have shown that, in poor environments, selection on survival can be stronger, but the amount of heritable genetic variation smaller (Charmantier and Garant 2005). Conversely, while there may be high heritable variation in good environments, selection may be relaxed in these conditions (Wilson et al. 2006), explaining why phenotypic trait means do not always correspond well to (directional) selection acting on them.

Unfortunately there are only a few studies investigating trade-offs in human life-history at both the phenotypic and genetic level, and none in men. This lack of information on the heritability and genetic constraints of reproductive traits in human populations has resulted in a limited understanding of whether the phenotypic selection documented could lead to evolutionary changes over time. The two exceptions include studies into the heritability of key life-history traits in contemporary Australian and rural historical Finnish women. In both populations, female life-history traits, such as age at menarche and menopause, reproductive rate and longevity, had significant additive genetic heritability, suggesting the possibility for a rapid evolutionary response to selection, and also 47% of the variance in fitness itself in Finns and 39% in Australians was attributable to additive genetic effects (Kirk et al. 2001; Pettay et al. 2005). Moreover, there were also detectable genetic constraints between reproductive traits and longevity (negative genetic trade-offs): genes related to capacity for a high birthrate appeared to also lead to reduced survival and shorter overall longevity of the mothers (Pettay et al. 2005). This supports the hypothesis that rate of reproduction should trade-off with longevity, and can maintain additive genetic variation in nature (Kirkwood 1977; Williams 1957). The fact that correlations between reproductive investment and longevity are often not present at the phenotypic level (see above) calls for further studies investigating such correlations at the underlying genetic level.

However, similarly to costs of reproduction, also heritabilities can be modified by age and environment. First, female fertility in humans shows clear changes with age, possibly affecting calculations of heritabilities. In young women, maternal effects may be important for successful reproduction, such as wealth of the parents, which correlates both with female body condition, and thus their age at menarche (reviewed in Voland 1998), and with marital success (Voland 1990). Furthermore, family help, such as the presence of grandmothers, may be an important determinant of female reproductive rate: daughters enjoying help from their post-reproductive mothers show reduced inter-birth intervals (Lahdenperä et al. 2004; Voland and Beise 2002) and increased breeding probability (Sear et al. 2003). Finally, female fertility also shows senescence with age: the natural conception rate falls rapidly already from the mid-30s onwards (Sievert 2001), and the risk of unsuccessful pregnancy (miscarriage) increases with age, while the quality of offspring, in terms of developmental and genetic problems, may decrease (Holman and Wood 2001). Consequently, a single over-lifetime estimate of heritability for traits may give too simplistic a view of their response to the documented selection pressures, because evolutionary theories of senescence predict that the additive genetic variance in fitness traits is age-dependent: aging leads to increase of additive genetic variance in late life (e.g. Charmantier et al. 2006).

Currently, the only evidence for the age-dependent changes in heritability of human life-history traits comes from a study by Pettay et al. investigating the historical pedigree records available for eighteenth and nineteenth century Finns (Pettay et al. 2008). A key female life-history trait, fecundity, had significant overall additive genetic heritability (0.31) across all ages, potentially permitting rapid evolutionary responses to selection (Pettay et al. 2005). However, this additive genetic variation in fecundity in women was age-dependent and increased with age, as suggested by the theories of senescence (Rose 1991). In contrast, family effects (nongenetic material inherited from parents) appeared important early in the reproductive career but diminished later on (Pettay et al. 2008). Contrary to many animals, women have high survival in later age classes (Hawkes 2004). The presence of age-dependent additive genetic variation suggests that the common practice of using a single estimate of heritability over all age classes may give an incomplete idea of whether and how selection can lead to an evolutionary change in trait mean values. It is also important in the light of the declining force of natural selection with age (Williams 1957). However, again, further studies on other populations living in differing ecological conditions and experiencing different age-specific reproductive rates and selection pressures would be helpful in clarifying the generality of these findings.

Heritabilities can also change across environments, and certain genotypes may be most successful in particular environments but not in others. Recently it has been increasingly demonstrated that environmental conditions can influence the genetic control of life-history traits such as survival and reproduction in laboratory populations (Sgrò and Hoffmann 2004). As such genotype–environment interaction can result in genetic correlations between life-history traits being environment-dependent (Wilson et al. 2006, 2008), trade-offs have to be examined across all environments experienced by a population. Such studies for human life-history traits are entirely lacking at the moment, although understanding life-history evolution requires examination of both the environmental and genetic relationships between fitness components across the ages and environments in which the traits are expressed.

9.7 Conclusions

In most animals, fitness is maximized by optimizing the trade-off between current and future reproduction, with the amount of selection on early reproduction relative to late reproduction influencing, in part, the rate at which individuals senesce and die. In contrast, in human women, menopause has been proposed to enable women to avoid reproducing at a time when the fitness costs begin to outweigh the benefits, while prolonged post-reproductive lifespan in turn offers an opportunity to increase one's overall genetic contribution to future generations by helping existing offspring to raise their families more successfully. Thus, by increasing current reproductive effort, mothers might not only reduce their future reproductive success but also their post-reproductive survival rates and ability to help their own adult sons and daughters in raising grand-offspring. Life-history theory consequently predicts that fitness in women should be governed by optimization of trade-offs both within the reproductive phase and between the reproductive and post-reproductive phases. We are only beginning to understand such processes, and the ultimate tests of this idea—that genetic trade-offs between rate of parity and post-reproductive lifespan increase with age—are yet to be conducted.

Understanding interactions between reproductive tactics, success, senescence, and lifespan in humans appeals to a large range of people both within and without the scientific community. First, recent changes in population age structure are a growing issue of concern in many "developed" societies, and current models have failed to predict demographic transitions in many developing countries. Second, although there have been dramatic gains in the survival rates of older people in Western countries over the last several decades, the scope for improvements remains an open question because our knowledge about the interaction of biology, behavior, and environmental conditions in determining rates of senescence and age-specific mortality rates is still limited. Moreover, it is likely that increasing knowledge of the human genome may lead to manipulations of genes and gene products, possibly resulting in boosts in lifespan. Better understanding of the forces that have evolved by natural selection to provide our species-typical cell and DNA repair mechanisms are important for these tasks, and this is provided by the life-history framework.

Acknowledgments Thanks to Palestina Guevara-Fiore for temporarily removing my personal costs of reproduction; Duncan Gillespie, Samuli Helle, Mirkka Lahdenperä, Jianghua Liu, Jenni Pettay, Ian Rickard, and Matthew Robinson for help with the literature; and the Royal Society of London for funding.

References

Ardia DR, Schat KA, Winkler DW (2003) Reproductive effort reduces long-term immune function in breeding tree swallows. Proceedings of the Royal Society of London B 270:1679–1683

Beeton M, Yule GU, Pearson K (1900) Data for the problem of evolution in man V: on the correlation between duration of life and number of offspring. Proceedings of the Royal Society of London B 67:159–179

Beise J, Voland E, Helle S, Lummaa V, Jokela J (2002) Effect of producing sons on maternal longevity in premodern populations. Science 298:317

Bourke AFG (2007) Kin selection and the evolutionary theory of aging. Annual Review of Ecology, Evolution and Systematics 38:103–128

Charmantier A, Garant D (2005) Environmental quality and evolutionary potential: lessons from wild populations. Proceedings of the Royal Society of London B 272:1415–1425

Charmantier A, Perrins C, McCleery RH, Sheldon BC (2006) Age-specific genetic variance in life-history trait in the mute swan. Proceedings of the Royal Society of London B 273:225–232

Côté K, Blanchard R, Lalumière ML (2003) The influence of birth order on birth weight: does the sex of preceding siblings matter? Journal of Biosocial Science 35:455–462

Daan S, Tinbergen JM (1997) Adaptations of life histories. In: Krebs JR, Davies NB (eds) *Behavioural Ecology: An Evolutionary Approach*. Blackwell, Oxford

Doblhammer G (2000) Reproductive history and mortality later in life: a comparative study of England and Wales and Austria. Population Studies 54:169–176

Doblhammer G, Oeppen J (2003) Reproduction and longevity among the British peerage: the effect of frailty and health selection. Proceedings of the Royal Society of London B 270:1541–1547

Dribe M (2004) Long-term effects of childbearing on mortality: evidence from pre-industrial Sweden. Population Studies 58:297–310

Ellison PT (2001) *Reproductive Ecology and Human Evolution*. Aldine de Gruyter, Hawthorne, NY

Faurie C, Russell AF, Lummaa V (2009) Middleborns disadvantaged? Testing birth-order effects on fitness in pre-industrial Finns. PLoS ONE 4(5):e5680. doi: 10.1371/journal.pone.0005680

Gagnon A, Smith KR, Tremblay M, Vezina H, Pare P-P, Desjardins B (2009) Is there a trade-off between fertility and longevity? A comparative study of women from three large historical databases accounting for mortality selection. American Journal of Human Biology 21:533–540

Gillespie DOS, Russell AF, Lummaa V (2008) When fecundity does not equal fitness: evidence of a quantity-quality trade-off in pre-industrial humans. Proceedings of the Royal Society of London B 275:713–722

Gillespie DOS, Russell AF, Lummaa V (2010) Older mothers produce less fit offspring. Submitted manuscript

Grundy E, Tomassini C (2005) Fertility history and health in later life: a record linkage study in England and Wales. Social Science & Medicine 61:217–228

Gustafsson L, Part T (1990) Acceleration of senescence in the collared flycatcher *Ficedula albicollis* by reproductive costs. Nature 347:279–281

Hamdoun A, Epel D (2007) Embryo stability and vulnerability in an always changing world. Proceedings of the National Academy of Sciences of the USA 104:1745–1750

Hamilton WD (1966) Moulding of senescence by natural selection. Journal of Theoretical Biology 12:12–45

Hanna CW, Bretherick KL, Gair JL, Fluker MR, Stephenson MD, Robinson WP (2009) Telomere length and reproductive ageing. Human Reproduction 24:1206–1211

Hawkes K (2003) Grandmother and the evolution of human longevity. American Journal of Human Biology 15:380–400

Hawkes K (2004) Human longevity—The grandmother effect. Nature 428:128–129

Helle S, Käär P, Jokela J (2002a) Human longevity and early reproduction in pre-industrial Sami populations. Journal of Evolutionary Biology 15:803–807

Helle S, Lummaa V, Jokela J (2002b) Sons reduced maternal longevity in preindustrial humans. Science 296:1085

Helle S, Lummaa V, Jokela J (2004) Accelerated immunosenescence in preindustrial twin mothers. Proceedings of the National Academy of Sciences of the USA 101:12391–12396

Helle S, Lummaa V, Jokela J (2005) Are reproductive and somatic senescence coupled in humans? Late, but not early, reproduction correlated with longevity in historical Sami women. Proceedings of the Royal Society of London B 272:29–37

Hill K, Boesch C, Goodall J, Pusey A, Williams J, Wrangham R (2001) Mortality rates among wild chimpanzees. Journal of Human Evolution 40:437–450

Holman DJ, Wood JW (2001) Ecology, reproduction, and human evolution. In: Ellison PT (ed) *Reproductive Ecology and Human Evolution*. Aldine de Gruyter, Hawthorne, NY

Hurt LS, Ronsmans C, Campbell OMR, Saha S, Kenward M, Quigley M (2004) Long-term effects of reproductive history and all-cause mortality among adults in rural Bangladesh. Studies in Family Planning 35:189–196

Jasienska G (2009) Reproduction and lifespan: trade-offs, overall energy budgets, intergenerational costs, and costs neglected by research. American Journal of Human Biology 21:524–532

Jasienska G, Nenko I, Jasienski M (2006a) Daughters increase longevity of fathers, but daughters and sons equally reduce longevity of mothers. American Journal of Human Biology 18:422–425

Jasienska G, Thune I, Ellison PT (2006b) Fatness at birth predicts adult susceptibility to ovarian suppression: an empirical test of the predictive adaptive hypothesis. Proceedings of the National Academy of Sciences of the USA 103:12759–12762

Käär P, Jokela J, Helle T, Kojola I (1996) Direct and correlative phenotypic selection on life-history traits in three pre-industrial human populations. Proceedings of the Royal Society of London B 263:1475–1480

Käär P, Jokela J, Merilä J, Helle T, Kojola I (1998) Sexual conflict and remarriage in preindustrial humans: causes and fitness consequences. Evolution and Human Behavior 19:139–151

Kawecki TJ (1993) Age and size at maturity in a patchy environment—fitness maximization versus evolutionary stability. Oikos 66:309–317

Kirk KM, Blomberg SP, Duffy DL, Heath AC, Owens IPF, Martin NG (2001) Natural selection and quantitative genetics of life-history traits in western women: a twin study. Evolution 55:423–435

Kirkwood TBL (1977) Evolution of ageing. Nature 270:301–304

Knodel JE (1988) *Demographic Behavior in the Past*. Cambridge University Press, New York, NY (Chap. 5)

Korpelainen H (2000) Fitness, reproduction and longevity among European aristocratic and rural Finnish families in the 1700s and 1800s. Proceedings of the Royal Society of London B 267:1765–1770

Kozlowski J (1992) Optimal allocation of resources to growth and reproduction: implications for age and size at maturity. Trends in Ecology and Evolution 7:15–19

Kumle M, Lund E (2000) Patterns of childbearing and mortality in Norwegian women. A 20-year follow-up of women aged 40–96 in the 1970 Norwegian census. In: Robine J-M, Kirkwood TBL, Allard M (eds) *Sex and Longevity: Sexuality, Gender, Reproduction, Parenthood*, pp. 117–128. Berlin: Springer-Verlag

Lahdenperä M, Lummaa V, Helle S, Tremblay M, Russell AF (2004) Fitness benefits of prolonged postreproductive lifespan in women. Nature 428:178–181

Le Bourg E (2007) Does reproduction decrease longevity in human beings? Ageing Research Reviews 6:141–149

Le Bourg E, Thon B, Légaré J, Desjardins B, Charbonneau H (1993) Reproductive life of French-Canadians in the 17–18th centuries: a search for a trade-off between early fecundity and longevity. Experimental Gerontology 28:217–232

Lindstrom J (1999) Early development and fitness in birds and mammals. Trends in Ecology and Evolution 14:343–348

Lummaa V (2001) Reproductive investment in preindustrial humans: the consequences of off-spring number, gender and survival. Proceedings of the Royal Society of London B 268:1977–1983

Lummaa V (2003) Reproductive success and early developmental conditions in humans: down-stream effects of pre-natal famine, birth weight and timing of birth. American Journal of Human Biology 15:370–379

Lummaa V, Clutton-Brock T (2002) Early development, survival and reproduction in humans. Trends in Ecology and Evolution 17:141–147

Lummaa V, Haukioja E, Lemmetyinen R, Pikkola M (1998) Natural selection on human twinning. Nature 394:533–534

Lycett JE, Dunbar RIM, Voland E (2000) Longevity and the costs of reproduction in a historical human population. Proceedings of the Royal Society of London B 267:31–35

Mace R, Sear R (1997) Birth interval and the sex of children in a traditional African population: an evolutionary analysis. Journal of Biosocial Science 29:499–507

Maklakov AA (2008) Sex difference in lifespan affected by female birthrate in modern humans. Evolution and Human Behavior 29:444–449

Martikainen P (1995) Mortality and socio-economic status among Finnish women. Population Studies 49:71–90

Medawar PB (1952) *An Unsolved Problem of Biology*. HK Lewis, London

Meij JJ, van Bodegom D, Ziem JB, Amankwa J, Polderman AM, Kirkwood TBL, de Craen AJM, Zwaan BJ, Westendorp RGJ (2009) Quality-quantity trade-off of human offspring under adverse environmental conditions. Journal of Evolutionary Biology 22:1014–1023

Menken J, Duffy L, Kuhn R (2003) Childbearing and women's survival: new evidence from rural Bangladesh. Population Development Reviews 29:405–426

Migliano AB, Vinicius L, Lahr MM (2007) Life history trade-offs explain the evolution of human pygmies. Proceedings of the National Academy of Sciences of the United States of America 104:20216–10219

Monaghan P, Charmantier A, Nussey DH, Ricklefs RE (2008) The evolutionary ecology of senescence. Functional Ecology 22:371–378

Moore SE, Cole TJ, Collinson AC, Poskitt EM, McGragor IA, Prentice AM (1999) Prenatal or early postnatal events predict infectious deaths in young adulthood in rural Africa. International Journal of Epidemiology 28:1088–1095

Müller H-G, Chiou J-M, Carey JR, Wang J-L (2002) Fertility and lifespan: late children enhance female longevity. Journal of Gerontology A 57:B202–B206

Nemoto S, Finkel T (2004) Ageing and the mystery at Arles. Nature 429:149–152

Nishida T, Corp N, Hamai M, Hasegawa T, Hiraiwa-Hasegawa M, Hosaka K, Hunt KD, Itoh N, Kawanaka K, Matsumoto-Oda A, Mitani JC, Nakamura M, Norikoshi K, Sakamaki T, Turner L, Uehara S, Zamma K (2003) Demography, female life history, and reproductive profiles among the chimpanzees of Mahale. American Journal of Primatology 59:99–121

Noorwijk AJ, de Jong G (1986) Acquisition and allocation of resources: their influence on variation in life history tactics. American Naturalist 128:137–142

Nussey DH, Coulson T, Festa-Bianchet M, Gaillard JM (2008) Measuring senescence in wild animal populations: towards a longitudinal approach. Functional Ecology 22:393–406

Nussey DH, Kruuk LEB, Donald A, Fowlie M, Clutton-Brock TH (2006) The rate of senescence in maternal performance increases with early-life fecundity in red deer. Ecology Letters 9:1342–1350

Nussey DH, Kruuk LEB, Morris A, Clutton-Brock T (2007) Environmental conditions in early life influence aging rates in a wild population of red deer. Current Biology 17:R1000–R1001

Partridge L, Harvey PH (1985) Evolutionary biology—costs of reproduction. Nature 316:20

Penn DJ, Smith KR (2007) Differential fitness costs of reproduction between the sexes. Proceedings of the National Academy of Sciences of the USA 104:553–558

Pettay JE, Charmantier A, Wilson AJ, Lummaa V (2008) Age-specific genetic and maternal effects in fecundity of pre-industrial Finnish women. Evolution 62:2297–2304

Pettay JE, Helle S, Jokela J, Lummaa V (2007) Natural selection on female life-history traits in relation to socio-economic class in pre-industrial human populations. Plos ONE 2(7):e606. doi: 10.1371/journal.pone.0000606

Pettay JE, Kruuk LEB, Jokela J, Lummaa V (2005) Heritability and genetic correlations of life-history trait evolution in preindustrial humans. Proceedings of the National Academy of Sciences of the USA 102:2828–2843

Réale D, Festa-Bianchet M, Jorgenson JT (1999) Heritability of body mass varies with age and season in wild bighorn sheep. Heredity 83:526–532

Rickard IJ (2008) Offspring are lighter at birth and smaller in adulthood when born after a brother versus a sister in humans. Evolution and Human Behavior 29:196–200

Rickard IJ, Lummaa V, Russell AF (2009) Elder brothers affect the life-history of younger siblings in preindustrial humans: social consequence or biological cost? Evolution and Human Behavior 30:49–57

Rickard IJ, Russell AF, Lummaa V (2007) Producing sons reduces lifetime reproductive success of subsequent offspring in pre-industrial Finns. Proceedings of the Royal Society of London B 274:2981–2988

Roff DA (2002) Life History Evolution. Sinauer Associates, Sunderland, MA

Rogers AR (1993) Why menopause? Evolutionary Ecology 7:406–420

Rose MR (1991) Evolutionary Biology of Aging. Oxford University Press, New York, NY

Sear R, Mace R (2008) Who keeps children alive? A review of the effects of kin on child survival. Evolution and Human Behavior 29:1–18

Sear R, Mace R, McGregor IA (2003) The effects of kin on female fertility in rural Gambia. Evolution and Human Behavior 24:25–42

Sgrò CM, Hoffmann AA (2004) Genetic correlations, tradeoffs and environmental variation. Heredity 93:241–248

Shanley DP, Sear R, Mace R, Kirkwood TBL (2007) Testing evolutionary theories of menopause. Proceedings of the Royal Society of London B 274:2943–2949

Sheldon BC, Verhulst S (1996) Ecological immunology: costly parasite defences and trade-offs in evolutionary ecology. Trends in Ecology and Evolution 11:317–321

Sievert LL (2001) Aging and reproductive senescence. In: Ellison PT (ed) *Reproductive Ecology and Human Evolution*. Aldine de Gruyter, Hawthorne, NY

Stearns SC (1992) *The Evolution of Life Histories*. Oxford University Press, New York, NY

Tracer DP (2002) Somatic versus reproductive energy allocation in Papua New Guinea: life history theory and public health policy. American Journal of Human Biology 14:621–626

Trivers RL (1972) Parental investment and sexual selection. In: Campbell B (ed) *Sexual Selection and the Descent of Man, 1871–1971*. Aldine, Chicago, IL

van de Pol M, Verhulst S (2006) Age-dependent traits: a new statistical model to separate within- and between-individual effects. American Naturalist 167:766–773

van de Putte B, Matthijs K, Vlietinck R (2004) A social component in the negative effect of sons on maternal longevity in pre-industrial humans. Journal of Biosocial Science 36:289–297

Voland E (1990) Differential reproductive success within the Krummhörn population (Germany, 18th and 19th centuries). Behavioral Ecology and Sociobiology 26:65–72

Voland E (1998) Evolutionary ecology of human reproduction. Annual Review of Anthropology 27:347–374

Voland E, Beise J (2002) Opposite effects of maternal and paternal grandmothers on infant survival in historical Krummhörn. Behavioral Ecology and Sociobiology 52:435–443

Walker R, Gurven M, Hill K, Migliano A, Chagnon N, De Souza R, Djurovic G, Hames R, Hurtado AM, Kaplan H, Kramer K, Oliver WJ, Valeggia C, Yamauchi T (2006) Growth rates and life-histories in twenty-two small scale societies. American Journal of Human Biology 18:295–311

Westendorp RGJ, Kirkwood TBL (1998) Human longevity at the cost of reproductive success. Nature 396:743–746

Williams GC (1957) Pleiotropy, natural selection and the evolution of senescence. Evolution 11:398–411

Wilson AJ, Charmantier A, Hadfield JD (2008) Evolutionary genetics of ageing in the wild: empirical patterns and future perspectives. Functional Ecology 22:431–442

Wilson AJ, Pemberton JM, Pilkington JG, Coltman DW, Mifsud DV, Clutton-Brock TH, Kruuk LEB (2006) Environmental coupling of selection and heritability limits evolution. PLoS Biology 4:1270–1275. doi: 10.1371/journal.pbio.0040216

Yi Z, Vaupel J (2004) Association of late childbearing with healthy longevity among the oldest-old in China. Population Studies 58:37–53

Chapter 10
Height and Reproductive Success

Is Bigger Always Better?

Rebecca Sear

Abstract Height is of great interest to the general public and academics alike. It is an easily observable and easily measurable characteristic, and one that appears to be correlated with a number of important outcomes, from survival to intelligence to employment and marriage prospects. It is also of interest to evolutionary biologists, as the end product of life history decisions made during the period of growth. Such decisions will depend at least partly on the payoffs to size in adulthood. This chapter surveys the costs and benefits of height during adulthood: what are the consequences of height in terms of mortality rate, mating success and fertility outcomes for each sex, and how much do these differ between environments? It is clear from this survey that relationships between height and fitness correlates show considerable variation between populations, suggesting that the costs and benefits of height depend on environmental conditions. If any tentative conclusion can be drawn it is that while short height is rarely advantageous, particularly for men, tall height is not universally beneficial, particularly for women. We can also conclude that height is clearly still important for fitness correlates in modern environments, thereby demonstrating that we have yet to leave our biological imperative behind.

10.1 Introduction

Height is a topic of great interest to academics and the general public alike. It is an easily observable and easily measurable characteristic, perhaps explaining its popularity in both academic research and popular culture. Websites abound on height. You can find out the height of your favorite celebrity (www.celebheights.com), discover average heights around the world (http://www.thegreatsleep.com/height.htm), calculate the percentage of Americans shorter than you (http://www.tallpeople.net/wiki/What_Percentage_of_People_are_Shorter_Than_Me), predict

R. Sear (✉)

Department of Social Policy, London School of Economics, London WC2A 2AE, UK

e-mail: r.sear@lse.ac.uk

U.J. Frey et al. (eds.), *Homo Novus – A Human Without Illusions*,
The Frontiers Collection, DOI 10.1007/978-3-642-12142-5_10,
© Springer-Verlag Berlin Heidelberg 2010

your child's ultimate height (http://www.bbc.co.uk/parenting/your_kids/toddlers_ heightcalculator.shtml), and find online support if you are particularly short (http://www.shortsupport.org/) or particularly tall (http://www.tallclub.co.uk/ index.asp). Height then is clearly salient to the internet-using general public. Academics are equally fascinated by the subject and their interest appears to be growing: Steckel reported earlier this year that approximately 325 publications on height have been published in the social sciences alone since 1995, and that the rate of publication on the topic had shown a four-fold increase over the previous 20 years (Steckel 2009). This academic research supports the lay view that height is of importance for many different outcomes, from intelligence and earnings (Case and Paxson 2008), to suicide rates (Magnusson et al. 2005), sexual orientation (Bogaert 1998), and jealousy (Buunk et al. 2008). Such academic research may be at least partly a matter of convenience, given the ready availability of data, including historical records as well as contemporary data. But to an evolutionary biologist, what does height actually mean?

Adult height is the end product of life history decisions made throughout the period of growth, from conception onwards. Life history theory is the branch of evolutionary biology that aims to understand how energy is allocated over the life course (Roff 1992). It is based on the principle of allocation, which states that energy used for one purpose cannot be used for another purpose. During their lifetimes, organisms continually make decisions about how to optimally allocate energy between functions such as growth, somatic maintenance (including immune function), and reproduction. An individual's final adult height results from the decisions it makes, and also the decisions its mother makes (Wells 2003), about how much energy to allocate to growth compared to maintaining body condition or reproduction. Both the speed and the timing of growth are relatively labile, so that individuals can speed up or slow down growth, or extend or shorten the period of growth to vary final adult height. An important decision in this process is the decision about when to stop growing and start reproducing, since humans, like many other species, separate out the periods of growth and reproduction given the costly nature of both. Height, then, is partly determined by the timing of the decision to stop allocating energy to growth and start allocating energy to reproduction instead. Any selection process affecting adult size will therefore act on the period of growth—the timing and speed of growth is the adaptation, rather than final adult size.

Such life history decisions will be affected by the amount of energy available to that individual, so that adult size differs between environments, but these cross-population size differentials are not simply the result of environmental constraints on growth. Individuals who are conceived and grow up in poor environments could grow a little faster during good periods (should they experience them) and/or continue growing a little longer in order to end up as larger adults. Both compensation mechanisms do in fact occur in poorly nourished populations. Children often experience periods of "catch-up" growth after stalls in growth due to episodes of disease or seasonal food shortages, for example (Martorell et al. 1994). Growth also tends to continue a little longer in poorly nourished populations, with a corresponding later age at reproductive maturity (Teriokhin et al. 2003). Growing fast and growing longer both have costs however. Fast growth appears to have adverse consequences in terms of higher mortality risk in later life (Rollo 2002). Growing longer delays the start to reproduction, which both increases the risk of dying before any offspring have been produced, and means fewer offspring are produced overall. Individuals must therefore tradeoff any potential benefits of large size against the costs of faster and longer growth. Recent research has begun to explore the potential costs and benefits of growth across populations, in order to understand these life history decisions in more detail. A cross-cultural analysis suggested that where juvenile mortality rates are relatively high, individuals stop growing sooner (and so end up as shorter adults), presumably to reduce the risk of dying before starting to reproduce (Walker et al. 2006). At the extreme end of the continuum, this may be at least part of the explanation for the very short height of pygmy populations in central Africa (Migliano et al. 2007).

The payoffs to investing in growth versus other functions are also likely to differ between environments because adult height may bring different costs and benefits in different environments. The aim of this chapter is to survey the costs and benefits of height during adulthood: what costs and benefits does height bring in terms of mortality rate, mating success, and fertility outcomes for each sex, and how much do these differ between environments? Large size, as a general rule, is frequently considered to be a good thing in fitness terms, bringing both survival and fecundity advantages (Harvey and Clutton-Brock 1985). Larger males tend to be more successful in the competition for mates, and larger females have greater energetic reserves to devote to reproduction. But large size is also energetically costly to maintain (Blanckenhorn 2000). Given the diversity of mating patterns, mortality rates, and energy access experienced by our geographically widespread species, it is worth considering exactly how height is related to the various correlates of reproductive success in diverse environments, before any conclusions can be drawn about the fitness benefits of size.

10.2 A Survey of Height and Correlates of Reproductive Success in Adulthood

10.2.1 Height and Adult Mortality

Mortality is one key component of fitness: to ensure any reproductive output, individuals must survive long enough to reproduce and to raise successfully any children produced. Of all the fitness components considered in this chapter, the literature on links between height and adult mortality is by far the largest. At a population level, height is clearly related to mortality rates—taller populations have lower mortality, and this holds for both women and men (Gage and Zansky 1995). Here, we are interested in whether this relationship holds at an individual, within-population level. Table 10.1 presents a summary of studies that have used longitudinal datasets and hazards analysis (the most appropriate statistical technique) to investigate the link between height and all-cause mortality at the individual level (several other

Table 10.1 Summary of studies analyzing the relationship between height and all-cause adult mortality (this page low mortality populations; next page high mortality populations)

Study population	Age	Sex	Height-mortality relationship	Comments	References
Norway	20+	Women Men	Negative Negative	Large population-based study; association weaker for tallest men and oldest ages	Waaler (1984)
Norway	20+	Women Men	U-shaped Negative	Similar to Waaler (1984) but larger sample and longer follow-up	Engeland et al. (2003)
Sweden	16+	Women Men	Negative Negative	Random sample of Swedish adults	Peck and Vagero (1989)
Sweden	18+	Men	Negative	Military conscripts	Allebeck and Bergh (1992)
Finland	25+	Women Men	Negative Negative	Population surveys in Eastern Finland	Jousilahti et al. (2000)
Finland	14+	Women	U-shaped	Fertile women recruited during childbearing years	Laara and Rantakallio (1996)
Finland	45+	Men	Negative	Helsinki business men	Strandberg (1997)
England	40+	Men	Negative	Whitehall study of civil servants; strength of association declined with length of follow-up	Leon et al. (1995)
Scotland	16+	Men	None	Male medical students	McCarron et al. (2002)
Scotland	45+	Women Men	Negative Negative	Renfrew/Paisley general population study	Davey-Smith et al. (2000)
US	<36	Women Men	None None	Framingham Heart Study	Kannam et al. (1994)
US	25+	Women Men	None None	NHANES 1	Liao et al. (1996)
South Korea	40+	Women Men	Negative Negative	Civil servants and their dependents	Song and Sung (2008) Song et al. (2003)

Table 10.1 (continued)

Study population	Age	Sex	Height-mortality relationship	Comments	References
Gambia	21+	Women Men	U-shaped None	Rural agriculturalists	Sear et al. (2004) Sear (2006b)
Bangladesh	10+	Women	None	Matlab study	Hosegood and Campbell (2003)
Belgium	50+	Men	Negative	Military conscript data; Birth cohort 1815–1828	Alter et al. (2004)
Belgium	50+	Men	None	Military conscript data; birth cohort 1829–1860	Alter et al. (2004)
US	55+ 56+	Men	Negative None	US Union Army records, nineteenth century; slightly different samples and methods yield different results	Costa (1993) Costa (2004)
US	20+	Men	None	Amherst College graduates 1834–1949	Murray (1997)

studies investigate the height–longevity link using other methods or in particularly biased samples: Samaras (2007) provides an excellent source of this literature).

Most of these studies have been conducted in relatively high income, and therefore low mortality, populations. Most, but not all, find a largely negative relationship between height and risk of death, though this relationship may be weaker at older ages, and for the particularly tall. And two of the studies find a U-shaped, rather than entirely linear relationship for women. Unfortunately there are very few studies on low income, high mortality populations, but the little evidence available suggests that the negative relationship between height and mortality may not hold so strongly here. Of the five relatively high mortality populations studied, in two there is evidence of a negative relationship between height and risk of death (nineteenth century US army veterans and East Belgians), but in both cases the results did not hold across the entire sample. In the other three populations—two in the contemporary developing world and one other historical US population—there is little evidence of any height–mortality relationship, except for Gambian women, in which the relationship is U-shaped.

The explanation for these variable results may lie in the fact that the relationship between height and mortality differs by cause of death. The broadly negative relationship between all-cause mortality and risk of death in high income countries is driven largely by lower risks of cardiovascular disease in tall individuals; the risk of death from cancer, particularly reproductive cancers, is actually higher among taller individuals (Davey-Smith et al. 2000; Lee et al. 2009; Okasha et al. 2002). The relationships between height and the main causes of death in high mortality populations, such as infectious or parasitic disease, have yet to be elucidated, given the lack of data on height and mortality in such populations. But since the main causes of death differ between populations, ages, and the sexes, we would not necessarily expect different populations to show identical relationships between height and mortality. Bearing in mind the caveat that we so far have little data on high mortality populations, some very tentative conclusions can perhaps be drawn from this survey. First, height seems to be more important in high income, low mortality populations, and less so in higher mortality populations, perhaps because of the clear link between height and cardiovascular disease, a significant cause of death in low mortality populations. Second, there are differences in the height–mortality link between men and women. For men, tall height seems to be broadly beneficial, or at least not detrimental, in that height seems to be either negatively related to mortality or not related at all. For women, several populations show a U-shaped relationship, so that tall height may sometimes bring disadvantages, as does short height.

10.2.2 Height and Mating Success

Once an individual has survived to reproductive age, the next step on the road to reproductive success is to find a partner. Anyone raised in western culture will be aware of the importance of height in the mating market. In the Anglophone world, women are traditionally supposed to seek "tall, dark, and handsome" men. Academic research backs up these perceptions. Evidence that height matters on the mate market can take one of two forms. We can indirectly test whether height matters by assessing whether individuals state preferences for a particular height, or we can more directly investigate whether height affects the choice of a mate in the real world of marriage. Research on mate preferences suggests that both men and women, at least in western populations, do prefer partners of a particular height. Lab tests of mate preferences and analysis of lonely hearts ads provides support for what social psychologists have described as the "cardinal rule of dating": both sexes prefer relationships in which the man is taller than the woman (Higgins et al. 2002; Pawlowski 2003; Pierce 1996), though women more so than men (Salska et al. 2008). There is consistent evidence that women do prefer tall men (Shepperd and Strathman 1989), though perhaps not very tall men (Hensley 1994). Men's preferences are sometimes less pronounced, but they do seem to have a preference for short (Shepperd and Strathman 1989) or average height women (Swami et al. 2008).

Studies of mate preferences are problematic, however. Such studies tend to have rather unrepresentative samples: college students and users of lonely hearts ads are not necessarily representative of all men or women. Studies of mate preferences have also been done almost exclusively in western populations. In addition, mate preferences are not necessarily converted into mate choice, which is ultimately what matters for reproductive success. Mate choices will result from mate preferences across a range of criteria, not just physical attractiveness; they will be affected by mate availability and one's own mate value, and perhaps also the preferences of one's parents and family. More convincing evidence that height matters for mating success would be research which found that individuals are actively choosing partners of particular heights as marriage partners.

Assortative mating for height is one indication that height may be of importance on the marriage market, and has been examined in many populations. A 1968 review found that a positive correlation between the heights of husbands and wives was relatively common among the 25 populations of European-origin included (Spuhler 1968). In a 1977 survey, 26 of 39 (67%) of populations that were European or of European-origin showed evidence of such assortative mating (Roberts 1977). Studies on non-European populations are less common, but the results are more mixed. The 1968 paper found no evidence for assortative mating in the 2 non-European populations it included, and in the 1977 review, only 2 of 10 (20%) non-European populations showed significant evidence of assortative mating for height. More recent research on non-European populations has found positive assortative mating for height in Bolivian forager-farmers (Godoy et al. 2008), Oaxaca, Mexico (Malina et al. 1983) and Pakistan (Ahmad et al. 1985), but not in Cameroon (Pieper 1981), Gambia (Sear 2006a), Korea (Hur 2003), or Hadza hunter-gatherers (Sear and Marlowe 2009). Overall, then, while positive assortative mating for height appears fairly common among human populations, it is by no means universal, and its frequency may be over-emphasized by the disproportionate number of studies on European populations, where assortative mating for height may be more common.

Assortative mating is the weakest evidence that height matters for mate choice. Alternative explanations are possible, for example, individuals could be assorting on a characteristic which is correlated with height (such as socio-economic status), or it could arise simply because different groups within a population are different in height. A better measure might be height-specific patterns of mating that are likely to be driven by actual preferences for particular heights. The male-taller norm is one such example. The proportion of marriages in which the female is taller than the male is considerably less than would be expected by random mating in populations in the US (Gillis and Avis 1980), and the UK (Sear et al. 2004). But in both Gambian agriculturalists and Hadza hunter-gatherers the proportion of female-taller marriages is exactly what would be expected by random mating, around 8–9% (Sear et al. 2004; Sear and Marlowe 2009).

A final piece of evidence that height matters would be analyses that showed individuals of particular heights were favored in terms of the probability or number of marriages (see Table 10.2). Most, but not all, of these studies find a positive relationship between height and marital success for men. There are fewer studies for women, and the results are a little more mixed: these studies variously show no relationship (Gambia, Hadza), a disadvantage for the shortest women (Bavaria), and a disadvantage for both the shortest and tallest women (UK).

This summary suggests both preferences and choices for height exist in both sexes in at least some populations, but not all. Women seem to choose either tall men, or have no preference at all; a marital advantage for short men is not seen. Men's choices may be somewhat more variable, and they may sometimes avoid shorter women (though such a pattern may also be driven by taller women

Table 10.2 Summary of studies analyzing the relationship between height and probability or number of marriages (top panel low mortality populations; bottom panel high mortality populations)

Study population	Dependent variable	Sex	Height-mortality relationship	Comments	References
UK	Probability and number of marriages	Women Men	∩-shaped Positive, except tallest men	Nationally representative sample (NCDS)	Nettle (2002b)
Poland	Probability of marriage	Men	Positive	Wroclaw	Pawlowski et al. (2000)
US	Number of wives	Men	Positive	West Point cadets	Mueller and Mazur (2001)
Gambia	Number of marriages	Women Men	None Positive	Rural agriculturalists	Sear et al. (2004) Sear (2006b)
Hadza	Number of marriages	Women Men	None None	Hunter-gatherers	Sear and Marlowe (2009)
Belgium	Probability of marriage	Men	Positive	Military conscript data; men born 1815–1860	Alter et al. (2004)
US	Probability of marriage	Men	Positive	US Union Army records, nineteenth century	Hacker (2008)
US	Probability of marriage	Men	Positive	Amherst College graduates 1834–1949	Murray (2000)
Bavaria	Probability of marriage	Women	Shortest women disadvantaged	Female prisoners, nineteenth century	Baten and Murray (1998)

attempting but failing to achieve their preferred taller partner). Such variability makes sense because what makes a potential mate attractive is likely to vary between populations. Height should confer different advantages and disadvantages in different ecologies and different subsistence strategies, perhaps growing in importance in agricultural and industrial populations, compared to hunter-gatherers, where large size may actually be a disadvantage in hunting game. As with the mortality research, a very tentative conclusion might be that height matters more for high income, developed country populations, compared to traditional societies, though again the data are far too limited to make this conclusion at all secure.

10.2.3 Height and Fertility Outcomes

Life history theory predicts that adult height will be negatively correlated with age at first reproduction, because of the tradeoff between growth and reproduction. Height could also plausibly be related to other fertility outcomes, such as reproductive rate in women, since taller women potentially have access to greater energetic reserves. There is relatively little research on relationships between height and fertility outcomes and what there is exists almost exclusively for women, rather than men. Male reproductive success is governed by a different set of factors to female: it is less constrained by the energetic ability to produce children and the time available in which to produce them, and more by the ability to attract mates. For these reasons, and because collecting accurate data on male fertility is more difficult, the study of male fertility is less well-developed than that of female. In this section we consider the evidence that height is correlated with the fertility outcomes of age at first birth and reproductive rate, including a discussion of how height affects child survival, for women only; in the final section we consider whether height is correlated with overall reproductive success for both sexes, measured by overall number of children or surviving children.

If there are any patterns between height and components of fitness that hold universally across populations, then one strong candidate is the relationship between height and age at reproductive maturity. The tradeoff between growth and reproduction results in a relatively consistent pattern of earlier maturity being correlated with shorter adult height across all types of populations. A number of studies in high or medium income countries—including Brazil (Gigante et al. 2006), two studies in the UK (Nettle 2002b; Ong et al. 2007), Scotland (Okasha et al. 2001), Copenhagen, Denmark (Helm et al. 1995), Greece (Georgiadis et al. 1997), and a comparative study of 9 European populations (Onland-Moret et al. 2005)—suggest that women who have a relatively early age at menarche are shorter as adults. This suggests that even in well-nourished populations women experience a tradeoff between devoting resources to growth and to reproduction. In natural-fertility societies an early menarche is likely also to lead to an earlier first birth, and one study of such a population in rural Gambia found the predicted tradeoff between age at first birth and adult height (Allal et al. 2004). In a similar vein, a study using a nationally representative Indian sample found that adult height was negatively related to the number of teenage births a woman had produced, likely to be correlated with her age at first birth (Brennan et al. 2005). The use of contraception might perhaps be expected to break the link between height and age at first birth, since women in such populations tend to delay births considerably beyond the age at which they could physiologically conceive, but in Finland women who were shorter as adults did have earlier first births, as well as earlier menarche (Helle 2008). These women didn't begin reproducing until their mid-1920s on average, but they may have been on a relatively fast life history track, despite the long lag between age at menarche and age at first birth. One partial exception to the earlier maturity–shorter height rule was demonstrated in a Guatemalan study, which found that both tall and particularly short women had delayed first births (Pollet and Nettle 2008), suggesting perhaps that no height–reproductive outcome relationship can be considered entirely universal in our species.

Whether height is correlated with reproductive rate, and the ability to conceive, once reproduction is underway is so far little studied. One study of a Gambian population found no relationship between height and the length of birth intervals (Sear et al. 2003). A complication with analyzing reproductive rate is that height is known to be correlated with other reproductive outcomes: taller women tend to have easier births (Cnattingius et al. 1998; Liljestrand et al. 1985), higher birthweight babies (Kirchengast et al. 1998), fewer stillbirths (Pollet and Nettle 2008), and frequently higher survival among their children. Higher survival among children will lengthen birth intervals, making a simple analysis of reproductive rate by height compli-cated. Higher survival amongst one's children is a fitness component itself, however, and may be one of the strongest determinants of fitness for women (Strassmann and Gillespie 2002). The majority of studies investigating whether maternal height is correlated with child survival show that tall maternal height brings benefits in terms of better child survival, including populations in the Gambia (Sear et al. 2004), Guatemala (Pollet and Nettle 2008), Bangladesh (Baqui et al. 1994), and Mozambique (Liljestrand et al. 1985). Despite one study in Peru showing shorter women have higher child survival (Frisancho et al. 1973), a recent comparative study using Demographic and Health Survey data from 42 developing countries has effectively demonstrated a clear positive relationship between maternal height and child survival that holds across varying levels of development (Monden and Smits 2008): this is one of the more convincing comparative studies indicating a clear advantage for tall height, at least for women in high mortality populations.

10.2.4 Height and Reproductive Success

The closest proxy for reproductive success in empirical studies is the number of surviving children, or simply number of children in societies with low child mortal-ity. There are a number of studies for both sexes on the relationship between height and overall number of children or surviving children, but these studies are perhaps the hardest of all to compare. The varying ages and control variables, if any, that these studies include in their analysis introduces considerable noise into the data. Analyzing only certain ages, for example, may introduce selection effects if height is related to mortality in the population under study. Many of these studies come from populations moving through the epidemiological and fertility transitions, again making not only comparisons difficult, but also an assessment of the effect of height in the population under study, if different cohorts have different fertility, mortality, or height.

Table 10.3 then needs to be interpreted with caution, but it draws together research that has investigated relationships between height and either total num-ber of children (for the low mortality populations) or number of surviving children (for the high mortality populations). These data tend to derive from rather unrepre-sentative samples so any conclusions must again be tentative, but we can see that for both women and men every possible relationship is seen between height and number of children—positive, negative, nonlinear, and no relationship. The only conclusion that can be drawn is that the diversity of these results suggests that the relationship between height and reproductive success is heavily dependent on environmental context.

Table 10.3 Summary of studies demonstrating the relationship between height and number of surviving children (top panel relatively low mortality populations; bottom panel high mortality populations)

Study population	Sex	Height-children relationship	References
Poland	Men	Positive	Pawlowski et al. (2000)
Finland	Women	None	Helle (2008)
UK	Women Men	∩-shaped None	Nettle (2002b) Nettle (2002a)
US	Women Men	None ∩-shaped	Mitton (1975)
US (West Point cadets)	Men	Positive	Mueller and Mazur (2001)
US college students 1880–1912	Women	∩-shaped	Vetta (1975)
US college students born 1912–1918	Women Men	None None	Scott and Bajema (1982)
Mexicans in US	Women Men	None None	Goldstein and Kobyliansky (1984)
Mexicans in US	Women Men	None None	Lasker and Thomas (1976)
China	Women Men	None None	Fielding et al. (2008)
Namibia hunter-gatherers	Women Men	Negative Positive	Kirchengast (2000)
Namibia urban	Men	Negative	Kirchengast and Winkler (1995)
Namibia horticultural pastoralists	Women Men	None Positive	Kirchengast and Winkler (1996) Kirchengast and Winkler (1995)
Gambia agriculturalists	Women Men	Positive None	Sear et al. (2004)
Colombia	Women Men	∩-shaped ∩-shaped	Mueller (1979)
India	Women	Negative	Devi et al. (1985)
Guatemala	Women	Positive	Martorell et al. (1981)
Guatemala	Women	Positive	Pollet and Nettle (2008)
Papua New Guinea	Women	∩-shaped	Brush et al. (1983)

10.3 Conclusion

This brief survey can only touch the surface of the costs and benefits of height in adulthood, not least because there is relatively little comparable data that can be used to assess the consequences of height across a range of environments. It should also be noted that even where methods and sampling strategies are similar, studies may not be directly comparable because average height varies between populations: a "tall" Gambian or Guatemalan, for example, will be considerably shorter than a "tall" westerner. Such empirical analysis investigating relationships between a single fitness outcome and height will also be misleading, since there may be interactions between fitness outcomes that may alter the association between height and overall fitness: age at menarche, for example, has been shown to be correlated with both later mortality (Jacobsen et al. 2007) and the birthweight of any children produced (Kirchengast and Hartmann 2000), so that even if an early menarche allows women to get a head-start on reproduction, it may come at a later cost of small babies and higher risk of death. Studies that try to holistically assess the relationship between height and multiple components of reproductive success, including age at first birth, probability of marriage and childlessness, number of children born and child survival within the same population, provide the best means for analyzing the height–reproductive success relationship (such as Nettle 2002b; Pollet and Nettle 2008; Sear et al. 2004), but even these will suffer from selection biases. Theoretical modeling of these relationships is likely to prove the most fruitful strategy for understanding these relationships, as has been done for Ache hunter-gatherers on the consequences of growth in terms of adult weight (Hill and Hurtado 1996).

If any conclusions can be drawn from this survey then, with appropriate caution, the following trends seem to appear. First, there are no obvious relationships between height and fitness outcomes that hold across all types of population, with the possible exceptions of a tradeoff between height and age at reproductive maturity for women, and the higher child survival of tall women in high mortality societies. Instead, the nature of almost all relationships discussed here clearly depends on ecological conditions, a finding that should not perhaps be surprising, given patterns of change in average human heights over our evolutionary history. Over the last few millennia, average height appears to have varied, rather than demonstrating a consistent increase. In fact, estimates suggest extant human populations are somewhat shorter and certainly lighter than ancestral species. Height did increase from australopithecines to *Homo*, but earlier *Homo* species may have been rather taller than modern *Homo sapiens* (Bogin 2001). Among modern humans, skeletal evidence suggests that human height may have been taller than the average today about 40,000 years ago, but declined over the next few millennia. The advent of agriculture is thought to have shrunk human populations further, after which average heights fluctuated until a steady increase in height began in economically developing countries over the twentieth century (Bogin 2001). Such variability suggests variation in the costs and benefits of investing in growth and in the payoffs to size in adulthood throughout our species' history.

Second, tall height may bring more benefits in adulthood to men than to women. Short height certainly brings no benefits to men, and tall height rarely seems costly, with the exception of a handful of the studies analyzing total number of children. Nonlinear relationships seem to be somewhat more common for women and are found across all outcomes: mortality, mating success, and fertility. These potentially negative effects of tall height for women may be related to the clear cost of a later start to reproduction, though this is counteracted in high mortality societies by higher survival amongst the children of tall women.

Finally, it is clear that the importance of height has not diminished in modern western societies. In fact, height may actually matter more in such societies than in more traditional populations, at least for mating success and mortality, if not overall number of children (though, given the low variance in number of children in such low fertility societies, a better measure of reproductive success might be the quality, rather than the quantity of these children, which is beyond the current review). Patterns of growth may therefore still have evolutionary importance today, with both mortality and mating success for men suggesting taller is better in own particular society. Does this mean we may ultimately become a species of Brobdingnagians: if we met our descendants, would they tower over us? This depends on whether this advantage of the tall is genuine, which will require more data. It will also depend on whether all populations will converge on this "bigger is better" pattern as they develop economically towards the developed world. That the very tall in our societies suffer adverse health and mating consequences, however, suggests that there may be limits to growth. Other factors may also come into play: two economics historians have recently suggested, somewhat controversially, that the obesity epidemic may cause heights to stagnate or even decline in the coming years (Komlos and Baur 2004). A confident prediction of what average heights will be across the world in a few decades time may be difficult but, given the large number of studies demonstrating correlations between height and fitness outcomes, it seems likely that height will still be important for reproductive success.

Acknowledgments   The author thanks Gert Stulp and Alexandre Courtiol for sharing unpublished work.

References

Ahmad M, Gilbert RI, Naqui AUN (1985) Assortative mating for height in Pakistani arranged marriages. Journal of Biosocial Science 17:211–214

Allal N, Sear R, Prentice AM, Mace R (2004) An evolutionary model of stature, age at first birth and reproductive success in Gambian women. Proceedings of the Royal Society of London, Series B 271:465–470

Allebeck P, Bergh C (1992) Height, body mass index and mortality: do social factors explain the association? Public Health 106:375–382

Alter G, Neven M, Oris M (2004) Height, wealth and longevity in XIXth century east Belgium. Annales de démographie historique 2:19–37

Baqui AH, Arifeen SE, Amin S, Black RE (1994) Levels and correlates of maternal nutritional-status in urban Bangladesh. European Journal of Clinical Nutrition 48:349–357

Baten J, Murray JE (1998) Women's stature and marriage markets in preindustrial Bavaria. Journal of Family History 23:124–135

Blanckenhorn WU (2000) The evolution of body size: what keeps organisms small? Quarterly Review of Biology 75:385–407

Bogaert AF (1998) Physical development and sexual orientation in women: height, weight, and age of puberty comparisons. Personality and Individual Differences 24(1):115–121

Bogin B (2001) *The Growth of Humanity*. Wiley-Liss, New York, NY

Brennan L, McDonald J, Shlomowitz R (2005) Teenage births and final adult height of mothers in India, 1998–1999. Journal of Biosocial Science 37:185–191

Brush G, Boyce AJ, Harrison GA (1983) Associations between anthropometric variables and reproductive performance in a Papua New Guinea highland population. Annals of Human Biology 10:223–234

Buunk AP, Park JH, Zurriaga R, Klavina L, Massar K (2008) Height predicts jealousy differently for men and women. Evolution and Human Behavior 29:133–139

Case A, Paxson C (2008) Stature and status: height, ability, and labor market outcomes. Journal of Political Economy 116:499–532

Cnattingius R, Cnattingius S, Notzon FC (1998) Obstacles to reducing cesarean rates in a low-cesarean setting: the effect of maternal age, height, and weight. Obstetrics and Gynecology 92:501–506

Costa DL (1993) Height, weight, wartime stress, and older age mortality: evidence from the Union Army records. Explorations in Economic History 30:424–449

Costa DL (2004) The measure of man and older age mortality: evidence from the Gould sample. Journal of Economic History 64:1–23

Davey-Smith G, Hart C, Upton M, Hole D, Gillis C, Watt G, Hawthorne V (2000) Height and risk of death among men and women: aetiological implications of associations with cardiorespiratory disease and cancer mortality. Journal of Epidemiology and Community Health 54:97–103

Devi MR, Kumari JR, Srikumari CR (1985) Fertility and mortality differences in relation to maternal body size. Annals of Human Biology 12:479–484

Engeland A, Bjorge T, Selmer RM, Tverdal A (2003) Height and body mass index in relation to total mortality. Epidemiology 14:293–299

Fielding R, Schooling CM, Adab P, Cheng KK, Lao XQ, Jiang CQ, Lam TH (2008) Are longer legs associated with enhanced fertility in Chinese women? Evolution and Human Behavior 29:434–443

Frisancho AR, Sanchez J, Pallarde. D, Yanez L (1973) Adaptive significance of small body size under poor socioeconomic conditions in Southern Peru. American Journal of Physical Anthropology 39:255–261

Gage TB, Zansky SM (1995) Anthropometric indicators of nutritional status and level of mortality. American Journal of Human Biology 7:679–691

Georgiadis E, Mantzoros CS, Evagelopoulou C, Spentzos D (1997) Adult height and menarcheal age of young women in Greece. Annals of Human Biology 24:55–59

Gigante DP, Horta BL, Lima RC, Barros FC, Victora CG (2006) Early life factors are determinants of female height at age 19 years in a population-based birth cohort (Pelotas, Brazil). Journal of Nutrition 136:473–478

Gillis JS, Avis WE (1980) The male taller norm in mate selection. Personality and Social Psychology Bulletin 6:396–401

Godoy R, Eisenberg DTA, Reyes-Garcia V, Huanca T, Leonard WR, McDade TW, Tanner S (2008) Assortative mating and offspring well-being: theory and empirical findings from a native Amazonian society in Bolivia. Evolution and Human Behavior 29:201–210

Goldstein MS, Kobyliansky E (1984) Anthropometric traits, balanced selection and fertility. Human Biology 56:35–46

Hacker JD (2008) Economic, demographic, and anthropometric correlates of first marriage in the mid-nineteenth-century United States. Social Science History 32:307–345

Harvey PH, Clutton-Brock TH (1985) Life history variation in primates. Evolution 39:559–581

Helle S (2008) A tradeoff between reproduction and growth in contemporary Finnish women. Evolution and Human Behavior 29:189–195

Helm P, Munster K, Schmidt L (1995) Recalled menarche in relation to infertility and adult weight and height. Acta Obstetricia et Gynecologica Scandinavica 74:718–722

Hensley WE (1994) Height as a basis for interpersonal attraction. Adolescence 29:469–474

Higgins LT, Zheng M, Liu Y, Sun CH (2002) Attitudes to marriage and sexual behaviors: a survey of gender and culture differences in China and United Kingdom. Sex Roles 46(3):75–89

Hill K, Hurtado AM (1996) *Ache Life History: The Ecology and Demography of a Foraging People.* Aldine de Gruyter, New York, NY

Hosegood V, Campbell OMR (2003) Body mass index, height, weight, arm circumference, and mortality in rural Bangladeshi women: a 19-y longitudinal study. American Journal of Clinical Nutrition 77:341–347

Hur Y-M (2003) Assortative mating for personality traits, educational level, religious affiliation, height, weight, and body mass index in parents of a Korean twin sample. Twin Research 6:467–470

Jacobsen BK, Heuch I, Kvale G (2007) Association of low age at menarche with increased all-cause mortality: a 37-year follow-up of 61,319 Norwegian women. Am J Epidemiol 166:1431–1437

Jousilahti P, Tuomilehto J, Vartiainen E, Eriksson J, Puska P (2000) Relation of adult height to cause-specific and total mortality: a prospective follow-up study of 31,199 middle-aged men and women in Finland. American Journal of Epidemiology 151:1112–1120

Kannam JP, Levy D, Larson M, Wilson PWF (1994) Short stature and risk for mortality and cardiovascular disease events: the Framingham heart study. Circulation 90:2241–2247

Kirchengast S (2000) Differential reproductive success and body size in !Kung San people from northern Namibia. Collegium Antropologicum 24(1):121–132

Kirchengast S, Hartmann B (2000) Association between maternal age at menarche and newborn size. Social Biology 47:114–126

Kirchengast S, Winkler EM (1995) Differential reproductive success and body dimensions in Kavango males from urban and rural areas in northern Namibia. Human Biology 67: 291–309

Kirchengast S, Winkler EM (1996) Differential fertility and body build in !Kung San and Kavango females from northern Namibia. Journal of Biosocial Science 28:193–210

Kirchengast S, Hartmann B, Schweppe KW, Husslein P (1998) Impact of maternal body build characteristics on newborn size in two different European populations. Human Biology 70:761–774

Komlos J, Baur M (2004) From the tallest to (one of) the fattest: the enigmatic fate of the American population in the 20th century. Economics and Human Biology 2(1):57–74

Laara E, Rantakallio P (1996) Body size and mortality in women: a 29 year follow up of 12,000 pregnant women in northern Finland. Journal of Epidemiology and Community Health 50:408–414

Lasker GW, Thomas R (1976) Relationship between reproductive fitness and anthropometric dimensions in a Mexican population. Human Biology 48:775–791

Lee CMY, Barzi F, Woodward M, Batty GD, Giles GG, Wong JW, Jamrozik K, Lam TH, Ueshima H, Kim HC, Gu DF, Schooling M, Huxley RR, for The Asia Pacific Cohort Studies C (2009) Adult height and the risks of cardiovascular disease and major causes of death in the Asia-Pacific region: 21,000 deaths in 510,000 men and women. International Journal of Epidemiology 38:1060–1071

Leon DA, Smith GD, Shipley M, Strachan D (1995) Adult height and mortality in London: early-life, socioeconomic confounding, or shrinkage. Journal of Epidemiology and Community Health 49(1):5–9

Liao Y, McGee DL, Cao G, Cooper RS (1996) Short stature and risk of mortality and cardiovascular disease: negative findings from the NHANES I epidemiologic follow-up study. Journal of the American College of Cardiology 27:678–682

Liljestrand J, Bergstrom S, Westman S (1985) Maternal height and perinatal outcome in Mozambique. Journal of Tropical Pediatrics 31:306–310

Magnusson PKE, Gunnell D, Tynelius P, Davey Smith G, Rasmussen F (2005) Strong inverse association between height and suicide in a large cohort of Swedish men: evidence of early life origins of suicidal behavior? American Journal of Psychiatry 162:1373–1375

Malina RM, Selby HA, Buschang PH, Aronson WL, Little BB (1983) Assortative mating for phenotypic characteristics in a Zapotec community in Oaxaca, Mexico. Journal of Biosocial Science 15:273–280

Martorell R, Delgado HL, Valverde V, Klein RE (1981) Maternal stature, fertility and infant mortality. Human Biology 53:303–312

Martorell R, Khan LK, Schroeder DG (1994) Reversibility of stunting: epidemiologic findings in children from developing countries. European Journal of Clinical Nutrition 48:S45–S57

McCarron P, Okasha M, McEwen J, Smith GD (2002) Height in young adulthood and risk of death from cardiorespiratory disease: a prospective study of male former students of Glasgow University, Scotland. American Journal of Epidemiology 155:683–687

Migliano AB, Vinicius L, Lahr MM (2007) Life history trade-offs explain the evolution of human pygmies. Proceedings of the National Academy of Sciences of the USA 104:20216–20219

Mitton J (1975) Fertility differentials in modern societies resulting in normalizing selection for height. Human Biology 47:189–201

Monden CWS, Smits J (2008) Maternal height and child mortality in 42 developing countries. American Journal of Human Biology 21:305–311

Mueller U, Mazur A (2001) Evidence of unconstrained directional selection for male tallness. Behavioral Ecology and Sociobiology 50:302–311

Mueller WH (1979) Fertility and physique in a malnourished population. Human Biology 51:153–166

Murray JE (1997) Standards of the present for people of the past: height, weight, and mortality among men of Amherst College, 1834–1949. Journal of Economic History 57:585–606

Murray JE (2000) Marital protection and marital selection: evidence from a historical-prospective sample of American men. Demography 37:511–521

Nettle D (2002a) Height and reproductive success in a cohort of British men. Human Nature 13:473–491

Nettle D (2002b) Women's height, reproductive success and the evolution of sexual dimorphism in modern humans. Proceedings of the Royal Society of London, Series B 269:1919–1923

Okasha M, McCarron P, Smith GD, McEwen J (2001) Age at menarche: secular trends and association with adult anthropometric measures. Annals of Human Biology 28:68–78

Okasha M, Gunnell D, Holly J, Smith GD (2002) Childhood growth and adult cancer. Best Practice and Research in Clinical Endocrinology and Metabolism 16:225–241

Ong KK, Northstone K, Wells JCK (2007) Earlier mother's age at menarche predicts rapid infancy growth and childhood obesity. PLoS Medicine 4:737–742

Onland-Moret NC, Peeters PHM, van Gils CH, Clavel-Chapelon F, Key T, Tjonneland A, Trichopoulou A, Kaaks R, Manjer J, Panico S, Palli D, Tehard B, Stoikidou M, Bueno-De-Mesquita HB, Boeing H, Overvad K, Lenner P, Quiros JR, Chirlaque MD, Miller AB, Khaw KT, Riboli E (2005) Age at menarche in relation to adult height: the EPIC study. American Journal of Epidemiology 162:623–632

Pawlowski B (2003) Variable preferences for sexual dimorphism in height as a strategy for increasing the pool of potential partners in humans. Proceedings of the Royal Society of London, Series B 270:709–712

Pawlowski B, Dunbar RIM, Lipowicz A (2000) Evolutionary fitness: tall men have more reproductive success. Nature 403:156

Peck AM, Vagero DH (1989) Adult body height, self perceived health and mortality in the Swedish population. Journal of Epidemiology and Community Health 43:380–384

Pieper U (1981) Assortative mating in the population of a German and a Cameroon city. Journal of Human Evolution 10:643–645

Pierce CA (1996) Body height and romantic attraction: a meta-analytic test of the male-taller norm. Social Behavior and Personality 24:143–149

Pollet TV, Nettle D (2008) Taller women do better in a stressed environment: height and reproductive success in rural Guatemalan women. American Journal of Human Biology 20:264–269

Roberts DF (1977) Assortative mating in man: husband/wife correlations in physical characteristics. Bulletin of the Eugenics Society of London Supplement 21–45

Roff DA (1992) *The Evolution of Life Histories*. Chapman and Hall, New York, NY

Rollo CD (2002) Growth negatively impacts the life span of mammals. Evolution and Development 4(1):55–61

Salska I, Frederick DA, Pawlowski B, Reilly AH, Laird KT, Rudd NA (2008) Conditional mate preferences: factors influencing preferences for height. Personality and Individual Differences 44(1):203–215

Samaras TT (2007) *Human Body Size and the Laws of Scaling*. Nova Science, New York, NY

Scott EC, Bajema CJ (1982) Height, weight and fertility among the participants of the 3rd Harvard growth study. Human Biology 54:501–516

Sear R (2006a) Height and reproductive success: how a Gambian population compares to the West. Human Nature 17:405–418

Sear R (2006b) Size-dependent reproductive success in Gambian men: does height or weight matter more?. Social Biology 53(3–4):172–188

Sear R, Marlowe FW (2009) How universal are human mate choices? Size does not matter when Hadza foragers are choosing a mate. Biology Letters 5(5):606–609

Sear R, Mace R, McGregor IA (2003) A life-history approach to fertility rates in rural Gambia: evidence for trade-offs or phenotypic correlations? In: Rodgers JL, Kohler HP (eds) *The Biodemography of Human Reproduction and Fertility*. Kluwer Academic, Boston, MA

Sear R, Allal N, Mace R (2004) Height, marriage and reproductive success in Gambian women. Research in Economic Anthropology 23:203–224

Shepperd JA, Strathman AJ (1989) Attractiveness and height: the role of stature in dating preference, frequency of dating, and perceptions of attractiveness. Personality and Social Psychology Bulletin 15:617–627

Song Y-M, Sung J (2008) Adult height and the risk of mortality in South Korean women. American Journal of Epidemiology 168:497–505

Song YM, Smith GD, Sung J (2003) Adult height and cause-specific mortality: a large prospective study of South Korean men. American Journal of Epidemiology 158:479–485

Spuhler JN (1968) Assortative mating with respect to physical characteristics. Eugenics Quarterly 15:128–140

Steckel RH (2009) Heights and human welfare: recent developments and new directions. Explorations in Economic History 46(1):1–23

Strandberg MDTE (1997) Inverse relation between height and cardiovascular mortality in men during 30-year follow-up. American Journal of Cardiology 80:349–350

Strassmann BI, Gillespie B (2002) Life-history theory, fertility and reproductive success in humans. Proceedings of the Royal Society of London, Series B 269:553–562

Swami V, Furnham A, Balakumar N, Williams C, Canaway K, Stanistreet D (2008) Factors influencing preferences for height: a replication and extension. Personality and Individual Differences 45:395–400

Teriokhin AT, Thomas F, Budilova EV, Guegan JF (2003) The impact of environmental factors on human life-history evolution: an optimization modelling and data analysis study. Evolutionary Ecology Research 5:1199–1221

Vetta A (1975) Fertility, physique, and intensity of selection. Human Biology 47:283–293

Waaler HT (1984) Height, weight and mortality: the Norwegian experience. Acta Medica Scandinavica Supplement 679:1–51

Walker R, Gurven M, Hill K, Migliano A, Chagnon N, De Souza R, Djurovic G, Hames R, Hurtado AM, Kaplan H, Kramer KL, Oliver WJ, Valeggia CR, Yamauchi T (2006) Growth rates and life histories in 22 small-scale societies. American Journal of Human Biology 18:295–311

Wells JCK (2003) The thrifty phenotype hypothesis: thrifty offspring or thrifty mother? Journal of Theoretical Biology 221:143–161

Illusion Number 4
The Past Doesn't Echo in Our Heads

Chapter 11
Developmental Psychology Without Dualistic Illusions

Why We Need Evolutionary Biology to Understand Developmental Psychology

Athanasios Chasiotis

Reality is that which, even when you stop believing in it, doesn't go away
(Philip K. Dick 1981)

Abstract This chapter starts with an epistemological introduction for two reasons: First, if you talk about a human without illusions—as the subtitle of this volume suggests—you refer to epistemological categories. Second, because this book is dedicated to a renowned evolutionary anthropologist and bio-philosopher, this chapter relies on works of other evolutionary theorists, and philosophers also (Bischof 2008; Gadenne 2004; Vollmer 1975; for a more psychological account see Chasiotis 2010 in press). After some epistemological prolegomena on the implicit dualism in psychology, an evolution-based developmental psychology is outlined and selected empirical findings based on this perspective are presented. The chapter concludes with (meta-)theoretical implications of an evolutionary account of developmental psychology. The seemingly paradoxical conclusion is that an evolutionary developmental psychology can help us to gain a more differentiated view of the concept of *environment* without abandoning a naturalistic and monistic view of reality.

11.1 The Implicit Dualism in Psychology

It is easier to defend human uniqueness if you divide reality into a hemisphere consisting of the triad of soul, human, and meaning and another hemisphere in which body, animal, and mechanistic causality is located. (Bischof 2008, p. 163).

Psychology, a science in a hybrid position between the humanities and the natural sciences, suffers from a subtle disease. Although it is mainly defined as being an empiricist science where all (mental and nonmental) phenomena are based on one

A. Chasiotis (✉)
Tilburg University Faculty of Social and Behavioural Sciences, 5000 LE, Tilburg, Netherlands
e-mail: achasiot@uvt.nl

U.J. Frey et al. (eds.), *Homo Novus – A Human Without Illusions*,
The Frontiers Collection, DOI 10.1007/978-3-642-12142-5_11,
© Springer-Verlag Berlin Heidelberg 2010

and the same physical substrate, this unfortunately does not mean that it presents itself as a monistic science, according to which the mind is just a part of the body. Historically, this problem can be traced back to the Aristotelian division of the world into body ("physis") and soul ("psyche"), and his postulate that the goal ("telos") each living creature has ("echein") lies in the soul, a vital force he thus called "entelechy". This Aristotelian dualism more or less dominated science—in biology at least until Darwin (Mayr 1984) and in psychology perhaps until today (see Bischof 1981, 2008). The mind–body dualism peaked in the distinction of the *res extensa* and the *res cogitans* by René Descartes in the seventeenth century. While the *res extensa* deals with all matter and can be described via deterministic mechanisms, the *res cogitans* is exclusively human and is based on nondeterministic semantics. Accordingly, because of this Cartesian heritage, Bischof calls the implicit dualism in psychology a "cartesian contamination" (Bischof 2008, p. 127). According to this notion, the link of an individual to its environment is either defined by attributing a teleological cause, a meaning, purpose, or intention to the subject without assuming any generalizable regularities with the environment or by ignoring introspection and locating all causes of behavior as being external to the individual. From that latter perspective, a stance held in Behaviorism, the dominating field of psychology in the first half of the twentieth century, everything there is to know lies in the environment and inner mental states are not considered at all. In Behaviorism, the organism is determined by an unspecified general drive, and the environment structures behavior via a virtually infinite set of reinforcing stimuli. Even after the so-called cognitive revolution in the 1950s, which set out to (re-)"establish meaning as the central concept of psychology" (Bruner 1990, p. 2), the dualistic stance did not disappear. In cognitive psychology, the organism is causally determined by so-called primitive, biological drives, while again the environment, now labeled society, offers all secondary or social motives, which are infinite in number and indefinite in quality. Ironically, while considered to be a counterpart to Behaviorism's notion, the reintroduction of meaning was not achieved by locating it within the individual, as in the traditional Aristotelian notion. Instead, information became synonymous with semantics, coming from outside the organism, although information is a quantitative term defining the frequency of a signal without any qualitative (semantic, meaningful) content. Even in contemporary cognitive science, the impressive results from neuroimaging often mask this hermeneutical leap one has to take to arrive at a meaningful interpretation of the colorful illustrations of activated brain areas (see e.g. Vul et al. 2009).

Where does this bias for dualistic conceptualizations come from? Obviously, we love symmetrical dichotomies, not only for aesthetic reasons, but also because of their seemingly exhaustive explanatory value: If it is not A, it is (i.e., is defined to be) non-A. Is that not all there is to know? But the problem with dichotomies, as philosophers of science know, is that they are irrefutable and therefore scientifically useless: Epistemologically speaking, if one is dissatisfied with the notion of God as an explanation for the evil in this world, introducing the notion of a devil does not make God any more plausible either (Vollmer 1988, 1995; see also Frey 2007).

To understand this affinity for symmetric aesthetics in our meta-theoretical models in psychology, it might be informative to draw on the work of Gestalt psychologist Wolfgang Metzger (Metzger 1954). He postulated that psychological concepts can be conceived in two different ways: The first, which he calls Eleatic (referring to the pre-Socratic school of Parmenides and Zenon in Elea, 500 BC), claims that our senses cannot be trusted and that every phenomenological experience should be scrutinized by rational reasoning to justify what intuitively has been perceived as true. The second he calls Phenomenological: here it is claimed that every psychological phenomenon should be described as it is experienced, regardless of how unusual, unexpected, or even illogical it might seem to be. It is important to note that these two conceptualizations are not just another dichotomy, because they are not opposites, but interchangeably used, sometimes we follow the one stance, while at another time, we argue according to the other (see Chasiotis 2008). One main difference concerning the validity of psychological concepts lies in this distinction between evidence and veridicality: While the Phenomenological School claims that everything that is evident is also true, the Eleatic perspective postulates that every evident phenomenon needs to be scrutinized to avoid misconceptions (see the motto of this chapter). For many scholars (e.g., Bischof 2008), the difference between the Eleatic and the Phenomenological stance still lies at the heart of many problems where interdisciplinarity is involved (see also van de Vijver and Chasiotis 2010). In empirical psychology, there are many terms describing these two differing views on our psyche: the Eleatic principle is more associated with behavioristic, syntactical, positivistic, or quantitative approaches, whereas the Phenomenological principle is associated with semantic, hermeneutical, and qualitative approaches. According to Bischof, they also differ in their heuristic principles: While the Phenomenological perspective has teleology as a heuristic guideline and searches for meaning and goals that in the final analysis can only be discovered within the individual, the Eleatic perspective uses aesthetics as a heuristic guideline (Bischof 1988, 2008). Here we re-encounter the numerous symmetric dichotomies in Psychology. There are dichotomies in emotions (positive ↔ negative affects), in social psychology (prosocial ↔ antisocial), in psychoanalysis (Eros ↔ Thanatos), and in cross-cultural psychology (Individualism ↔ Collectivism, see Chasiotis 2010 in press). Finally, there is a dichotomous view in mainstream developmental psychology in which endogenetic factors are identified with biological determinism and maturation while exogenetic factors (society, culture) provide all meaning required for ontogenesis. It is to this psychological discipline we will turn now.

11.2 Evolutionary Developmental Psychology as an Environmentalist Discipline

Organisms did not evolve to survive, but to reproduce: Life is finite and our psychological makeup is ultimately not aimed at merely surviving or well-being but on reproductive success. This perspective helps us in explaining not only the psychological, proximate causes of a behavior (e.g. by trying to show *how* we pursue a happy life), but also *why* we strive to obtain one certain state of mind and not another, that is, why some things, like having children (and grandchildren, see Voland et al. 2005), make us happy while others do not.

The Darwinian concept of adaptation is crucial to understanding why and how individual traits fit environmental conditions, and thus have ultimately resulted in reproductive success. Adaptations carry environmental information that has become represented in phenotypes during evolution because it helped organisms to (survive in order to) reproduce. Accordingly, there are no organism-independent environmental factors: without an organism, there is no environment. It is important to note that this is not a solipsistic stance in which reality does not exist at all, but the evolutionary epistemological stance (Vollmer 1975): We can only know something because it reflects something that is in an adaptive relation to reality. Contrary to the common miscomprehension of evolutionary biology as fully deterministic (*if it is in the genes, it cannot be changed*), the epigenetic view of development is bidirectional: if a gene is switched on, its genetic activity is a cause for the development of an organism, but the expression of the involved genes during ontogenesis is also influenced by the ontogenetic experiences (i.e., maturational processes and behavior; see Bischof 2008; Bjorklund and Pellegrini 2002; Gottlieb 1991). From this perspective, the phenotype is the result of epigenetic processes during development: genes interact epigenetically with the environment to produce the behavior we study in psychology. So there is neither a "pure" genetic nor "pure" environmental determination of behavior but an environmentally mediated, epigenetic relation between a genotype and a phenotype.

But if one claims that the nature–nurture dualism is not useful because there are no pure genetic or environmental effects, does it mean that we cannot say anything about the interaction of genes and environment? Of course not, quite to the contrary: if we abandon the traditional dichotomy of nature versus nurture, we only discard the extremes of the continuum lying between what we call "genes" and "environment". What is gained instead, though, is a much more differentiated, epigenetic description of the interactions between these postulated poles. As I will show in the following, the dualistic misconception is not due to an all-encompassing and therefore useless conceptualization of biology or nature, but an undifferentiated view of *environment*.

To know more about an organism, we need an environmental theory explaining the species-specific epigenetic effects of the environment on the organism. According to Bischof, there are three different segments of the environment determining the relations of genotype and phenotype: *selection*, *alimentation*, and *stimulation* (Bischof 2008).

1. *Selection* is responsible for the finality (or semantic quality) of an organism. Finality is a function of a system that acts as if it had an interest in adapting or a goal to adapt to the environment. This gives the system a semantic quality, i.e. meaning. To distinguish it from the metaphysical notion of teleology, this naturalistic goal-directedness is called *teleonomy* (see Mayr 1984; Bischof 1995, 2008). The notion of teleonomy provides us with a meaning of a behavior:

A behavior is shown because it had an adaptive value leading to reproductive success in the evolutionary past. Hence, selection deals with the *phylogenetic* development. Its adaptational pressure is aimed at reproductive success.

2. *Alimentation* is the term subsuming all intra- and extra-uterine environmental influences that lead to a macroscopic development of the genetic code. The phenotype, thus, is the result of the interaction of a genotype with alimentative aspects of the environment. Alimentation deals with the *ontogenetic* development, its adaptational pressure aims at survival and its typical developmental mechanism is maturation, mainly affecting the morphology of the organism. A striking example of environmental alimentative pressure is monozygotic twins raised apart in dissimilar environments (see Tanner 1978).

3. *Stimulation* from the environment does not alter the phenotype, but affects the behavior of an organism. The organism is evolutionary prepared (by selection, see the next section) to detect stimuli from the environment and to react accordingly. Environmental stimuli are telling us something about the current state of the selective environment, and thus lead finally (but only ephemerally) to the psychological adaptation of well-being. Its typical developmental mechanism is learning.

With these specifications, the epigenetic perspective on which evolutionary developmental psychology is based can be formulated as the interplay of stimulation and alimentation leading to selection. The underlying process can be described as if the genetic adaptation copies the learned adaptation: if the environment is stable enough, there will be a genetic adaptation irrespective of the flexibility/plasticity of the organism's learning capacity. The proportion of fixed genetic programs in the behavioral output of the organism increases with the stability of the environment (a process coined *obligatory genocopy*, see Lorenz 1965; see also Dennett's notion of "genetic learning" 1995).

11.3 Childhood as a Sensitive Period

One of the most obvious manifestations of the just presented epigenetic interplay is the evolution of life spans. This view implies that different developmental stages are not transitory phases toward adulthood but evolutionary end-products per se, because many features of childhood can be considered *preparations for adulthood* (Alexander 1987; Bjorklund 1997): if environmental change is slow compared to an individual lifespan, the optimal mode of adaptation is to establish sensitive learning situations early in life as preparations for adulthood that guide later development (Draper and Harpending 1988). These sensitive learning situations are characterized in our terms as stimulative alimentation, where certain environmental stimuli during a certain sensitive period are needed additionally to alimentative processes (e.g. the presence of a parental figure in imprinting, see Lorenz 1965).

This evolutionary perspective fits with empirical evidence in the psychological literature and in mainstream developmental psychology, in which the first 6 years of childhood are considered as psychologically the most important for individual development (Lamb and Sutton-Smith 1982). Every child is reared in a unique environment characterized by contextual variables such as number of siblings, specific birth order, and socioeconomic conditions. According to his/her ordinal position within the family, the child receives a specific form of parental treatment (Toman 1971; see also Moore et al. 1997). The ordinal position that thus shapes the developmental context has been shown to explain a huge array of phenomena, ranging from differences in personality traits to scientific discoveries and political revolutions (Sulloway 1996). Extensive value surveys in sociology (Inglehart 1997) and cross-cultural psychology (Allen et al. 2007) provide evidence for the importance of socioeconomic factors for developmental conditions: For example, the financial situation during childhood has been found to be a better predictor of the endorsement of values in adulthood than the current economic situation of the adult respondent. Such effects are typically summarized under the notion of "economic determinism" to refer to the impact of the economic situation on psychological outcomes. In the following, recent empirical evidence will be presented regarding these two building blocks of childhood context, birth order and socioeconomic status during childhood, and their explanatory power for cultural differences in pubertal timing, parenting motivation, social values, and autobiographical memory.

11.3.1 Pubertal Timing

Evolutionary developmental psychology offers a theoretical framework to conceptualize the influence of resource availability in childhood on consequent somatic, psychological, and reproductive development (Belsky et al. 1991; Chisholm 1993): Psychosocial contextual stressors such as inadequate resources or unstable employment lead to marital discord and foster insensitive parental behavior, which in turn induces behavioral problems in the child. If ecological conditions are largely held constant, as is the case with the majority of citizens in industrialized countries, the theory arrives at the critical prediction that aversive childhood experiences accelerate sexual maturation.

In a research project aimed at investigating the social changes in family development that occurred after the reunification of Germany in 1989, my colleagues and I provided support for this perspective (Chasiotis et al. 1998; Chasiotis 1999; Chasiotis et al. 2003). We confirmed the importance of birth order and its interaction with socioeconomic status in childhood by predicting somatic as well as psychological developmental outcomes in a comparison of samples from Osnabrück (West Germany) and Halle (East Germany). In one study, we used the subsample of all mother–daughter dyads from West and East Germany to test the assumption that the onset of puberty is affected by childhood experiences (Chasiotis et al. 1998). A comparison of the two samples of mother–daughter dyads showed that what seems to be inherited is not the timing of puberty per se, but the sensitivity for the prepubertal childhood context. The consideration of social status and birth order in other subsamples of the same research project led to the assumption that childhood context variables could also determine the East–West differences in intergenerational context continuity. Results of a reanalysis showed that birth order displayed significant and (mainly) expected effects of childhood variables on the age at menarche for women who do not have younger siblings (i.e. only children or later-borns) (Chasiotis et al. 2003). In contrast, participants with younger siblings (i.e. first-borns and middle-borns), showed no such effects. In the previous study differences in intergenerational context continuity between the parental and filial generations in East and West Germany were interpreted as being caused by different socio-cultural milieus prevalent in the former Federal Republic of Germany and the German Democratic Republic (Chasiotis et al. 1998). The reanalysis of the data revealed that the intergenerational context discontinuity affecting the onset of puberty was primarily due to different childhood experiences of last-born daughters and their mothers. It seems that the absence or existence of younger siblings influences the age at menarche, and not the "cultural" origin of the subjects.

11.3.2 Parenting Motivation

Parenthood constitutes an investment in genetic offspring as a part of reproductive effort while at the same time transmitting cultural values and practices between generations. Although much contextual and cultural variation in parenting behavior has been reported (Keller 2007), the motivational roots of this culturally divergent parenting behavior are barely known. Chasiotis, Hofer, and Campos proposed that interactive experiences with younger siblings should be considered an important factor for the emergence of parenting motivation (Chasiotis et al. 2006). Taking a cross-cultural, developmental perspective, they suggested that the presence of younger siblings triggers prosocial, nurturant motivations and caretaking behaviors. In turn, this implicit parenting motivation results in positive, loving feelings towards children on a conscious level, which finally leads to parenthood. Using structural equation modeling, they demonstrated that this developmental pathway is verifiable in both male and female participants, and in all cultural samples from Germany, Costa Rica, and Cameroon.

A further investigation of the relationship was warranted because implicit parenting motivation showed cultural variation and was associated with the existence of younger siblings—which was different across cultures. To investigate the impact of this childhood context variable on cultural differences (aiming to "peel the onion called culture", Poortinga et al. 1987), implicit parenting motivation was first regressed on the variable "younger siblings". In the next step, the unstandardized residual of implicit parenting motivation of that regression analysis was reentered in an analysis of variance (ANOVA) with culture as predictor. The ANOVA with the residual of implicit parenting motivation as dependent variable and culture as predictor showed a remarkable decrease in effect size of culture (in statistics, an effect size is a measure of the strength of the relationship between two variables in a statistical population), which meant that 62% of the original effect size of culture on implicit parenting motivation could be traced back to sibling effects. This impressive effect was replicated in three additional samples from Cameroon, Costa Rica, and Germany (Chasiotis and Hofer 2003), in which the effect size of "culture" decreased to 50%, and with three recent samples from Cameroon, Germany, and PR China, in which the reduction even approached 100% (Bender and Chasiotis 2010 in press).

11.3.3 Social Values

Building on these results of previous studies on implicit prosocial (parenting) motivation, we investigated whether explicit prosocial values are also influenced by childhood context variables. In two studies, data on social value orientations were collected (Chasiotis and Hofer 2003, Bender and Chasiotis 2010 in press). The first study with the Schwartz Value Survey (SVS, Schwartz 1994), and samples from Cameroon, Costa Rica, and Germany, reveals that 36% of the cultural differences of social values constituting the higher order value type of conservation (consisting of the subscales tradition, conformity, and security) can be traced back to sibling effects. After combining the effect of siblings with that of socioeconomic status in childhood (i.e. paternal profession), the amount of explained variance in conservation even increases to 55%. Analogous to the findings on economic determinism by Inglehart (1997) and Allen et al. (2007), present occupation was not related to conservation value orientation. In the second study (Bender and Chasiotis 2010 in press), the importance of sibling effects for social value orientations was further corroborated in samples from Germany and Cameroon: measuring conservation with the Portrait Values Questionnaire (PVQ, Schwartz et al. 2001), the number of siblings explains 72% of the cultural variance in conservation. These strong sibling effects only occur in scales in which intimate relationships with close relatives are almost explicitly mentioned (see, e.g., the definition of the Benevolence scale, Schwartz 2009: *the welfare of people with whom one is in frequent personal contact*), but not in scales dealing with more individualistic, autonomous social values such as self-direction and achievement.

11.3.4 Autobiographical Memory

Storytelling and narratives represent human universals (Brown 1991), and autobiographical narratives are a "natural kind" of human cognition (Bruner 1990). At age 3–4, children start to participate actively in "memory talk" with their parents (Nelson 2005), a process by which children begin learning to refer to themselves in the past. This emergence of autobiographical memory (AM) in the preschool years is an important event in human development: it is considered a unique feature of the psychological endowment of the human primate on the ultimate phylogenetic level (Bischof 2008), and on the proximate psychological level it seems to mediate the relationship between the development of implicit motives and theory of mind, thus constituting an important building block for a culture-specific development of the self (see Chasiotis et al. 2010).

An increasing number of studies have found differences in the content and structure of AM across cultural contexts, which have been traced back to different parental socialization practices (for an overview, see Nelson and Fivush 2004; see also Chasiotis et al. 2010). There are some indications that not just parents, but siblings (or their absence), may play a crucial role in the formation of AM, which are corroborated by recent findings from a study in Cameroon, PR China, and Germany (Bender and Chasiotis 2010 in press). Two of the most widely investigated variables in cross-cultural research on AM, namely the age at which the earliest memory took place, and the specificity of the mnemonic account, have been related to childhood contextual variables. These measures were supplemented with a measure for cognitive complexity, which allows the identification of "separated" (differentiation) and "connected" (integration) ways of processing autobiographical information also in cross-cultural samples (see Chasiotis et al. 2010). The number of siblings had a substantial effect: while 30% of cultural differences in the age of the earliest childhood recollection can be explained through the number of siblings, for cognitive complexity (99%) and specificity (99%) the sibling effect even renders cultural group membership insignificant (Bender and Chasiotis 2010 in press).

11.3.5 Childhood Context Explains Cultural Differences

These results on childhood context effects on diverse psychological variables across cultures imply that the family context during childhood is a powerful tool for explaining cross-cultural differences in developmental outcomes. Context variables such as socioeconomic status during childhood, birth order, or number of siblings can be expected to exert similar influences on somatic, psychological, and reproductive developmental trajectories across different cultural contexts. On the basis of the explanatory power of these childhood context variables for cultural differences in such highly diverse areas as pubertal timing, implicit motivation, social value orientations, and autobiographical memory, it can be suggested that many psychological characteristics that are typically attributed to cultural differences may reflect systematic variations in family constellations across cultural contexts. For example, differences in self-construals, which are interpreted as due to culture-specific socialization (Markus and Kitayama 1991), could be at least partially dependent on relevant characteristics shared by participants from cultural samples such as systematic biases due to having (or not having) siblings.

11.4 Conclusion: A Developmental Psychology Without Dualistic Illusions

With these reflections on an evolutionary developmental psychology in mind, what are the implications of a naturalistic and monistic view for environmental effects on human development? In this concluding section, some main themes revolving around an evolutionary view of environmental effects on human development will be reconsidered to illustrate the strengths of this approach. In short, the seemingly paradoxical conclusion is that an evolutionary view of development helps us to obtain a more differentiated view of the somewhat shallow concept of *environment* without abandoning a monistic and naturalistic view of reality.

Epistemological distinctions reconsidered. The presented distinction between the Eleatic and the Phenomenological perspective would only then entail methodological consequences for conducting research if the two perspectives were based on an actually inherent dualism of our psychological apparatus, in which the Eleatic perspective just dealt with bodily sensations and the phenomenological perspective with the soul. However, if one takes the monistic stance—as is the case in the naturalistic worldview of modern evolutionary theory—these two perspectives are just two sides of the same coin: what can be *described* eleatically, is *experienced* in a phenomenological way (see also the concept of "qualia" in the philosophy of mind literature, e.g., Dennett (1995) and Gadenne (2004)): What we experience as meaningful can be described as goal-directed at the same time (in system theory, see Bischof (1995)). Thus, psychology, as an empirical science, cannot ignore our phenomenological experiences and should treat them seriously by describing them as they are, but should also distinguish between their evidence (face value) and their veridicality (validity) without adding any metaphysical attributes to them.

Psychological functionalism reconsidered. A nonevolutionary view of our biological heritage considers stimulation and alimentation at most: if we talk of biological (or so-called primal) needs, we often refer to alimentative processes (like hunger) that evolved for survival purposes. If we only consider stimulation and alimentation and ignore selection as a driving environmental force, this is as far a "biological" view of the psyche goes: its function seems to be confined to guaranteeing survival (alimentation) and well-being (stimulation), but not reproduction.

Criticism of sociobiology reconsidered. Exactly the opposite can be observed in some sociobiological notions: an epigenetic perspective also clarifies why the sociobiological view of selection as the driving force of the adaptational efforts of the organism is often criticized as being too "unpsychological": it often deals only with the selective segment of the environment, although adaptations occur in the alimentative and on the stimulative part of the environment as well (Bjorklund and Pellegrini 2002).

Behavioral genetics. What behavioral genetics measure are therefore not genetic or environmental effects per se, but genotypic or phenotypic variance based on alimentation (for a more elaborate discussion on the relation of behavioral genetics and evolutionary theory see Chasiotis 2006, 2007).

Gender differences. The more stable the considered environmental features are over time, the more probable is a genetic fixation: The intrauterine conception in mammals leading to obvious gender differences in parental investment first occurred 400 million years ago. That is why it is very unlikely that gender differences in behavior such as competitive aggressiveness and risk proneness are only learned during differential socialization (see Chasiotis and Voland 1998).

Biological versus cultural transmission. Another example of a misleading dichotomy between nature/biology and nurture/culture is the distinction of the modes of informational transmission. One way of clarifying the difference between biogenetic and tradigenetic transmission (Boyd and Richerson 1985) is by contrasting it with individual learning (stimulation): learning allows for fast adaptations to changes in immediate actual-genetic circumstances. While genetic changes (genetic "learning", see Dennett 1995) need at least hundreds of generations, sociocultural changes (via stimulative alimentation during childhood (see above) and individual learning) are normally observed within a generation (Chasiotis 1999; Chasiotis et al. 2003; Voland et al. 1997). If we dichotomize these learning rates, we reify the underlying processes and implicitly assume different units of transmission, while the unit of information that is intergenerationally transmitted is still the same, namely the gene (Chasiotis 2007).

The environment of evolutionary adaptedness (EEA) reconsidered. The EEA is often mentioned in debates on evolutionary approaches in psychology and is often misconceived as the specific environment of the Pleistocene (Daly and Wilson 1999). This misconception is rightly regarded as one of the most important weaknesses of the evolutionary approach in psychology (Panksepp and Panksepp 2000). Properly considered, the EEA is neither a habitat nor a phylogenetic period, but a statistical term for all stimulus-relevant environmental features of our phylogenetic past (Tooby and Cosmides 1990). Interestingly, this conceptualization of the EEA by Tooby and Cosmides (1990) is synonymous to Bischof's notion of the "inborn environment":

> If a genotype builds a phenotype via alimentation which reacts to stimulation of the environment in such a way that selection does not have to change the genotype, this natural environment can be labeled in an almost paradoxical way the inborn environment. (Bischof 2008, p. 153f)

From that perspective, it is not surprising that the adaptivity of human reproductive behavior is historically and culturally far less restricted as a vulgar understanding of the EEA as just the Pleistocene period might suggest (see also Chasiotis 2006, 2007). On the contrary: empirical evidence in evolutionary anthropology suggests that not only in foraging peoples, but also in agrarian, pre-industrial, and historical societies until the nineteenth century at least, that is, before the demographic transition of a society, adaptive mechanisms have still been at work at least until very recently, if not even until today (Voland 1998, 2000, 2009).

References

Alexander RD (1987) The biology of moral systems. Aldine, New York
Allen MW, Ng SH, Ikeda K, Jawan JA, Sufi AH, Wilson M, Yang KS (2007) Two decades of change in cultural values and economic development in eight East Asian and Pacific Island nations. Journal of Cross-Cultural Psychology 38:247–269
Bender M, Chasiotis A (2010 in press) Number of siblings in childhood explains cultural variance in autobiographical memory in PR China, Cameroon, and Germany. Journal of Cross-Cultural Psychology

Belsky J, Steinberg L, Draper P (1991) Childhood experience, interpersonal development, and reproductive strategy: an evolutionary theory of socialization. Child Development 62:647–670

Bischof N (1981) Aristoteles, Galilei, Kurt Lewin – und die Folgen. In: W Michaelis (ed) *Bericht des 32. Kongresses der Deutschen Gesellschaft für Psychologie, Zürich, 1980.* Hogrefe, Göttingen

Bischof N (1988) Ordnung und Organisation als heuristische Prinzipien des reduktiven Denkens. In: H Meier (ed) *Die Herausforderung der Evolutionsbiologie.* Piper, Munich

Bischof N (1995) *Struktur und Bedeutung. Eine Einführung in die Systemtheorie für Psychologen, Biologen und Sozialwissenschaftler.* Huber, Bern

Bischof N (2008) *Psychologie. Ein Grundkurs für Anspruchsvolle.* Kohlhammer, Stuttgart

Bjorklund DF (1997) The role of immaturity in human development. Psychology Bulletin 122:153–169

Bjorklund DF, Pellegrini A (2002) *The Origins of Human Nature: Evolutionary Developmental Psychology.* American Psychological Association, Washington, DC

Boyd R, Richerson P (1985) *Culture and the Evolutionary Process.* University of Chicago Press, Chicago, IL

Brown D (1991) *Human Universals.* Temple University Press, Philadelphia, PA

Bruner J (1990) *Acts of Meaning.* Harvard University Press, Cambridge, MA

Chasiotis A (1999) *Kindheit und Lebenslauf. Untersuchungen zur evolutionären Psychologie der Lebensspanne.* Huber, Bern

Chasiotis A (2006) Evolutionsbiologische Ansätze in der Psychologie. In: Pawlik K (ed) *Handbuch Psychologie.* Springer, Berlin

Chasiotis A (2007) Evolutionstheoretische Ansätze im Kulturvergleich. In: Trommsdorff G, Kornadt H-G (eds) *Kulturvergleichende Psychologie, Band I, Theorien und Methoden,* Enzyklopädie der Psychologie. Hogrefe, Göttingen

Chasiotis A (2008) Über (die Illusion der) Betreuungsalternativen und den Preis der Freiheit— Evolutionsbiologische und entwicklungspsychologische Aspekte frühkindlicher Erziehung. In: Rietmann S, Hensen G (eds) *Tagesbetreuung im Wandel. Das Familienzentrum als Zukunftsmodell.* VS Verlag, Wiesbaden

Chasiotis A (2010 in press) An epigenetic view on culture: what evolutionary developmental psychology has to offer for cross-cultural psychology. In: Breugelmans S, Chasiotis A, van de Vijver F (eds) *Fundamental Questions in Cross-Cultural Psychology.* Cambridge University Press, Cambridge, MA

Chasiotis A, Bender M, Kiessling F, Hofer J (2010) The emergence of the independent self: autobiographical memory as a mediator of false belief understanding and motive orientation in Cameroonian and German preschoolers. Journal of Cross-Cultural Psychology 41:368–390

Chasiotis A, Hofer J (2003) Die Messung impliziter Motive in Deutschland, Costa Rica und Kamerun. Research report to the German Research Foundation (DFG)

Chasiotis A, Hofer J, Campos D (2006) When does liking children lead to parenthood? Younger siblings, implicit prosocial power motivation, and explicit love for children predict parenthood across cultures. Journal of Cultural and Evolutionary Psychology 4:95–123

Chasiotis A, Keller H, Scheffer D (2003) Birth order, age at menarche, and intergenerational context continuity: a comparison of female somatic development in West and East Germany. North American Journal of Psychology 5(2):153–170

Chasiotis A, Scheffer D, Restemeier R, Keller H (1998) Intergenerational context discontinuity affects the onset of puberty: a comparison of parent-child dyads in West and East Germany. Human Nature 9:321–339

Chasiotis A, Voland E (1998) Geschlechtliche Selektion und Individualentwicklung. In: Keller H (ed) *Lehrbuch Entwicklungspsychologie.* Huber, Bern

Chisholm JS (1993) Death, hope, and sex: life-history theory and the development of reproductive strategies. Current Anthropology 34:1–24

Daly M, Wilson M (1999) Human evolutionary psychology and animal behavior. Animal Behaviour 57:509–519

Dennett D (1995) *Darwin's Dangerous Idea: Evolution and the Meanings of Life*. Simon & Schuster, New York, NY

Dick PK (1981) *Valis*. Vintage, New York, NY

Draper P, Harpending H (1988) A sociobiological perspective on human reproductive strategies. In: MacDonald KB (ed) *Sociobiological Perspectives on Human Development*. Springer, New York, NY

Frey U (2007) *Der blinde Fleck. Kognitive Fehler in der Wissenschaft und ihre evolutionsbiologischen Grundlagen*. Ontos, Frankfurt

Gadenne V (2004) *Philosophie der Psychologie*. Huber, Bern

Gottlieb G (1991) Experiental canalization of behavioral development. Developmental Psychology 27:4–13

Inglehart R (1997) *Modernization and Post-modernization: Cultural, Economic, and Political Change in 43 Nations*. Princeton University Press, Princeton, NJ

Keller H (2007) *Cultures of Infancy*. Erlbaum, Mahwah, NJ

Lamb M, Sutton-Smith B (eds) (1982) *Sibling Relationships: Their Nature and Significance Across the Lifespan*. Erlbaum, Mahwah, NJ

Lorenz K (1965) *Evolution and Modification of Behavior*. University of Chicago Press, Chicago, IL

Markus HR, Kitayama S (1991) Culture and the self: implications for cognitions, emotion, and motivation. Psychological Review 98:224–253

Mayr E (1984) *The Growth of Biological Thought*. Harvard University Press, Cambridge, MA

Metzger W (1954) *Psychologie. Die Entwicklung ihrer Grundannahmen seit der Einführung des Experiments*. Steinkopff, Darmstadt

Moore G, Cohn J, Campbell S (1997) Mother's affective behavior with infant siblings: Stability and change. Developmental Psychology 33:856–860

Nelson K (2005) Evolution and development of human memory systems. In: Ellis BJ, Bjorklund DF (eds) *Origins of the Social Mind: Evolutionary Psychology and Child Development*. Guilford, New York, NY

Nelson K, Fivush R (2004) The emergence of autobiographical memory: a social cultural developmental theory. Psychological Review 111:486–511

Panksepp J, Panksepp JB (2000) The seven sins of evolutionary psychology. Evolution and Cognition 6:108–131

Poortinga YH, van de Vijver FJR, Joe RC, van de Koppel JMH (1987) Peeling the onion called culture: a synopsis. In: C Kağitçibaşi (ed) *Growth and Progress in Cross-Cultural Psychology*. Swets & Zeitlinger, Lisse

Schwartz SH (1994) Beyond individualism and collectivism: new cultural dimensions of values. In: Kim U, Triandis HC, Kağitçibaşi C, Choi SC, Yoon G (eds) *Individualism and Collectivism: Theory, Method and Applications*. Sage, Thousand Oaks, CA

Schwartz SH (2009) Values: cultural and individual. In: Breugelmans S, Chasiotis A, van de Vijver F (eds) *Fundamental Questions in Cross-Cultural Psychology*. Cambridge University Press, Cambridge, MA

Schwartz SH, Melech G, Lehmann A, Burgess S, Harris M, Owens V (2001) Extending the cross-cultural validity of the theory of basic human values with a different method of measurement. Journal of Cross-Cultural Psychology 32:519–542

Sulloway F (1996) *Born to Rebel: Birth Order, Family Dynamics, and Creative Lives*. Pantheon Books, New York, NY

Tanner J (1978) *Foetus into Man: Physical Growth from Conception to Maturity*. Open Books, London

Toman W (1971) The duplication theorem of social relationships as tested in the general population. Psychological Review 78:380–390

Tooby J, Cosmides L (1990) The past explains the present: adaptations and the structure of ancestral environments. Ethology and Sociobiology 11:375–424 van de Vijver F, Chasiotis A (2010) Making methods meet: mixed designs in cross-cultural research. In: Brown M & Johnson T (eds) *Survey Methods in Multinational, Multiregional, and Multicultural Contexts*. Wiley, New York, NY, pp 1–9

Voland E (1998) Evolutionary ecology of human reproduction. Annual Review of Anthropology 27:347–374

Voland E (2000) Natur oder Kultur?—Eine Jahrhundertdebatte entspannt sich. In: Fröhlich S (ed) *Kultur—Ein interdisziplinäres Kolloquium zur Begrifflichkeit*. Landesamt für Archäologie, Halle/Saale

Voland E (2009) *Grundriss der Soziobiologie*, 3rd ed. Springer, Berlin

Voland E, Chasiotis A, Schiefenhövel W (eds) (2005) *Grandmotherhood—The Evolutionary Significance of the Second Half of Female Life*. Rutgers University Press, Piscataway, NJ

Voland E, Dunbar RIM, Engel C, Stephan P (1997) Population increase and sex biased parental investment in humans: evidence from 18th and 19th century Germany. Current Anthropology 38:129–135

Vollmer G (1975) *Evolutionäre Erkenntnistheorie*. Hirzel, Stuttgart

Vollmer G (1988) Das alte Gehirn und die neuen Probleme. In: Vollmer G (ed) *Was können wir wissen? Band 1: Die Natur der Erkenntnis*. Hirzel, Stuttgart

Vollmer G (1995) Bin ich Atheist? Orientierungshilfen für ernsthafte Zweifler. In: Vollmer G (ed) *Auf der Suche nach der Ordnung. Beiträge zu einem naturalistischen Welt- und Menschenbild*. Hirzel, Stuttgart

Vul E, Harris C, Winkielmann P, Pashler H (2009) Voodoo Correlations in fMRI Studies of Emotion, Personality, and Social Cognition. Perspectives on Psychological Science 4:274–290

Chapter 12
The Psychology of Families

Harald A. Euler

Abstract Two illusions about the nature of human families are the Illusion of Gender Sameness and the Illusion of Family Socialization. The assumption of gender sameness is critically evaluated and found to be deficient. The lack of sex differences on many variables is not disputed, but on variables where sex-specific past selection pressures can be assumed, the differences are considerable. Neglecting to cut nature at its joints, to use Socrates' butcher metaphor, and using effect size estimates averaged over "wrongly cut" areas gives a mistaken impression of the absence of sex differences. Moreover, socially important sex differences may appear as variance differences. How evolutionarily designed sex differences invade mating, parenting, grandparenting, and extended family relationships, and produce asymmetries in family life, is exemplified, particularly with respect to grandparenting. The Illusion of Family Socialization denotes the belief that the human adult personality is formed by parenting practices. Robust data from behaviorial genetics attest that the shared environment, and thus family-specific socialization practices, does not—with exceptions—account for the variances in personality. Considerations from evolutionary theory, particularly life history theory and parent–child conflict, deliver plausible reasons why parents are not able to mold permanently their offspring's personality. A human evolutionary behavioral science is well equipped to expose and debunk these illusions.

Modern men are no longer like the patriarchs of past times. The "new men" touted in the media presumably take an equal share in housework and childcare, or at least try to, especially if they have an academic education, are open-minded, inclined to left or liberal views, and participate in enlightened discourses. Such men participate in birth preparation courses, are empathetically present at delivery, push prams in public, and are willing to take an equal share of domestic chores, or even the role of house husband if necessary.

H.A. Euler (✉)
Institut für Psychologie, Universität Kassel, 34109 Kassel, Germany
e-mail: euler@uni-kassel.de

U.J. Frey et al. (eds.), *Homo Novus – A Human Without Illusions*,
The Frontiers Collection, DOI 10.1007/978-3-642-12142-5_12,
© Springer-Verlag Berlin Heidelberg 2010

Good intentions and honest promises are cheap, but their consequences costly. Do men with egalitarian gender role attitudes, which are especially prominent among the young of university faculties, stick to their ideals? Steven Rhoads, a professor of Public Policy at the University of Virginia, initiated a nationwide study of how male and female faculty members used parental leave (Rhoads 2004; Rhoads and Rhoads 2004). His research team conducted lengthy interviews about infant care with 184 male and female assistant professors who were trying or had tried to obtain tenure while at the same time raising a child under age two. These are the type of people we may assume to be at the very frontier of gender equality.

The statement "Families usually do best if the husband and wife share equally in childcare, household work, and paid work" was answered affirmatively by 75% of the female professors, 10% disagreed. Of the men, 55% agreed and 33% disagreed. This is not very egalitarian, but let us note that a majority of both genders supported equal roles and shares in childcare. The actual performance, however, did not match the professors' attitudes. Whereas 67% of eligible female professors took the available paid leave, only 12% of the males did. The participants were interviewed about 25 childcare tasks, namely whether each task was always or usually done by the respondent, by the spouse, or by both equally. The tasks covered the whole gamut of childcare that can be split between both parents, such as basic tasks (e.g., changing diapers), logistics (e.g. bringing child to day care), consulting and planning (e.g., seeking advice about childcare), recreation (e.g., playing with child), and emotional involvement (e.g., comforting the child).

The female academics did all 25 tasks significantly more often than the male academics, for all but two tasks even highly significant. The gender difference remained extremely large when male and female leave takers were compared on the one hand, and those who did not on the other. The men who had taken the leave had ample opportunity to engage in childcare, and they did more of it than the males who had not taken the leave, but they still did significantly less than the women in either group, those who took leave and those who did not. Even if only those men and women were compared who had stated that childcare should be shared equally, a large difference remained. From the whole sample, less than 3% of the males said they did more care than their spouses, whereas 96% of the females said they did more. Asked about how much they liked doing the baby care tasks, the gender difference was the same: the women liked almost all of the tasks more than the men did. For example, more women liked to change diapers than disliked it, whereas the reverse was true for men.

The findings about this gender gap are recent, but not new. A similar story has been told about the kibbutzim (Spiro 1979). The kibbutz movement in Israel tried to overturn the traditional social order, abolish sustainably any division of labor between the sexes, and disburden women from the "yoke of childcare". These ideals were not externally imposed but were generally adopted wholeheartedly by the first-generation kibbutz members. As it turned out, the third generation of female kibbutz members engaged in a sort of counterrevolution. These women demanded emphatically to have the right to take care of their children themselves and to have them around all the time, especially at night. They placed emphasis on their female appearance, expressed preferences for domestic work, and valued household chores and childcare as satisfactory. As to the traditional three German K's of gender segregation (*Küche*, *Kinder*, *Kirche*), they fought their way back to kitchen and kids by their own desire.

This is a chapter on prevalent illusions about the psychology of family matters. A modern Western illusion is the one exemplified above, the Illusion of Gender Sameness. This illusion is so basic that it pervades various aspects of mating, partnership, parenting, grandparenting, and solidarity in extended families. Another major illusion may be called the Family Socialization Illusion, the widespread belief that a child's personality is molded lastingly by parenting practices. Both illusions will be dealt with in some detail. It goes without saying that a human evolutionary behavioral science is best equipped to expose and debunk these illusions, and has the duty to inform the public. Expressly, the aim is not to turn the wheel of progress back, but to help avoid pitfalls and dead ends in science, unrealizable hopes in private family life, and inefficient public policies.

12.1 The Illusion of Gender Sameness

The Illusion of Gender Sameness comes in many forms and degrees. It originated in equity feminism and quickly developed into gender feminism with a central claim about human nature: the differences between men and women are not primarily biological but are socially constructed due to traditional male dominance (e.g. Fausto-Sterling 2000). The task is not to empirically investigate and support the claim but to "deconstruct".

Moderate forms of this position can be subsumed under the banner of the gender similarity hypothesis (Hyde 2005; Eagly 1987): males and females are similar on most psychological variables. There are admittedly some genetically mediated somatic sex differences, especially men's greater strength and size and women's childbearing and lactation, which interact with shared cultural beliefs and economic demands and thus lead to gender role assignments that constitute the sexual division of labor. Any psychological sex differences are thus secondary, are gender differences.

The proponents of the gender similarity hypothesis take laudable pains to present empirical support, typically with meta-analyses. In a second-order meta-analysis, Hyde reviewed 128 meta-analyses and found that most psychological gender differences were close to zero ($d \leq 0.10$) or small ($0.11 < d < 0.35$), a few a moderate, and very few are large ($0.66 \leq d \leq 1.00$) or very large ($d > 1.00$) (Hyde 2005). Let us take a critical look at these results.

Human evolutionary behavioral scientists would fully agree that on many variables, if not most, both sexes differ not at all or just slightly. If there has been no evolutionary history of sex-specific selection pressures, no sex differences have been formed. For example, both sexes equally like a person with a likeable personality for a long-term mate (Buss et al. 1990) because a nice partner is of equal fitness interest for both sexes. Both sexes equally seek acceptance from peers because rejection from the group has been equally harmful. Both sexes seem to get jealous about equally often and equally strongly, but the stimuli that provoke jealousy are quite different for the two (Buss et al. 1992).

We have to cut nature at its joints, to use Socrates' butcher metaphor. If, for example, we find evidence that shows men to have better spatial abilities (Gaulin and Hoffman 1988), we did not cut at the right joint location because spatial ability is too heterogeneous to be lumped together. Men get higher scores on many spatial abilities, such as spatial orientation, spatial visualization, and mental rotation, but not on all. Women perform better in tasks of object-location memory (Silverman and Eals 1992), especially if the objects are or have been relevant in the life of a gatherer, such as plants (Neave et al. 2005) or food items (New et al. 2007; Spiers et al. 2008). So the questions whether one sex is more "emotional" and the other more "rational", or whether the sexes differ in "mathematical" or "musical" ability are badly stated and as wrong as the question of whether one sex is more intelligent than the other. Evolutionary selection pressures make finer differentiations. Thus, the meta-analytic findings of no gender difference in sexual satisfaction (Oliver and Hyde 1993) or in life satisfaction and happiness (Pinquart and Sörensen 2001; Wood et al. 1989) come as no surprise to an evolutionary scientist (Lippa 2005), who instead would predict differences in attitudes about casual sex and masturbation, where they are indeed large (Oliver and Hyde 1993). The overall gender difference in aggression is moderate (Hyde 1984; Eagly and Steffen 1986), but higher if specific types of aggression are considered (e.g., physical aggression, assertive aggression), or reversed with female-specific types of aggression (e.g., indirect aggression).

Natural and sexual selection do not pick out serially one characteristic after another, but target individual systems. It is system configurations that get selected. Sex-specific selection pressures change the frequencies of trait constellations, not of isolated traits. Therefore, the sexes differ in the profiles of their characteristics. The standard procedure in estimating the magnitude of the difference in central tendencies of a distribution, for example Cohen's d, is applied to one characteristic at a time. More appropriate for estimating sex differences is a multivariate effect size measure that quantifies properly the difference in the constellation of characteristics, the Mahalanobis distance D. Del Giudice recalculated two meta-analyses with this procedure and reported that on the Big Five personality traits the average Cohen's d was just 0.28, whereas the Mahalanobis D was 0.84 (Del Giudice 2009). In a meta-analysis of aggression, a similar difference between the univariate and the multivariate effect size was found.

Moreover, evolutionarily meaningful sex differences may show up in distribution parameters other than differences in central tendencies. Intelligence tests, for example, are mostly constructed to be gender-fair by the choice of tasks; males and females do not differ in averages. But males tend to show higher variances than females (Feingold 1992; Irwing and Lynn 2005). A similar variance difference is found, among others, with respect to height (Bell et al. 2002), academic high-school success (Nowell and Hedges 1998), and verbal competence (Gallagher et al. 2000). The phenomenon is also seen in everyday life. Men are overrepresented among Nobel laureates, great artists, and workaholics, but also among those with severe intellectual disabilities, among junkies, criminals, and losers (cf. Pinker 2008). This variance difference makes evolutionary sense. Because of higher male than female reproductive potential and thus over-proportionate fitness returns with an optimal male, nature tries out more variability in the construction of male than female phenotypes (Euler and Hoier 2008). The variance differences are generally rather small, but at the tails of the distribution, and thus among those persons who are particularly noticed because they are "outliers", the difference may become socially meaningful.

Even if sex differences turn out to be small in controlled individual tests, this does not mean that they remain small in the dynamics of social interactions. Maccoby noted that in social interactions between girls and boys a plethora of obvious gender differences could be observed (Maccoby 1990) and that her conclusion in her and her colleague's seminal book on sex differences (Maccoby and Jacklin 1974) of only a few gender differences had been artifactual because the old gender difference studies were typically done with individual tests. The evolutionary heritage of our psychological sex differences does not show up primarily in performance scores of achievement tests, but in thresholds, inclinations, preferences, and openness for learning. The context matters. For example, men can and can learn to take care of a baby as well as women do. But if the baby wakes up at night and starts to fuss, the chances are high that the mother wakes up before the father does, that she starts to feel uneasy before he does, and, if she likes to take care of the baby just a bit more than he does, she will in the end be the one to get up and comfort the baby. The baby herself will react to the sex of the more frequent and thus most reliable caregiver and want mother to come, not father, display this preference and thus amplify the originally slight sex differences into cemented divisions of labor, into sex stereotypes with normative appeal.

Everyone who has raised kids of different sexes knows that boys and girls differ. Everyone who looks at the data about human behavior and does not commit the moralistic fallacy "for ... that which must not, cannot be" (Christian Morgenstern, transl. by Max Knight) finds sex differences with impact on family relationships. Everyone who realizes that the exclusive and isolated look at just current human behavior is anthropocentric tunnel vision knows that a view which incorporates somatic and life history traits in a wider cross-species, cross-cultural and historical perspective (e.g. Geary 1998; Lippa 2005; Low 2000; Mealey 2000; Voland 2000) is incompatible with the gender sameness position.

We shall now turn to the impact sex differences have on family matters. Mating and parenting will be dealt with only cursorily. I emphasize grandparental and extended family relationships because this is the area of my particular expertise.

12.1.1 Sex Differences in Mating

The sex differences in mating are so stunning that the obvious might be overlooked, namely sexual orientation. Males and females differ as to which sex arouses them, with an extremely large effect size. If biological sex did not matter, why would homosexuality and heterosexuality not be equally frequent? If sex were a historical construction due to patriarchal oppression by men, why is the prevalence of homo-sexuality so stable over diverse cultures? If sex were a social construction, why have serious past attempts at individual deconstruction of homosexuality (e.g., psy-chotherapy) all been to no avail? A theory must also explain the obvious, not only the rare and the subtle. Confucius says: "The common man marvels at uncommon things; the wise man marvels at the commonplace."

Apart from sexual orientation, the findings about sex differences in mating, many of them culturally universal, are so rich and recurrent that only a selected sample can be presented here (please excuse my popular vernacular): The girl wants a guy, and the guy wants sex. The woman thinks about love, the man thinks about sex. For her, sex may be the consequence of love; for him, love may be the consequence of sex. He unconsciously looks for signs of fertility, she is impressed by indicators of resources (Marilyn Monroe in *Gentlemen Prefer Blondes*: "Don't you know that a man being rich is like a girl being pretty?"). He courts, she audits. She tends to infidelity if she is dissatisfied with her partner; he does so if it is just another woman ("If I can't be near the woman I love, I love the woman I'm near"). She wants to wait with the first sex, for him it cannot be soon enough. She looks for the one Mr. Right, he counts the sheer number of his sexual conquests. She wishes to get the attention of a particular tall man, he wishes to impress all young women. She wants long-term exclusive commitment, he dodges and delays. At the wedding she thinks that she can change him; he thinks that she will stay the same (they are both wrong). He consumes pornography, she reads love stories. His display of intelligence is an asset, hers only if she can keep it hidden from him (Mae West). He tends to look at his partner in a more negative light after meeting a single, attractive woman; she is likelier to appreciate her partner more after meeting an available, attractive man (Lydon et al. 2008). If all this sounds like obsolete stereotypes from shallow party chitchat, here are a few of the many scientific references: Buss 2003, 2008; Buss and Schmitt 1993; Chasiotis and Voland 1998; Symons 1979; Townsend 1998.

The importance of sex differences in mating may be finally exemplified by a classical study (Clark 1990; Clark and Hatfield 1989, 2003) on responses to sexual offers. Male and female US-American student confederates of average attractive-ness approached unfamiliar members of the opposite sex and asked, after a few friendly and complimentary sentences, one of three questions: (1) "Would you go out with me tonight?"; (2) "Would you come over to my apartment tonight?"; (3) "Would you go to bed with me tonight?". Combining the data from all three studies ($N = 144$) gave the following stunning results. The females agreed to the propos-als with the following percentages: 50% (date), 7% (apartment), and 0% (sex). The males, however, agreed with 56% (date), 63% (apartment), and 71% (sex). While 29% of the males declined the proposal for sex, consider that many of these males came up with excuses ("I have no time tonight, how about tomorrow night?") and that about 5% of males are gay!

As far as I know, the experiment has not been replicated more recently (it is dif-ficult to find males to play the confederate), but the last question, slightly changed ("Would you like to have sex with me, as soon as possible?"), was part of a German live TV game show in 2005 (RTL, *Typisch Frau—Typisch Mann*, presenter: Günther

Jauch) where I served as the "expert". The over 30-year-old moderately good-looking female confederate got a positive answer within a few minutes from every male she asked, whereas the somewhat younger, very handsome, tall, and friendly male confederate got nothing but rebuffs, and most women reacted with outrage and anger. Many of the reactions of both males and females were hilariously funny.

Sex differences like the one talked about so far can be seen as a consequence of a basic mammalian sex difference: The obligatory minimal investment into a single reproduction is considerably larger for females with their internal gestation and postpartum lactation than for males (Trivers 1972). The reproductive potential is, therefore, higher for males than for females. Because organisms are reproductive strategists that try to maximize their genetic replication, the higher male than female reproductive potential leads to sex-specific tradeoffs between parental effort and mating effort. The difference in fitness payoff between mating effort (maximizing mates) and parental effort (maximizing care for offspring) is generally more in favor of males than of females. This asymmetry in reproductive costs (Voland 2007) and thus in reproductive strategy between males and females pervades all kinds of family matters, in addition to a second asymmetry due to internal fertilization and the consequent paternal uncertainty, and can be the cause of all kinds of conflicts (Buss 1989).

A frequently heard objection against an evolutionary analysis of human behavior is that nowadays people are no longer interested in their genetic replication. People use birth control methods and may remain childless by choice. The reply is that the biological imperative to maximize reproduction is instilled into organisms by the design of their motivational structure and not by a conscious will. Even if reproduction is no longer the expressed aim, people still want and do things that in the ancestral past ensured reproduction. They like sex, strive for status, react positively to cute little kids, seek the company of friends, and so on. The continued power of our ancestral motive structure is well illustrated by an example from Tooby and Cosmides (2005): Men pay for the service of a prostitute, although they know that the prostitute uses birth control or even hope she will, but they get paid for a semen donation to a sperm bank. People are no longer reproduction maximizers, but they are still adaptation executors.

12.1.2 Sex Differences in Parenting

The human sex differences in parenting go far beyond the example presented at the beginning of this chapter, and the ultimate reasons for these differences are by now mostly well understood (Buss 2008; Daly and Wilson 1983; Geary 2005, 2008; Keller and Chasiotis 2007). Of the many differences, only a few shall be mentioned here. Paternal investment, in humans and in many other species, is facultatively expressed, whereas—in mammals at least—maternal investment is obligatory. If paternity certainty is high, if paternal investment improves offspring survival, and if other mating opportunities are rare or costly, paternal investment can be expected.

For human males this means that they are on average better fathers if they can be certain of their paternity. If there are cues to paternity certainty, such as exclusive sexual access to the woman during time of conception or phenotypic similarity between putative father and child (Burch et al. 2006), they tend to be more caring fathers, whereas the same is not true for mothers. If social norms (e.g., religion) makes male extra-pair mating effort more costly, for example with loss of reputation or stairways down to hell, fathers tend to stay home instead of philandering, whereas mothers are already by nature less inclined to stray.

12.1.3 Sex Differences in Grandparenting

In parenting, the sex differences of two generations are involved, in grandparenting there are three generations, with the parent as a mediating link. The sex of the parent is a strong determinant of grandparental care, as is the grandparent sex, stronger than the child's sex. In the past, social science theorists usually lumped grandparents into one category, thus ignoring sexual asymmetries. But there are grandfathers and grandmothers, and both can be either matrilineal or patrilineal. This makes for four kinds of grandparent, and when it comes to how much they do for grandchildren, the kinds of grandparent differ considerably owing to the sex asymmetry in reproductive strategy and paternity certainty. If we want to cut nature at its joints, the differentiation along sex *and* lineage is crucial. What is good for the goose may be good for the gander, but humans are not geese. (Geese are monogamous and thus show little sex dimorphisms, humans are not monogamous.) Grandparental investment cannot be understood adequately without consideration of its fitness consequences. The particular consequences are not always straightforward, but are mediated by socioecological circumstances, such as subsistence conditions, mating and kinship system, division of labor, residence pattern, lineality, resource control, and inheritance rules (Holden et al. 2003; Leonetti et al. 2005; Voland and Beise 2002, 2005).

Grandparents can still be reproductive, in the sense that they can do things that increase their own genetic replication. They can engage in extraparental nepotistic effort (Alexander 1987; Voland 2000) and thus increase their own inclusive fitness. Grandparents can assist their adult offspring in the offspring's parental effort by transferring resources either to their offspring and/or their grandoffspring. Families may thus be seen as a joint enterprise for reproductive profit of the participating entrepreneurs.

If grandparents are set to maximize their own inclusive fitness, it pays to allocate their resources preferentially. Mothers carry a higher burden with childcare than do fathers, and they do good to muster all the help they can get. According to the rule that "the squeaky wheel gets the grease", grandparents do good to prefer—all else being equal—to help their daughter and invest in her children rather than to help their son and his children. The son has a wife to whose parents the same logic applies. We may therefore expect maternal grandparents to exhibit more grandparental investment than paternal grandparents.

The grandparental biological relationship certainty varies between the four kinds of grandparents. The maternal grandparent can be certain that her grandchildren are her biological grandchildren. The maternal grandfather and the paternal grandmother each have one link of paternity uncertainty, and the paternal grandfather has two links. He can neither be certain of the paternity of his son nor of his son's children. All else being equal, the maternal grandmother can therefore be expected to be the most caring grandparent, the paternal grandfather the least caring.

In our first study on preferential grandparental care (Euler and Weitzel 1996), we asked participants in a wide age range (16–80 years) how much each grandparent had cared for them (*gekümmert*) up to the age of 7 years. From the total sample, those 603 cases were selected for analysis whose four grandparents were all still alive when the participant was 7 years old. The maternal grandmother was rated as having been the most caring, followed by the maternal grandfather, the paternal grandmother, and the paternal grandfather. Maternal grandparents were significantly more caring than were the paternal grandparents, and grandmothers significantly more than grandfathers. The effect sizes (partial η^2) were 0.11 for the lineage effect (maternal vs. paternal) and 0.17 for the effect of sex of grandparent. Both effects together account for a sizable proportion of the variance.

Of special interest is the finding that the maternal *grandfather* cared more than the paternal grandmother. If grandparental care-giving were solely determined by the social role of women as childcare-givers, both types of grandmothers should provide more care than both grandfathers. The social role argument should apply particularly to the older study participants, whose grandparents presumably subscribed more to traditional gender roles than grandparents of younger participants. However, the difference between more care from the maternal grandfather and less care from the paternal grandmother was even more pronounced for the older (40 years or more) than for the younger participants.

This same pattern of preferential grandparental solicitude has been found in various Western countries (for a summary see Euler and Michalski 2007), but the pattern is not a cemented, unalterable law of nature. In traditional pastoral societies with patrilocality, patrilinearity, and corresponding inheritance rules, paternal grandparents seem to get more involved in care for grandoffspring than in our environments.

Care is one type of investment of the caregiver, but there are other kinds of investment. It thus comes as no surprise that a plethora of indicators of investments have been found to differ between grandparents in the abovementioned gradation: emotional closeness, time spent together and interaction frequencies, gifts for grandchildren, grandparental mourning after a grandchild's death, naming favorite grandparents, adoption of grandchildren, and so on (Euler and Michalski 2007). No scientific data about bequests from grandparents seem to be available, but it might be predicted that inter vivos transfers are preferentially directed to children of daughters than to children of sons.

Even the informal terms of address of grandparents reflect the salient position of the maternal grandmother, who is most often of all grandparents addressed with an endearing or diminutive name. For example, in Germany the maternal grandmother might be called "my dear granny" (German: "die liebe Oma", or "Omi-lein"), whereas the paternal grandmother is often just called "the other grandmother" or "the grandmother from Hannover" (Euler et al. 1998).

Could it be that paternal grandparents and grandfathers lose out because they do not invest less but invest differently, that is, do different things with grandchildren than maternal grandparents and grandmothers do, and that this possibility is not adequately reflected in the investment measures reported so far? I asked 230 students (171 females, 59 males) in a rating scale questionnaire with a list of 55 activities that grandparents can do with their grandchild (e.g., "picked me up from school", "played games with me", "was proud of me") how often each of the four grandparents had done that activity. For 49 of the 55 items, the standard pattern held significantly: grandmothers more than grandfathers and maternal grandparents more than paternal grandparents. Only two activities were done more by grandfathers than by grandmothers, namely "taught me skills, like cycling or swimming" and "did repairs for me". But in both cases, the maternal grandfathers outdid the paternal grandfathers. Nonsignificant differences were seen only with activities that occurred so rarely that a floor effect did not allow differences to appear. In accord with sex-specific activity preferences, grandmothers tend to do the time-consuming, empathic, caring, and consoling activities, whereas grandfathers tend to do the repairs, teach skills (except for cooking), and spend money.

It goes without saying that the type of grandparent is not the sole determinant of grandparental investment, but it accounts for a sizable share of the variance (about 28%). Residential distance between grandparents and grandchildren understandably influences heavily how much grandparents care, but the distance between grandparent and grandchild reduces the investment of the maternal grandmother the least, the investment of the paternal grandfather the most. If the grandparents live separated or are divorced, they care less—except for the maternal grandmother. Her solicitude is the most obligatory one. Grandfathers, however, drastically reduce their concern for grandchildren if they are separated or divorced. Old male age does not seem to protect from the folly of mating effort.

The sex of the grandchild does not seem to matter, and there is no evolutionary reason why it should. Girls and boys alike ought to take what they can get. The number of grandkids a grandparent calls his or her own, however, understandably has a big effect on how much investment can be allocated to each single grandchild.

Of particular interest in the evolutionary sciences is a particular feature of the female human reproductive lifecycle, namely the menopause and its evolutionary explanation ("grandmother hypothesis"). The menopause frees the woman from the demands of newborns, so that she can devote time to raise her existing children into adolescence and help in the care of newborn grandoffspring. The grandmother hypothesis shall not be elaborated here because very good scholarly works about it are available (Hrdy 1999, 2009; Voland 2007; Voland et al. 2005).

A final question, however, might be raised. Even if modern humans are no longer reproduction maximizers and the reproductive benefit of grandmothers in terms of grandchild survival can only be shown in natural fertility populations (Sear and Mace 2008), how about psychological benefits from grandparenting? Are grandparents with their devotion to grandchildren just thoughtless adaptation executors in an environment of mismatch to the ancestral environment, like men consuming pornography and women spending their spare time viewing Hollywood or Bollywood love story films, or does modern grandparenting translate, if no longer into grandchild survival, into the grandchildren's cognitive, verbal, and social abilities, its mental health and well-being? There are only a few studies which give some answers, but there is evidence that under conditions of risk, such as teenage pregnancy, maternal depression, and postpartum depression, grandparents can indeed provide support that helps to safeguard their children and grandchildren against adverse risks (Coall and Hertwig, 2010).

12.1.4 Sex Differences in In-Laws

One and the same grandparent can be a parent and a parent-in-law, and the adult offspring a son or daughter, or an in-law. Mothers have a positive image, mothers-in-law a negative one, apparently across cultures. The same discrimination does not apply, at least not as pronounced, to fathers vs. fathers-in-law. Evolutionary adaptations to family life are clearly relationship-specific (Daly et al. 1997) and thus frequently sex-specific. The reasons for the unfair stereotypes can be explained well from an evolutionary vantage point (Euler et al. 2009). There the maternal and paternal families each have different fitness interests in the reproductive potential of the mother (Voland and Beise 2005). A daughter-in-law is replaceable, a daughter is not. The maternal grandmother has invested massively in her daughter, but for the paternal mother-in-law the same woman is a newcomer to the family without much previous investment. The fitness interest of the maternal grandmother is thus the conservation of her daughter's maternal resources, whereas the fitness interest of the paternal grandmother is the exploitation of her daughter-in-law's maternal resources. Maternal grandparents, especially maternal grandmothers, thus add to grandchild survival, whereas paternal grandparents facilitate birth rate (Mace and Sear 2005). The intrafamilial conflict between the matriline and the patriline can even be traced down to the genetic level, where maternally imprinted genes may inhibit and paternally imprinted genes stimulate fetal growth (Burt and Trivers 2006). At the psychological level, however, the hypothesis of matrilineal conservation vs. patrilineal exploitation awaits empirical verification.

12.1.5 Sex Differences in Extended Family Relationships

The impact of sex differences in family relationships reaches well into extended family relationships. Aunts and uncles differ, as do matrilateral or patrilateral aunts/uncles. The analysis of investment of aunts and uncles has the advantage of avoiding co-residence as a confound. Brothers and sisters of a parent, unlike grandmothers and grandfathers, typically do not live together and thus act relatively independently of each other. Due to paternity uncertainty, more investment is put into the offspring of sisters than into the offspring of brothers and more by aunts than by uncles (Gaulin et al. 1997; Hoier et al. 2001). In the same vein, female twins from same-sex pairs express greater closeness toward their nieces/nephews than male twins from same-sex pairs (Segal et al. 2007). And finally, the relationship with cousins from the mother's side tends to be closer than with those from the father's side (Jeon and Buss 2007).

12.2 The Illusion of Family Socialization

For almost 100 years everybody has known that behavioral differences between people are brought about by socialization and that the most effective socialization agent is the family. The baby is born with her mind still empty ("blank slate"), and the social environment makes more or less permanent engravings into the slate (Pinker 2002). The notion of the family as the prime and most enduring socialization agent is intuitively appealing because the family is the first social environment with the opportunity to write on the slate and thus has the most slate space. There seemed to be ample evidence for this assumption, and even more eminence: Freud had taught that adult behavior problems ("neuroses") can be traced back to traumatic experiences in early childhood. The nature–nurture controversy was understood in a dualistic way: nature is for the body and nurture is for the mind. Even the Vatican could accept this position at the end of the twentieth century when the pope declared that evolution could explain the body, but the soul was of divine source. The genes have done their work when the baby is born; from then on comes cultural influence. So what you are in personality and intelligence is so because your parents made you that way. "As the twig is bent, so grows the tree".

The data in favor of this socialization assumption were overwhelming. Socialization research found solid parent–child correlations in many traits and behaviors. The sovereignty of interpretation was in the hands of environmental theories, so no one thought, well, was allowed to assume that behavior could also be genetically determined. There was already twin research which clearly showed that intelligence was also heritable, but during my student time it was not "scientifically correct" (the term "politically correct" did not yet exist) to consider these data, let alone to consider the possibility that personality traits could be heritable. The correlations were interpreted causally in the zeitgeist of dominant milieu theories.

Behavior genetics shattered these culturalistic beliefs. The robust findings of behavior genetics may best be summarized with Turkheimer's Three Laws of Behavior Genetics (Turkheimer 2000): "(1) All human behavioral traits are heritable. (2) The effect of being raised in the same family is smaller than the effect of genes. (3) A substantial portion of the variation in complex human behavioral traits is not accounted for by the effects of genes or families" (p. 161).

The first law of the omnipresence of heritability in human behavioral traits comes, on first sight, as no surprise to evolutionary behavioral scientists because genetic inheritance is a necessary ingredient for evolution to occur. On second sight, however, the law contradicts Fisher's Fundamental Theorem of Natural Selection, according to which traits that have undergone a natural selection process should show only small genetic variability. The heritability of many human behavioral traits, however, is usually rather high, generally roughly half of the total variance. This contradiction is a topic that has been discussed for years in evolutionary behavioral sciences and has been pretty much resolved. The discussion shall not be summarized here (see Euler and Hoier 2008).

The Turkheimer's second law reveals the family socialization belief to be an illusion and makes, together with the first law, the conventional findings of family socialization research uninterpretable. The second law is worded cautiously when it states that the family effect is "smaller" than the genetic effect. In most behavior genetic studies, the effect of the family, the so-called shared-environment component, is close to zero or even straight zero for personality traits other than intelligence (Plomin et al. 2001; Rowe 1994). Living together in the same family does not make siblings more similar to each other compared to children from other families. Put in popular terms: If you, the reader, and I, the writer, had exchanged our families in childhood, you having grown up in mine and I in yours, we would both be the same today as we are now. You might not sit there and read my chapter and I might not write it right now, but in terms of overall personality we would be the same. With general intelligence the matter is a bit different: Good or poor intellectual family socialization might be an asset or a handicap, which, however, does not last lifelong but disappears after at most a few decades.

A few caveats about the shattered family socialization illusion are necessary. Behavior genetic studies are typically carried out with samples within the normal range of family life in Western industrialized countries. Their findings can, therefore, not be generalized to exceptional family contexts, such as maltreatment of children and the like. Secondly, personality as assessed in these studies, usually with the Big Five personality dimensions, is much but not all of it. There are some traits that do have a sizeable family socialization component, namely general cognitive abilities, love style, vocabulary, musicality, and a few more (see Euler 2002). Third, Turkheimer's second law applies to adults, who generally do not co-reside with their parents, not to children or adolescents. Therefore, it should state "the effect of having been raised" instead of "the effect of being raised". A recent publication (Burt 2009) indeed indicates that there is some effect of shared environment in adolescent psychopathology.

Why are parents unable to mold the personality of their children durably, with all the possibilities of influence they have over many childhood years? The answer is found in life history theory. Childhood is a time of somatic effort, the acquisition of reproductive resources to be used for later reproduction, that is, for mating, parental, and nepotistic effort (Voland 2000). If new life efforts are required, such as mating effort after puberty and parental effort some time later, behaviors and interests change, in many species even the phenotype (Alexander 1987). The caterpillar spends all its time in somatic effort: eat and avoid being eaten. After pupation the very same animal, in a completely different form, has a new interest, namely mating. What the butterfly learned as a caterpillar is of little help now.

Who is the best teacher or model for adolescents when they begin with mating effort, when many models are concurrently available? It would be inefficient to utilize only the vertical cultural transmission mode (learning from parents) irrespective of age and sex of the learner, stability of the environment, and content of the transmission (McElreath and Strimling 2008). Better is to use context-dependently also oblique transmission (learning from adults outside the family) and best is horizontal transmission (learning from peers). To take parents as models has several disadvantages. Within the family the task is to organize family interactions, not to learn about flirting and sex. As Harris remarked:

> A child's goal is not to become a successful adult, any more than a prisoner's goal is to become a successful guard. A child's goal is to be a successful child. (Harris 1998, p. 198)

The adolescent's peers, however, constitute the players in the mating market, and the peers are much more up-to-date on what is currently "in" than the old-fashioned parents, who don't even know what twittering is.

We tend to think that parents want the best for their children, but this is not always true and turns out to be another illusion that has been dispelled decades ago (Alexander 1974; Trivers 1974). Parents and offspring share half of the alleles, but only half. The fitness interests of the parents and those of the offspring are not identical. Parents may unconsciously manipulate their offspring for the service of parental fitness. It is in the deep interest of children and adolescents to evade this parental manipulation by not letting their parents shape their personality.

If no other models are available, as in early childhood, and when intra-familial behaviors are concerned, we can expect parental influences to show effects. When the offspring leaves the natal family, the parental investment is taken along as seed capital, but the parental norms are left behind. Parental influence may crop up again later, when the offspring has founded his or her own family (Euler 2002) and now does things the way the parent did and taught, like how to fold a man's shirt. Data on this conjecture are not available, because such particular family customs and practices are of little interest in mainstream psychology.

If an individual personality is determined only in part by genes, and not by family-specific influences, by which influences then? These influences are hidden in the non-shared component of environmental influences. The problem is that the non-shared part of the total variance is a left-over component, and it is unclear what it comprises. Harris has a prime candidate, namely the peers (Harris 1995, 1998). Harris presents convincing data and a group socialization theory to explain how this peer influence happens at the level of social-psychological group processes. However, environment is not only the social environment, but everything outside the genome, from the intracellular environment over the uterine environment to the social environment, including the internal representations of the world with all their dramas played out on the theater stage of one's own imagination. Ontogenetic development is a complex process in which there is ample room for chance events that may lead in a process of self-organization to individual differences (Molenaar et al.

1993). Not only evolution, also ontogenesis is to some extend open-ended. Fate may play a role in it.

The illusion of family socialization grew out of several mistaken ideas about children (Harris 1998): The nuclear family is a new invention, as is the private nature of family life. In the past it was more often a village that raised the child. Socialization is not something that parents do to kids; children socialize themselves and are guided in this process by a multitude of time-tested evolutionary psychological adaptations. Behavior is shown to be highly situation-specific; the play performed within the family is a different theater production than the acts shown outside the family. The reach of genetic influence into adulthood has been underestimated tremendously.

The illusion of family socialization is not just academic but has large societal consequences, even if not typically splashed out in news headlines. Even good parents may have problem kids, and problem parents may have children who turned out amazingly well. If a child has behavioral problems of any kind, parents, especially mothers, blame themselves and asked what they did wrong. Friends and other relatives find plausible causal attributions for the child's problem in past parenting procedures. The school teacher who talks to the mother with the problem boy during the yearly consultation hour inquires about past parent–child interactions, and if not, the mother surmises that the teacher will think that she was not a good mother. In psychoanalytically oriented psychotherapies the question of early parent–child interactions comes up again and again, so that the patient is induced to blame the parents. If we can trash the conventional illusion of family socialization, mothers can stop blaming themselves.

On the other hand, parents may be upset if they are told that what they do with the child will have no long-term effect. This is not justified, for two reasons. Parents can certainly influence their children, for example by influencing their children's peer contact: living in certain neighborhoods, preferring one school to another, mixing selectively with certain other couples with children, and so on. But most important: The best parents can do is not to try to form a certain child personality of their own preference, but to assure that the child has a happy and fulfilled childhood. The poet and philosopher Khalil Gibran says in his poem *On Children* (Gibran 1923):

> You may give them your love but not your thoughts,
> for they have their own thoughts.
> You may house their bodies but not their souls,
> for their souls dwell in the house of tomorrow,
> which you cannot visit, not even in your dreams.
> You may strive to be like them,
> but seek not to make them like you.
> For life goes not backward nor tarries with yesterday.

12.3 Conclusion

I have tried to expose two widely held beliefs about human nature, namely the negligible differences between man and woman and the power of parents to shape their child's personality permanently by the way they treat the child, as illusions with major impact on the psychology of the family. Both beliefs, together with the blank slate illusion, I consider as the most important ones, but am not convinced that these illusions will be rejected soon. It may rather take generations before people will look back at the twentieth century and wonder how these beliefs could have been so well engrained and be maintained for so long.

There are a few other illusions about family interactions of minor importance, one of which might be mentioned briefly, namely the illusion of reciprocal exchange in families. Reciprocity regulates interactions between individuals outside the family unless modern economic markets determine prices and thus transactions (Fiske 1992). Family transactions, however, are characterized by communal sharing. Family membership alone entitles one to use the family resources without balancing gives and takes ("Family is where everyone can take things out of the fridge without asking"). Fairness disputes are therefore largely absent in the intergenerational transfer of resources. A mother does not expect her child to pay her investments back at a later time. She may be disappointed if Mother's Day was forgotten, but does not expect continuous thankfulness. I venture the hypothesis that this absence of reciprocity expectation is less obvious in nonbiological parent–child relationships, as with step-, foster-, and maybe even adoptive children, but supportive evidence seems unavailable.

To conclude, if human—as well as nonhuman—families are seen as a joint enterprise for reproductive profit (Davis and Daly 1997; Emlen 1995), the varieties of family structures can become highlighted and illusions instilled by an ideology of political correctness and by obsolete eminence-based assumptions can be exposed and debunked.

References

Alexander RD (1974) The evolution of social behavior. Annual Review of Ecology and Systematics 5:325–383

Alexander RD (1987) *The Biology of Moral Systems*. Aldine de Gruyter, New York, NY

Bell AC, Adair LS, Popkin BM (2002) Ethnic differences in the association between Body Mass Index and hypertension. American Journal of Epidemiology 155:346–353

Burch RL, Hipp D, Platek SM (2006) The effect of perceived resemblance and the social mirror on kin selection. In: Platek SM, Shackelford TK (eds) *Female Infidelity and Paternal Uncertainty*. Cambridge University Press, Cambridge, MA

Burt SA (2009) Rethinking environmental contributions to child and adolescent psychopathology. Psychological Bulletin 135:608–637

Burt A, Trivers R (2006) *Genes in Conflict*. Belknap, Cambridge, MA

Buss DM (1989) Conflict between the sexes. Journal of Personality and Social Psychology 56: 735–747

Buss DM, 53 co-authors (1990) International preferences in selecting mates. A study of 37 cultures. Journal of Cross-Cultural Psychology 21:5–47

Buss DM (2003) *The Evolution of Desire*, 2nd ed. Basic, New York, NY

Buss DM (2008) *Evolutionary Psychology*, 3rd ed. Pearson, Boston, MA

Buss DM, Larsen RJ, Westen D, Semmelroth J (1992) Sex differences in jealousy: evolution, physiology, and psychology. Psychological Science 3:251–255

Buss DM, Schmitt DP (1993) Sexual strategies theory. Psychological Review 100:204–232

Chasiotis A, Voland E (1998) Geschlechtliche Selektion und Individualentwicklung. In: Keller H (ed) *Lehrbuch Entwicklungspsychologie*. Huber, Bern

Clark RD (1990) The impact of AIDS on gender differences in willingness to engage in casual sex. Journal of Applied Social Psychology 20:771–782

Clark RD, Hatfield E (1989) Gender differences in receptivity to sexual offers. Journal of Psychology and Human Sexuality 2:39–55

Clark RD, Hatfield E (2003) Love in the afternoon. Psychological Inquiry 14:227–231

Coall DA, Hertwig R (2010) Grandparental investment: past, present, and future. Behavioral and Brain Sciences 33:1–19

Daly M, Salmon C, Wilson M (1997) Kinship: the conceptual hole in psychological studies of social cognition and close relationships. In: Simpson JA, Kenrick DT (eds) *Evolutionary Social Psychology*. Erlbaum, Mahwah, NJ

Daly M, Wilson M (1983) *Sex, Evolution, and Behavior*, 2nd ed. Wadsworth, Belmont, CA

Davis JN, Daly M (1997) Evolutionary theory and the human family. The Quarterly Review of Biology 72:407–435

Del Giudice M (2009) On the real magnitude of psychological sex differences. Evolutionary Psychology 7:264–279

Eagly AH (1987) *Sex Differences in Social Behavior*. Erlbaum, Hillsdale, NJ

Eagly AH, Steffen V (1986) Gender and aggressive behavior. Psychological Bulletin 100:309–330

Emlen ST (1995) An evolutionary theory of the family. Proceedings of the National Academy of Sciences of the USA 92:8092–8099

Euler HA (2002) Verhaltensgenetik und Erziehung: Über "natürliche" und "künstliche" Investition in Nachkommen. Bildung und Erziehung 55:21–37

Euler HA, Hoier S (2008) Die evolutionäre Psychologie von Anlage und Umwelt. In: Neyer FJ, Spinath FM (ed) *Anlage und Umwelt*. Lucius & Lucius, Stuttgart

Euler HA, Hoier S, Pölitz E (1998) Kin investment of aunts and uncles. Paper at the 21st Annual Meeting of the European Sociobiological Society, Russian State University for the Humanities, Moscow, May 31–June 3

Euler HA, Hoier S, Rohde P (2009) Relationship-specific intergenerational family ties. In: Schönpflug U (ed) *Cultural Transmission*. Cambridge University Press, Cambridge, MA; New York, NY

Euler HA, Michalski R (2007) Grandparental and extended kin relationships. In: Salmon CA, Shackelford TK (eds) *Family Relationships*. Oxford University Press, Oxford

Euler HA, Weitzel B (1996) Discriminative grandparental solicitude as reproductive strategy. Human Nature 7:39–59

Fausto-Sterling A (2000) *Sexing the Body*. Basic, New York, NY

Feingold A (1992) Sex differences in variability in intellectual abilities. Review of Educational Research 62:61–84

Fiske AP (1992) The four elementary forms of sociality. Psychological Review 99:689–723

Gallagher A, Bridgeman B, Cahalan C (2000) The effect of computer-based tests on racial/ethnic, gender, and language groups. ETS Research Report 00-8. Educational Testing Service, Princeton, NJ

Gaulin SJC, Hoffman HA (1988) Evolution and development of sex differences in spatial ability. In: Betzig L, Borgerhoff Mulder M, Turke P (eds) *Human Reproductive Behaviour*. Cambridge University Press, Cambridge, MA

Gaulin SJC, McBurney DH, Brakeman-Wartell SL (1997) Matrilateral biases in the investment of aunts and uncles. Human Nature 8:139–151

Geary DC (1998) *Male, Female*. APA, Washington, DC

Geary DC (2005) Evolution of paternal investment. In: Buss DM (ed) *The Handbook of Evolutionary Psychology*. Wiley, Hoboken, NJ

Geary DC (2008) Evolution of fatherhood. In: Salmon C, Shackelford TK (eds) *Family Relationships*. Oxford University Press, Oxford

Gibran K (1923) The prophet. *Alfred a Knopf*, New York, NY

Harris JR (1995) Where is the child's environment? Psychological Review 102:458–489

Harris JR (1998) *The Nurture Assumption*. Free Press, New York, NY

Hoier S, Euler HA, Hänze M (2001) Diskriminative verwandtschaftliche Fürsorge von Onkeln und Tanten. Zeitschrift für Differentielle und Diagnostische Psychologie 22:206–215

Holden CJ, Sear R, Mace R (2003) Matriliny as daughter-biased investment. Evolution and Human Behavior 24:99–112

Hrdy SB (1999) *Mother Nature*. Pantheon, New York, NY

Hrdy SB (2009) *Mothers and Others*. Belknap, Cambridge, MA

Hyde JS (1984) How large are gender differences in aggression? Developmental Psychology 20:722–736

Hyde JS (2005) The gender similarity hypothesis. American Psychologist 60:581–592

Irwing P, Lynn R (2005) Sex differences in means and variability on the progressive matrices in university students. British Journal of Psychology 96:505–524

Jeon J, Buss DM (2007) Altruism towards cousins. Proceedings of the Royal Society of London, Series B 274:1181–1187

Keller H, Chasiotis A (2007) Maternal investment. In: Salmon C, Shackelford TK (eds) *Family Relationships*. Oxford University Press, Oxford

Leonetti DL, Nath DC, Hemam NS, Neill DB (2005) Kinship organization and the impact of grandmothers on reproductive success among the matrilineal Khasi and patrilineal Bengali of northeast India. In: Voland E, Chasiotis A, Schiefenhövel W (eds) *Grandmotherhood*. Rutgers University Press, New Brunswick, NJ

Lippa RA (2005) *Gender, Nature, and Nurture*, 2nd ed. Erlbaum, Mahwah, NJ

Low BS (2000) *Why Sex Matters*. Princeton University Press, Princeton, NJ

Lydon JE, Menzies-Toman D, Burton K, Bell C (2008) If-then contingencies and the differential effects of the availability of an attractive alternative on relationship maintenance for men and women. Journal of Personality and Social Psychology 95:50–65

Maccoby EE (1990) Gender and relationships. American Psychologist 45:513–520

Maccoby EE, Jacklin CN (1974) *The Psychology of Sex Differences*. Stanford University Press, Palo Alto, CA

Mace R, Sear R (2005) Are humans cooperative breeders? In: Voland E, Chasiotis A, Schiefenhövel W (eds) *Grandmotherhood*. Rutgers University Press, New Brunswick, NJ

McElreath R, Strimling P (2008) When natural selection favors learning from parents. Current Anthropology 49:307–316

Mealey L (2000) *Sex Differences*. Academic, San Diego, CA

Molenaar PCM, Boomsma DI, Dolan CV (1993) A third source of developmental differences. Behavior Genetics 23:519–524

Neave N, Hamilton C, Hutton L, Tildesley N, Pickering AT (2005) Some evidence of a female advantage in object location memory using ecologically valid stimuli. Human Nature 16:146–163

New J, Krasnow MM, Truxaw D, Gaulin SJC (2007) Spatial adaptations for plant foraging: women excel and calories count. Proceedings of the Royal Society of London, Series B 274:2679–2684

Nowell A, Hedges LV (1998) Trends in gender differences in academic achievement from 1960 to 1994. Sex Roles 39:21–43

Oliver MB, Hyde JS (1993) Gender differences in sexuality. Psychological Bulletin 114:29–51

Pinker S (2002) *The Blank Slate*. Penguin Putnam, New York, NY

Pinker S (2008) *The Sexual Paradox*. Scribner, New York, NY

Pinquart M, Sörensen S (2001) Gender differences in self-concept and psychological well-being in old age. Journal of Gerontology: Psychological Sciences 56B:P195–P213

Plomin R, DeFries JC, McClearn GE, McGuffin P (2001) *Behavioral Genetics*, 4th ed. Worth, New York, NY

Rhoads SE (2004) *Taking Sex Differences Seriously*. Encounter Books, San Francisco, CA

Rhoads SE, Rhoads CH (2004) Gender roles and infant/toddler care: the special case of tenure track faculty. Paper at the Annual Meeting of the MidWest Political Science Association, April 16. http://faculty.virginia.edu/sexdifferences/NewFiles/paper1.pdf. Accessed 23 July 2009

Rowe DC (1994) *The Limits of Family Influence*. Guilford, New York, NY

Sear R, Mace R (2008) Who keeps children alive? Evolution and Human Behavior 29:1–18

Segal NL, Seghers JP, Marelich WD, Mechanic MB, Castillo RR (2007) Social closeness of MZ and DZ twin parents toward nieces and nephews. European Journal of Personality 21:487–506

Silverman I, Eals M (1992) Sex differences in spatial abilities. In: Barkow JH, Cosmides L, Tooby J (eds) *The Adapted Mind*. Oxford University Press, New York, NY

Spiers MV, Sakamoto M, Elliott RJ, Baumann S (2008) Sex differences in spatial object-location memory in a virtual grocery store. CyberPsychology and Behavior 11:471–473

Spiro ME (1979) *Gender and Culture: Kibbutz Women Revisited*. Duke University Press, Durham, NC

Symons D (1979) *The Evolution of Human Sexuality*. Oxford University Press, New York, NY

Tooby J, Cosmides L (2005) Conceptual foundations of evolutionary psychology. In: Buss DM (ed) *The Handbook of Evolutionary Psychology*. Wiley, Hoboken, NJ

Townsend JM (1998) *What Women Want—What Men Want*. Oxford University Press, Oxford

Trivers RL (1972) Parental investment and sexual selection. In: Campbell B (ed) *Sexual Selection and the Descent of Man 1871–1971*. Aldine, Chicago, IL

Trivers RL (1974) Parent–offspring conflict. American Zoologist 14:249–264

Turkheimer E (2000) Three laws of behavior genetics and what they mean. Current Directions in Psychological Science 5:160–164

Voland E (2000) *Grundriss der Soziobiologie*, 2nd ed. Spektrum, Heidelberg

Voland E (2007) *Die Natur des Menschen*. Beck, Munich

Voland E, Beise J (2002) Opposite effects of maternal and paternal grandmothers on infant survival in historical Krummhörn. Behavioral Ecology and Sociobiology 52:435–443

Voland E, Beise J (2005) "The husband's mother is the devil in the house". In: Voland E, Chasiotis A, Schiefenhövel W (eds) *Grandmotherhood*. Rutgers University Press, New Brunswick, NJ

Voland E, Chasiotis A, Schiefenhövel W (eds) (2005) *Grandmotherhood*. Rutgers University Press, New Brunswick, NJ

Wood W, Rhodes N, Whelan M (1989) Sex differences in positive well-being. Psychological Bulletin 106:249–264

Illusion Number 5
Moral, Religion and Culture
Are Social Constructions

Chapter 13
Moral Normativity Is (Naturally) Grown

Kurt Bayertz

Abstract This chapter suggests an approach to a "naturalistic" explication of moral "ought" that, at the same time, avoids reductionism. The central thesis states that "X ought to do something" means nothing but Y wants X to do it, with "X" and "Y" being any individuals or groups of individuals with the only conditions being that (a) they want something from somebody else and (b) they can understand that "want". Thus, normativity is a part of the real world. This thesis goes beyond mere biological principles by understanding the moral "ought" as a process of translating volitions that can be described as "exteriorization". By means of scripture, these exteriorized volitions are institutionalized by existing members of a community, thus outlasting the particular individuals. Since they become part of the cultural heritage, they can be re-interiorized by new members of the community. Interiorized norms are a part of the social world, which every new generation will find and will have to internalize (even if it changes them).

13.1 Introduction

> I fully subscribe to the judgement of those writers who maintain that of all the differences between man and the lower animals, the moral sense or conscience is by far the most important. This sense ... is summed up in that short but imperious word 'ought', so full of high significance. (Darwin 1879, p. 120)

Darwin's thesis of the central significance of the short but imperious word "ought" will be the starting point of the following reflections. As a matter of fact the phenomenon of "ought" is no marginal phenomenon of morality, but the key to its adequate understanding. Still, what does it mean, "somebody *ought* to do something"? Moral philosophy has had problems in dealing with the subject.

K. Bayertz (✉)
Philosophisches Seminar, Westfälische Wilhelms-Universität, 48143 Münster, Germany
e-mail: bayertz@uni-muenster.de

U.J. Frey et al. (eds.), *Homo Novus – A Human Without Illusions*,
The Frontiers Collection, DOI 10.1007/978-3-642-12142-5_13,
© Springer-Verlag Berlin Heidelberg 2010

Apart from a few exceptions it has been marginalized or avoided altogether; hence, theoretical interpretations of naturalistic and non-naturalistic approaches alike remain in short supply and of unsatisfactory quality. The former are troubled by the (meanwhile generally accepted) prohibition of concluding from "is" to "ought". By this prohibition "ought" seems to be radically cut off from the real world and condemned to a mystical existence in some supernatural sphere. Some theoreticians have concluded that "to ought" (if it can't be derived from "to be") doesn't exist at all. In this chapter I should like to make a first argument by plausibly showing that this conclusion is premature: Normativity is a part of the real world. My second argument will be to illustrate which part of the real world it belongs to.

13.2 From "Volition" to "Ought"

My thesis is that "X ought to do something" means nothing other than "Y wants X to do it", with "X" and "Y" being any individuals or groups of individuals, with the only conditions being that (a) they want something from somebody else and (b) they can understand that "want". Thus, the essence of what we mean by saying "ought" seems identified: "ought" is the "complying to volition"; if Y *wants* X to do something, than X *ought* to do it.

Of course, this explication is incomplete; yet, its advantage lies in its independence from any supernatural instances. According to the explication, the "ought" is a part of our common reality. Embedded in the context of evolutionary theoretical considerations it can be added that X and Y can also be animals, if they fulfill the previously stated conditions. The analysis of phylogenic connections between the "ought" of primates and the "ought" of human beings is not blocked by definitions; the explication virtually invites such analyses. According to Flack and de Waal "prescriptive rules" can also be found in primates, especially in chimpanzees, understood as "expectations about how others should behave", although this assessment seems to be controversial (Flack and de Waal 2000, pp. 8–9). Nevertheless there are serious objections to consider. The first says that "to ought" is derived from a fact (the volition) and thus a violation of Hume's Law. I will not address this objection directly; nevertheless a good part of the following remarks may be taken as an indirect response. Second, at first glance the connection with morality if Y wants X to do something is unclear. Even if it is correct that in certain contexts "X ought" means nothing but "Y wants X to do something", this would be no explication of the *moral* "ought"; and that is the problem we are dealing with here.

This second objection is appropriate and leads us to the fact that an occasional or isolated "want" by Y doesn't suffice for a moral "ought". We have to add further conditions. One of them is *generality*. We only speak of a moral "ought" if it is not only applicable to certain occasions or individuals, but to all situations of a specific type. Hence: Whenever this situation occurs, Y wants X to do something specific. Of course this condition isn't hard to meet, so that we can assume for any individual (animal or human) living in groups, that they (a) want something from each other and (b) know about it. On this small basis there will emerge a relatively stable net of mutual ought-relations within the group, on which the individuals can orientate their actions. The advantages for all involved individuals are obvious, since such an orientation helps to avoid conflicts and to establish cooperation. Hence, it is in all individuals' interest that such a net of ought-relations exists and provides orientation.

The "ought" thus belongs to the social reality: It is to be considered as a social relation between individuals of a group that on the one hand is (relatively) temporally stable and on the other hand is detached from "supporting" individuals. If a new member enters the group (for instance by birth), the net is already waiting for him; likewise, if a certain individual leaves the group (for instance by death), the relation remains undisturbed. The relation surely depends on a *critical mass* of individuals who stand in relation; nevertheless, it exists independently from *certain* individuals. To summarize: We are dealing with a process that begins with volition "within" the individuals and then leads to a relation "between" individuals, which is partially detached from the individuals.

Initially, this "ought" is implicit. It consists of "mute" mutual behavior expectations (note the normative/descriptive ambiguity of "expectation"). At a certain level of their development, human beings start to verbalize these expectations. By means of language the implicit norms are turned into explicit ones, by which important factors come into play to make possible the development of morality. On the one hand, the generality of "ought" is enhanced, since it is one of the special abilities of language to provide *general* expressions (descriptive as well as normative). Linguistically expressed contents may have a (relative) independence from their respective contexts, as well as from the speaking individuals. The meaning of words or phrases doesn't first come into being in the context of their utterance; furthermore, we use expressions of our language according to their already established meaning within our community. Hence, this meaning remains beyond the context of utterance as well as the immediately involved individuals. The situational independence of linguistic utterances gives the opportunity to talk about absent (for instance past or future) as well as abstract subjects. Thus, the verbalization of abstract norms and values (for instance "justice") becomes possible.

Furthermore, it is one of the characteristics of verbal expressions that they can be uttered with a claim of correctness or validity; and that claim can be doubted, questioned, or denied. Unlike the receivers of an implicit norm, the receivers of an explicit norm not only have the possibility not to comply, but also have the possibility to give their own opinion. Linguistically expressed norms are an invitation to talk, to discussion, to dispute: to what is nowadays characterized as "discourse". Within this discourse a second layer becomes evident because now there are *reasons* that inevitably come into play, for discourses don't simply consist of allegations countered by other allegations, but of arguments (backed up by reasons) that are defended or criticized. Thus, from the beginning, norms exist in a dual space: on the one hand in a space of causes (as far as their genesis can be explained causally) and on the other hand in a space of reasons. The existence and validity of norms without a possibility to reason is unthinkable. At least partially, the discourse takes place on a meta-layer: while norms deal with actions, reasons deal with norms. Such questions about (rational) reasons for the validity of norms do not depend on the examination of their (empirical) causes.

If the practice of reasoning is sufficiently initialized within a community, it will develop beyond daily contexts and will be performed in a way of methodical reflection. So, philosophical texts can emerge, which home in on the rational reasoning of the currently valid norms. Even more: such theories can formulate reasons for or against the validity of norms and even ask for criteria for the adequacy of reasons.

13.3 Digression on Moral Reasoning

Talking about a "space of reasons" might sound a bit philosophical. In the real world, aren't causes the only things that matter? In order to show that this is not the case, that—when we start to discuss and try to convince somebody else of something—reasons become inevitable, I would like to turn to a prominent protagonist of "evolutionary ethics". In his book *The Biology of Moral Systems*, Richard D. Alexander proposes two theses, namely that (a) all moral and ethical questions derive from conflicts of interest and (b) ultimately, all interests are of a biological nature.

> In terms of evolutionary history the ultimate interests of organisms, including humans, are in maximizing the likelihood of survival of their genetic material through reproduction; we expect organisms to find pleasure in, seek out, or be satisfied by activities that in the environments of history would accomplish this end. (Alexander 1987, p. 139)

This is a social biologic determination of the causes of human action and of the conflicts emerging within this action, which shall not be discussed here in detail. More interesting for our context is the fact that Alexander did not keep to categorical and programmatic statements, but dealt with concrete moral problems of our times and proposed *solutions*. One of these problems deals with the rights of embryos and the moribund. Following Alexander's point of view, even these problems emerge from conflicts of interests, for example between the embryo and its parents. In the course of his search for a biologically well-grounded solution, Alexander first outlines the prevalent positions of the abortion debate. In order to answer the question about the rights of embryos, he then reverts to a definition of brain death, previously drafted in the same chapter. A moribund person may be removed from life support systems and be committed to transplantation, if irreversible loss of all functions of the entire brain is diagnosed. Alexander now translates this principle of "postconsciousness" by analogy to the question of the moral status of embryos:

> Suppose we now take the same approach as appears to be currently adopted for determining the rights of the moribund—that of postconsciousness. We can ask if the offspring is conscious or preconscious—i.e. whether or not it has acquired any kind of self-knowledge equivalent to that we regard as existing in ourselves. (Alexander 1987, p. 208)

Since apparently this isn't the case, an absolute right for life of the embryo can't reasonably be defended.

An evaluation regarding the content of this solution and its implications cannot be offered here. Yet, the kind of reasoning applied by Alexander is of interest. First, it is striking that he *doesn't* argue biologically or evolutionarily in a specific sense. As far as he refers to biological facts, he treats them as empirical constraints, which are not constitutional for the finally stated proposal for a solution. Alexander has acknowledged that himself. His answer to the question about the rights of embryos and moribund (on his own account) is

> disappointingly nonradical and (I hope) commonsense, distinguished only by some aspects of attitude in no way restricted to biologists, and by the weighting in of certain kinds of information, especially about conflicts of interest, that are usually not considered ... My opinion stems not from some kind of direct application of biological knowledge, but from a playing of the question of the interests of one individual (or individuals) against those of another (in this case the embryo). Even if others disagree with how I have done it in this case, I can see no alternative to some kind of balancing of interests. (Alexander 1987, p. 212)

This biological abstinence of an evolutionary biologist is no coincidence. For it is unimaginable how such a balancing of interests could be performed on the basis of biological facts alone, i.e., without any assessment. Sound moral reasoning necessarily goes beyond biology and includes kinds of concepts, arguments, and principles that are—*horribile dictu*—nonempirical.

This is clearly recognizable in Alexander's considerations, for they are based on coherence arguments of the following form: "If our arguments in one case are such-and-such, we have to argue in the same way in a (in all relevant aspects identical) second case and we also have to apply the same approach to a solution!" Such arguments are frequently used in daily ethical discussions and are widespread in moral philosophy. Although their premises may have empirical content, they still base their validity on nonempirical, i.e., rational principles, for instance, the principle that equal cases are to be treated in the same way. This fact of equality of two cases serves as the *reason* to act the same way in both cases. So, Alexander *had to* enter the "space of reason" while trying to convince his reader of the validity of his solution. There is no moral reasoning beyond this space.

13.4 Exteriorization and Interiorization

Let us turn back to the moral "ought". My brief draft should at least give an idea that a natural explanation of the moral ought is possible: "natural" in the sense that it doesn't need to revert to supernatural instances; but also "natural" in the sense that its phylogeny can be tracked back to the realm of animals. Furthermore, this draft should have pointed out that the biological roots of normativity are still no conclusive argument for norms being "nothing but biology". Although the moral normativity emerges from biological causes, it grows beyond them. One characteristic mechanism rendering this possible has already been hinted at in my draft: the exteriorization of interior functions or abilities.[1]

In order to clarify this important mechanism, I will first look at the example of the development of tools. The first "tools" our ancestors had at their disposal were organs, extremities and parts of their bodies. For instance, they could dig with their hands. Later, they learned to perform this action with tools, such as sticks, bones, or shovels crafted especially for this purpose. These tools took over the functions that had previously been executed by the hands; the function was translated to the tool. It is exactly this process of translation that can now be described in a double sense as "exteriorization": on the one hand we have a translation from the "interior" to the "exterior"; on the other hand we have a translation of biologically predetermined functions and abilities to external material, generally nonbiological objects.

Of course, this exteriorization is especially attractive because significant enhancements of effectiveness and efficiency of a given action become possible: even with a stick, and even more with a shovel, digging is easier than with the bare hands. Again even more so, if in later stages of development a shovel is combined with an engine so that a excavator comes into being. In this way the separation of practical function from biological endowment is advanced. For the function of a shovel stays bound to an individual, who conveys his power. The excavator conveys the power automatically and only needs to be controlled; until even this function is assumed by a computer. Hence, each step of exteriorization is a step towards the *de-biologization* of the current function: biological mechanisms are replaced by technological mechanisms. Even if the excavator is ultimately employed for biological interests (for instance in order to build a water reservoir), it doesn't operate under biological but other laws, for instance physical. A conspicuous indication of this de-biologization is the fact that the development of exteriorized functions (in this case: tools and their technology) take place in wholly different temporal dimensions as the development of nonexteriorized functions.

Not only material, but also intellectual functions and abilities can be exteriorized. The genesis of norms is one example. In the beginning of the process, there is a psychic condition: the volition of human or animal individuals. Other individuals of the same community perceive this volition and anticipate it, so that stable behavior expectations emerge. Perhaps the genesis of independent ought-relations *between* individuals can already be seen as a first step of exteriorization, for these relations result from a translation from the interior of an individual to an inter-individual level. At any rate, the verbalization is such a step of exteriorization: the linguistically formulated norm is the exteriorized volition. Within the verbalization, the initially organism-internal function of influencing others' behavior is projected outwards and objectified; it has left a biologically constituted space and entered a socially constituted one, in which the biological mechanisms aren't actually disabled, but interfered with by other mechanisms.

---

[1] I adopt this term from André Leroi-Gourhan (Leroi-Gourhan 1980).

Anyway, the exteriorization process is not finished yet. A further important step is introduced by the genesis of writing. By their recording, norms gain a new degree of independence of both the concrete situation from which they initially emerged and the individuals who initially formulated them. Written norms can still exist and claim validity long after these situations and individuals have ceased to exist. It is not possible to discuss here to what extent certain social institutions such as religion, jurisdiction, and states have to be considered as exteriorizations of "ought". Although these institutions cannot simply be classified as "moral", they still overlap more or less with morality. A comprehensive "moral culture" is established, the development of which is—as the development of technology—not primarily ruled by biological determinants. The biological determinants do not disappear; instead, they define a corridor that is broad enough to allow space for many different developments.

Before we go on, the counterpart of exteriorization should briefly be mentioned: *interiorization*. If human history consists of outward projections of initially interior biological functions and abilities in a considerable way; and thus, if these exteriorized functions and abilities gain a growing independence of their biological origins; then, by extrapolating this trend, the idea of a situation in which all exteriorized functions and abilities have become *completely* independent from human beings as biological organisms becomes conceivable. Then, we would be dealing with a free-floating, absolutely autonomous culture (as Karl Popper's "World 3"). I do not intend to speculate on the probability or even possibility of such a world; there is little room for it and there won't be any in the foreseeable future, anyway. The trees of de-biologization don't rise up to the sky, so there's only one thing left to note: the last basis of even a relatively independent culture is and will be the human being as a biological organism. Yet, that means: those in the course of history increasingly exteriorized functions and abilities appear to individuals of each following generation as external, given reality and thus have to be adopted and internalized, i.e. "interiorized". Understanding and mastery of technical artifacts isn't given to us at birth; we have to labor and learn in an ever more complicated process. There is no comparable growth of complexity in the moral norm systems; but it still applies that they are not given to us from the beginning in a biological way, but have to be learnt and adopted. So, human beings in the course of their ontogenesis have to (re-)internalize extra-biologically existing norms. In the long run, there can be no exteriorization without interiorization.

It is, therefore, no mere coincidence that human beings are biologically equipped in a special and unique way for this process of interiorization. As a matter of fact, many empirical findings point out that the specific difference between human beings and biologically closely related primates lies in their special ability to cooperate with other human beings. The crucial difference is

an adaptation for participating in collaborative activities, involving shared intentionality—which requires selection during human evolution for powerful skills of intention reading as well as for a motivation to share psychological states with others. In ontogeny, these two components—the understanding of intentional action and the motivation to share psychological states with others—intermingle from the beginning to produce a unique development pathway for human cultural cognition, involving unique forms of social engagement, symbolic communication, and cognitive representation. Dialogic cognitive representations, as we have called them, enable older children to participate fully in the social–institutional–collective reality that is human cognition. (Tomasello et al. 2005, p. 16)

It is not their superior intelligence, but their *social* competence that enables human beings to accomplish their unique achievements. Because of this competence individuals have the ability to adopt the achievements of many thousands of years of human beings' cultural development within an extraordinary short period of time, i.e., internalizing the functions and abilities exteriorized in the course of history.

For the current argument it is crucial that interiorized norms are a part of the social world, which every new generation finds and will have to internalize (even if it will change them). Thus, norms will be interiorized as a part of cultural heritage. Yet, this means that it is characteristic and even essential for human beings to live in a dual world, as already mentioned above: on the one hand this is the world of biological determinants of behavior, in which all other animals live; and on the other hand this is a world of cultural determinants of behavior, with norms being a part of those determinants. By being interiorized, these norms become a part of a "second nature", which human beings are apparently especially disposed to adopt. In a word: We're dealing with an extra-biological factor that individuals can allow themselves to be guided by.

13.5 Do Norms Matter?

At this point, at the latest, the objection may come up that moral norms are without effect. From a biological perspective, the thesis seems likely that human beings— as any other animal—follow evolutionary evolved motives and impulses that serve to maximize fitness. Moral norms can either be congruent to those motivations or run contrary to them. In the former case, the actions comply with them; yet, they do so only because they command what human beings are determined to do at any rate. In the latter case, they are without effect, because the determinants evolved in evolution will rule them out.

A weak variant of this objection states that in the case of a divergence between moral norms and biological determinants, individuals have a *tendency* to prefer the latter. There is little to say against this variant, since it would indeed be illusionary to ignore the existence of biological determinants and/or the power they wield over our actions. Anyway, nobody who wants to be taken seriously will claim that we could lay down our biological constitution as we could remove our clothes. Even Immanuel Kant (to name an especially stringent protagonist of anti-biological ethics) assumed that human beings like to follow "inclinations" born from their "sensibility". It is exactly for this reason that moral norms are necessary. A *purely* rational creature, having no potentially irrational and immoral "inclinations", resulting from its "sensible nature", would have no need for such norms. Hence, the normativity of morality is an expression of that tendency of sensible, yet rational creatures to follow their "inclinations". Therewith Kant presumes that acting *against* these "inclinations" is possible at least. Even if human beings are subject to a biological determination, they still have the chance to deprive themselves of it, if the moral norms so demand. Whatever the (biological) basis of this ability may consist of, at any rate it has expanded the human beings' breadth of actions compared to animal primates, "so that we human beings are equipped with a measure of flexibility and plasticity that is unequalled in the animal kingdom." (Voland 2007, p. 26) If this opportunity/chance is denied, we are dealing with the strong variant of the objection, according to which biological determinants leave *no* room for individuals to follow norms if they diverge from the determinants. This thesis shall not be criticized here. It suffices to say: If this thesis applies, then there can be only one biology of human behavior, and no ethics, no "evolutionary ethics", either.

References

Alexander RD (1987) *The Biology of Moral Systems*. Aldine de Gruyter, New York, NY

Darwin C (1879/2004) *The Descent of Man and Selection in Relation to Sex*. Penguin Books, London

Flack JC, de Waal FBM (2000) "Any animal whatever". Darwinian building blocks of morality in monkeys and apes. Journal of Consciousness Studies 7(1–2):1–29

Leroi-Gourhan A (1980) *Hand und Wort. Die Evolution von Technik, Sprache und Kunst.* Suhrkamp, Frankfurt am Main

Tomasello M, Carpenter M, Call J, Behne T, Moll H (2005) Understanding and sharing intentions: the origins of cultural cognition. Behavioral and Brain Sciences 28:675–691

Voland E (2007) Seine Kultur ist des Menschen Natur. In: Eibl K, Mellmann K, Zymner R (eds) *Im Rücken der Kulturen*. Mentis, Paderborn

Chapter 14
The Origins of Symbolic Culture

Chris Knight

Abstract Symbolic culture is a realm of patently false signals. From a Darwinian standpoint, it is not easy to explain how strategies of reliance on such signals could have become evolutionarily stable. The archaeological record shows evolving modern humans investing heavily in cosmetics, with a particular emphasis on ochre pigments matching the color of blood. This chapter discusses the Female Cosmetic Coalitions model of the origins of symbolic culture in the context of hypotheses sometimes considered to be alternative explanations. It is shown that these various hypotheses are not genuine alternatives. Many are not Darwinian, while others either fail to address the question of symbolism or address it but make no reference to details of the archaeological record. It is concluded that the Female Cosmetic Coalitions model offers the most testable and parsimonious way of integrating these different perspectives.

> Symbolic culture ... requires the invention of a whole new kind of things, things that have no existence in the 'real' world but exist entirely in the symbolic realm. Examples are concepts such as good and evil, mythical inventions such as gods and underworlds, and social constructs such as promises and football games. (Philip Chase 1994, p. 628)

From a Darwinian standpoint, "symbolic culture" is an unsettling notion. Modern science became established in opposition to the idea that culturally accepted fictions can be equated with facts. Yet the concept of symbolic culture requires us to grasp just that paradoxical possibility. Long before the late twentieth century invention of the Internet, evolution allowed humans to flit between two realms, reality on the one hand, virtual reality on the other. Symbolic culture is an environment of objective facts—whose existence depends entirely on subjective belief. In this chapter, I attempt to bridge the gap between Darwinism and the human sciences by providing a materialist account of our species' puzzling reliance on moral, religious, and other cultural illusions.

C. Knight (✉)
Comenius University, Bratislava, Slovakia
e-mail: Chris.Knight@live.com

U.J. Frey et al. (eds.), *Homo Novus – A Human Without Illusions*,
The Frontiers Collection, DOI 10.1007/978-3-642-12142-5_14,
© Springer-Verlag Berlin Heidelberg 2010

14.1 Two Kinds of Fact

"Brute facts", in the terminology of John Searle (1996, p. 27), are facts that are true anyway, regardless of human belief. Suppose you don't believe in gravity: jump over a cliff and you'll still fall. Natural science is the study of facts of this kind. "Institutional facts" are fictions accorded factual status within human social institutions. Monetary and commercial facts are fictions of this kind. The complexities of today's global currency system are facts only while we believe in them: suspend the belief and the facts correspondingly dissolve. Yet although institutional facts rest on human belief, that doesn't make them mere distortions or hallucinations. Take my confidence that these two 5-lb banknotes in my pocket are worth 10 lb. That's not merely my subjective belief: it's an objective, indisputable fact. But now imagine a collapse of public confidence in the currency system. Suddenly, the realities in my pocket dissolve.

For scholars familiar with Rousseau, Marx, or Durkheim, none of this is especially surprising or difficult to grasp. Some facts are true anyway, irrespective of human belief. Others subsist in a virtual realm of hallucination or faith. For de Saussure (1983 [1915], p. 8), it was the parallel between linguistic meanings and currency values—all in some sense hallucinatory—which made a scientific linguistics so problematical:

> Other sciences are provided with objects of study given in advance, which are then examined from different points of view. Nothing like this is the case in linguistics . . . The object is not given in advance of the viewpoint: far from it. Rather, one might say that it is the viewpoint adopted which creates the object.

It was in rebellion against such troubling notions that Noam Chomsky (2000, pp. 106–133) redefined "language and similar phenomena" as "elements of the natural world, to be studied by ordinary methods of empirical enquiry". Linguistics, within Chomsky's new paradigm, ceased to be social and became instead a natural science. Ideologically hostile to Marx, Durkheim, and what they termed "Standard Social Science", a generation of Darwinians (Tooby and Cosmides 1992, 1995; Pinker 1994) embraced Chomsky's naturalistic approach. Somehow, language had now to be explained as an innate cognitive module without any animal precursor. Its emergence had also to be explained if possible without reference to social selection pressures (e.g., Hauser et al., 2002; Fitch et al., 2005). The consequence of all this was to render language's very existence an insoluble mystery (Knight 2004, 2009). Far from yielding to Darwinian explanation, the evolutionary emergence of language is nowadays considered "the hardest problem in science" (Christiansen and Kirby 2003).

14.2 Four Positions on the Origins of Symbolic Culture

Within the past 15 years, archaeological revelations from the African Middle Stone Age have transformed our picture of the timing of symbolic culture's emergence. Until the early 1990s, the prevailing view of the "human revolution" (Mellars and Stringer 1989) was notably Eurocentric, focused on the Upper Paleolithic Revolution as humanity's "Great Leap Forward". Recent discoveries from Africa have at least doubled the time-depth of acknowledged and accepted evidence of symbolic activity. This has left us with four main positions concerning the timeline for symbolic culture's emergence:

1. *Francesco d'Errico*. Multispecies transition across Africa and Eurasia. Symbolic capacities already in place with *Homo heidelbergensis* 300,000–400,000 years ago. Sporadic behavioral expressions of symbolism among ancestors of both Neanderthals and ourselves (d'Errico 2003).
2. *Sally McBrearty* and *Alison Brooks*. Down with the revolution! African ancestors of modern humans undergo gradual, sporadic build-up of modern cognition and behavior spanning 300,000 years. Symbolism presents no special theoretical difficulties, emerging as part of the package of modern, flexible, creative behaviors within Africa (McBrearty and Brooks 2000; McBrearty 2007).
3. *Christopher Henshilwood* and *Ian Watts*. The human revolution occurred as part of modern human speciation in Africa. Evidence for symbolism in the form of cosmetics and personal ornamentation is the archaeological signature of this transition. Symbolism was not an optional extra—life following the transition became fundamentally organized through symbols (Henshilwood and Dubreuil 2009; Watts 2009).
4. *Richard Klein*. Recent interpretations of the African Middle Stone Age record are wrong; the original "human revolution" theory remains correct. Middle Stone Age humans evolving in Africa may appear anatomically modern, but did not become cognitively modern until the Late Stone Age/Upper Paleolithic. Symbolic culture emerged some 50,000 years ago, caused by a genetic mutation that rewired the brain (Klein 1999; Klein and Edgar 2002).

14.3 The Archaeological Evidence

In the African archaeological record, the earliest persuasive evidence for symbolic culture includes certain engraved pieces of ochre (Henshilwood et al., 2002) associated with marine pierced shells (Henshilwood et al., 2004; d'Errico et al., 2004). Dated to around 70,000 years ago, these were recovered from Middle Stone Age levels at Blombos Cave, South Africa. Mounting evidence for symbolic behavior at still earlier dates includes a South African coastal site (Pinnacle Point) yielding mollusk remains, bladelets, and red ochre pigments dating to at least 164,000 years ago (Marean et al., 2007). Use of ochre pigments extends back between 250,000 years (ky) and 300 ky at some sites in the tropics; regular and habitual use dates back to the time of modern speciation (Watts 1999, 2009).

Most archaeologists now accept that the shells and pigments were used for personal ornamentation. Often, the shells were strung together to form a necklace. Traces of red pigment have been found on a set of 82,000 year-old perforated shells from the Grotte des Pigeons in North Africa, suggesting that the wearer's body was perhaps already ochred (d'Errico and Vanhaeren 2009, plate 2). Traces of red ochre pigment have similarly been found on some shells from Blombos in South Africa (d'Errico et al. 2004). At Blombos, several modified pieces of ochre have a sharp beveled edge, as if designed to produce a clear outline of color on a surface (Watts 2009, Plate 4). Ochres yielding the most saturated dark reds—especially "blood" reds—were subjected to the greatest intensity of grinding and use (Watts 2009). Pinnacle Point nearby yields similar "crayons" dated to 164,000 ky (Marean et al., 2007). Geometric engravings found on Blombos pieces (Henshilwood et al., 2002) add to the suggestion that many of these delicately shaped "crayons" were used to produce abstract designs, probably on the human body (Watts 2009). This cultural tradition can be traced back to at least a 100,000 years ago (Henshilwood et al. in press). Such evidence suggests that cultural traditions involving body painting were already being established with the speciation of *Homo sapiens*.

14.4 Explanatory Scenarios

To Christopher Henshilwood and Benoit Dubreuil, the cosmetic evidence indicates that Middle Stone Age people were capable of symbolic communication (Henshilwood and Dubreuil 2009). For individuals to wear cosmetics or a necklace, they must care about how they look. To adorn oneself appropriately, it is necessary to imagine one's appearance from the standpoint of others. The requisite capacities for multiple perspective taking are distinctively "modern" and underlie all symbolic communication, including language. Henshilwood and his colleagues on that basis conclude that the producers of the Blombos pigments and ornaments already had language-ready minds.

Ian Watts arrives at similar conclusions concerning language, but on different theoretical grounds (Watts 2009). Since my own theoretical position converges closely with that of Watts, and since we both support Camilla Power's Female Cosmetic Coalitions model (see discussion below), I will avoid repetition at this point and turn directly to Klein, who is the main archaeological opponent of the idea that African Middle Stone Age findings from sites such as Blombos have anything to do with symbolism.

The argument for a mutation generating language and then triggering symbolic culture (e.g., Klein 1999; Chomsky 2005) has little to recommend it. We should be suspicious when a puzzle regarding our own species is addressed using "special" methods—methods without parallel elsewhere in evolutionary science. No specialist in, say, elephant or social insect communication would invoke a single mutation to explain its evolution. We would be equally astonished at an appeal to elephant or honeybee psychology fixed by an "environment of evolutionary adaptedness" (Tooby and Cosmides 1992, 1995) in the remote past. Evolution is not driven by mental phenomena. In the case of any natural species, we explain cognition and communication by reference to reproductive strategies, foraging strategies, and other behavioral adaptations to environmental and social conditions as these fluctuate and change over evolutionary time. We need a theory of the emergence of *Homo sapiens* faithful to the methods of behavioral ecology that have proved so successful in addressing problems elsewhere in the living world.

It might be thought that by now we would have a number of theoretical attempts in this direction. Sadly, this is not so. If we are looking for hypotheses which are (a) based in behavioral ecology, (b) focused on the emergence of symbolism, and (c) testable in the light of relevant archaeological data, the range of suggestions is limited. Camilla Power's Female Cosmetic Coalitions model (see discussion below) meets all three conditions. But before presenting it, I will survey an array of models that meet at least some of these basic preconditions.

14.5 Costly Versus Cheap Signals: Cooperation Between Strangers

1. *Philip Chase: Symbolism enforces cooperation between strangers.* During the later phases of human evolution, humans began to invent entities lacking any existence in the real world—intangibles such as underworlds, promises and totems. Symbolic culture arose because its coercive rituals and associated belief systems provided the only mechanisms of punishment and reward capable of enforcing cooperation between strangers, in turn a prerequisite for the establishment of institutional facts. The term "cooperation between strangers" means cooperation on a scale transcending the limits of Darwinian kin-selection or reciprocal altruism (Chase 1994, 1999).
2. *Richard Sosis: Costly ritual enforces cooperation between strangers.* Religious communities are networks of "strangers" held together by costly ritual. The supernatural entities that help to inspire allegiance don't spontaneously replicate in human brains: they must be coercively installed. Painful ordeals such as initiation rites perform this function. The only way to reliably demonstrate religious commitment is to undergo rituals so demanding of personal sacrifice that the benefits of subsequent defection are likely to be outweighed by the costs (Sosis 2003).
3. *Merlin Donald: Mimesis.* Symbolic culture became established as *Homo erectus* came under communicative pressure to exercise cognitive control over previously hard-to-fake, emotionally expressive body language. Mimetic culture took the form of learned, culturally transmitted, simulated versions of such body language. Through dance, song, pantomime and ritual, evolving humans bonded with one another and became increasingly equipped to express in public their emotional and cognitive states (Donald 1991, 2001).
4. *Dan Sperber: To qualify as symbolic, a signal must be false.* To determine whether a signal or statement is "symbolic", a simple rule can be applied. Is it patently false? If so, it may qualify as a symbol. Falsehood is intrinsic to symbolism. Linguistic utterances are symbolic to the extent that they are patent falsehoods serving as guides to communicative intentions. Metaphor, irony,

sarcasm, and humor illustrate the principle. Language began to evolve when humans started reciprocally faking and falsifying in communicatively helpful ways (Sperber 1975, 2005; Sperber and Wilson 1986).

5. *Roy Rappaport: In the beginning was the Word.* Words are cheap and unreliable. Costly, repetitive and invariant religious ritual is the antidote. At the apex of a religious cosmology is an "ultimate sacred postulate"—an article of faith beyond possible denial. Words may lie, so it is claimed, but "the Word" emanates from a reliable source. Without such public confidence upheld by costly ritual, faith in the entire system of interconnected symbols would collapse. During the evolution of humanity, the crucial step was the establishment of rituals capable of upholding the levels of trust necessary for linguistic communication to work (Rappaport 1999).

6. *Jerome Lewis: Hunting, mimicry, and play.* Antelopes, monkeys, and other animals hunted by Central African forest people treat vocal signals as intrinsically reliable. Forest hunter-gatherers routinely exploit such gullibility, faking animal cries to lure their targets within range. When people subsequently recall a particular hunting episode, they act out the story drawing on the same sophisticated capacities for faking, mimicry, and pantomime. Story-telling, ritual, play, and religion in such societies is the in-group, cooperative, and correspondingly honest redeployment of capacities for deception initially deployed in the forest. This converges with the people's own view that their words, songs, and rituals echo the voices and sounds of the forest (Lewis 2009).

14.6 Symbolism: Puzzles and Paradoxes

Turning now to a review of these ideas, archaeologist *Philip Chase* asserts that Darwinism alone cannot explain cooperation between strangers. He also reminds us that symbolic culture enforces just this kind of cooperation. But how did symbolic culture itself emerge? Having posed the question with admirable clarity, he leaves the evolutionary emergence of symbolic culture unexplained.

Behavioral ecologist *Richard Sosis* does offer a Darwinian model in which individual strategies of alliance-building enforce cooperation between strangers. To explain the mechanisms at work, Sosis relies on costly signaling theory (Zahavi 1975; Zahavi and Zahavi 1997). Religious communities hold themselves together by insisting that everyone must pay admission and continued membership costs so heavy as to deter free-riding. The threshold of costs will be set by the probability of social defection. This explains why rituals of initiation are so often painful, and potentially why there should be variability in costliness. A ritual involving no hardship or sacrifice cannot signal commitment: it would allow free-riders to flourish.

Sosis has done his main studies on contemporary or recent historic religious communities, who are already immersed in symbolic culture. In principle, however, the model can apply to the evolutionary emergence of ritual and religion. Indeed Alcorta and Sosis discuss the African Middle Stone Age archaeological record, mainly the ochre evidence, in relation to this model (Alcorta and Sosis 2005). The value of this work is that it suggests a bridge between animal signaling and symbolic cultural display: the same body of theory can be applied in both domains. But why exactly must hard-to-fake ritual generate what Chase (1994) terms "things that have no existence in the 'real' world"? Hunter-gatherer ritual and religious landscapes are populated by animal spirits, tricksters, and other such fictional entities. What is the connection between these two apparently incompatible properties of ritual—its intrinsic reliability on the one hand and its trickery on the other?

In stark opposition to the hard-to-fake costly signal model stand *Merlin Donald* and *Dan Sperber*. For symbolism to evolve, if we accept their positions, evolving humans had to stop probing signals for their reliability and instead collude with patent fakes. At first sight, this seems wholly incompatible with Sosis's argument that symbolically constituted communities hold themselves together by resorting to signals whose reliability is underwritten by their costs. If Donald and Sperber are correct, symbolism seems to presuppose signals that are not just unreliable but patently false. But perhaps the cheap signals and the costly ones perform distinct functions, operating on quite different levels?

This is essentially the argument of *Roy Rappaport*, a social anthropologist who rejected modern selfish-gene Darwinism but independently converged on the costly signaling idea. Social acceptance of symbols presupposes high levels of trust already in place. Sosis in fact follows Rappaport's argument that costly ritual is designed to generate trust where none existed before. Integrating these lines of reasoning, we might conclude that ritual is needed to cement bonds sufficiently trusting to permit communication on the basis of cheap fakes.

Let me put this in another way. A distinction can be drawn between signaling costs of two kinds (cf. Grafen 1990; Guilford and Dawkins 1991). One alternative is that the signaler must generate trust signal by signal, using intrinsically convincing features to do so. Where this is the case, the costs involved in eliminating perceptual ambiguity won't suffice: added costs will have to be incurred to ensure reliability as well. A strong case can be made that all animal signals fall into this category, such that both kinds of costs ("efficacy costs" plus "strategic costs") are always involved. The reason for this is that animal signals must always carry at least some of the burden of generating the trust necessary for communication to work.

But what if the signaler doesn't have to generate trust at all? Trust could be assumed, leaving the signaler free to concentrate only on perceptual discriminability. If it were possible to reduce the strategic cost of proving reliability to zero, all signaling effort could be poured into efficacy. Carried to its conclusion, this should permit digital signaling—the cheapest and most efficient kind of communication. We know that human language is in fact digital on a number of levels, both phonological and semantic, and that this is one of its most remarkable and biologically unprecedented features (Burling 2005, pp. 25–27, 53–55). Animal signaling is never like this for the same reason that it doesn't have the luxury of being patently false or fictional. Costly signals of any kind can only be evaluated on an analog scale. Putting all this together, it seems that language is digital for the same reason that it consists of social fictions. Signals of this kind are acceptable only under highly unusual conditions—such as those internal to a ritually bonded community whose members are not tempted to lie.

Combining the insights of Chase, Sosis, Donald, Sperber, and Rappaport, we might summarize by defining symbolic culture as a domain of transparent false-hoods whose social acceptance depends on levels of trust generated through the performance of costly ritual. We might add that once such fictions are accepted, they qualify as "institutional facts" (Searle 1996). Human social institutions perpet-uating facts of this kind evolved associated with the uniquely human phenomenon of "cooperation between strangers". But it remains to be explained just how and why.

Following Maynard Smith and Harper (2003, p. 3), we may define a "signal" as any act or structure which alters the behavior of other organisms, which evolved because of that effect, and which is effective because the receiver's response has also evolved. If one animal pushes another away, that is not a signal. If one animal bares its teeth and the other retreats, it's a signal because the response depends on evolved properties of the brain and sense organs of the receiver. The signal must carry information of interest to the receiver. This need not always be correct, but it must be correct often enough for the receiver to be selected to respond to it. Krebs and Dawkins (1984) view signal evolution as an "arms race" between signalers as "manipulators" and receivers as "mind-readers". Zahavi proposed "the handi-cap principle" to explain why signal selection favors extravagance and apparent wastefulness as opposed to utilitarian efficiency (Zahavi 1975). Receivers on guard against deception force signalers to compete in producing signals so costly that they cannot be fakes.

The problem is that by these standards, conventional signals such as linguis-tic signs appear to be *theoretically* impossible—a point explicitly made by Zahavi (Zahavi 1993). Machiavellian primates can produce tactical deceptions, but these are frequency-dependent: they only work if most signals are honest. To explain the emergence of human cultural symbolism, we need a theory that addresses this dif-ficulty: How can we imagine fakes becoming so prevalent as to dominate social life? How can we imagine Machiavellianly evolving humans, by definition resistant to deception, allowing themselves to become immersed in whole realms of patent fiction and illusion?

Here is a possible solution. Whether a given signal is deceptive or reliable, costly or cheap, analog or digital depends on one's perspective. We need to know who is doing the evaluating and from what standpoint. Imagine a coalition of individuals cooperatively aiming deceptive signals at an external target. Viewed from inside the coalition, such deceptions will have positive value. Instead of being resisted, *from this standpoint* they should be celebrated and embraced. To quote de Saussure (1983 [1915], p. 8) once again: "The object is not given in advance of the viewpoint: far from it. Rather, one might say that it is the viewpoint adopted which creates the object."

Drawing on his work with the Mbendjele forest people of Central Africa, social anthropologist *Jerome Lewis* offers a proposal along similar lines, rooting human vocal deception capacities in hunting. Human volitional control over vocal signaling, he suggests, did not evolve initially in contexts of human *social* inter-action. Instead, it was used initially to deceive prey animals who would prove vulnerable again and again to such fakes. Humans cooperating with one another to deceive *external* targets would be predicted not to resist one another's decep-tions but on the contrary to echo and amplify them. In Lewis's account, vocal simulations redeployed internally within the community laid the basis for vocal humor, children's games, choral singing, narrative fiction, metaphor, religion, and so forth. Humans successfully "deceived" the forest and then constructed the symbolic domain as that forest's own echo, now directed back into the human social world.

We now need to consider how hunter-gatherer strategies of this kind might have become established in the evolutionary past.

14.7 Counterdominance, Egalitarianism, and Collective Intentionality

1. *Michael Tomasello: The cultural origins of human cognition.* Cultural evolution can proceed rapidly, a fact helping to explain the accelerated pace of evolu-tion associated with the emergence of *Homo sapiens.* It presupposes the "ratchet effect", in which innovations are preserved and accumulated down through the generations. This would have been fostered by cooperative strategies in which individuals subordinated their private purposes to collective goals. Apes are not capable of this kind of cooperation, which explains why they don't even point. Declarative pointing presupposes "we"-intentionality: a shared subjectivity ren-dering things interesting or relevant "for us". It involves a triadic structure of representation in which signaler and receiver share the same focus of attention. If ape cognition is poorly adapted to such tasks, the explanation is ultimately that these animals are by nature just too competitive (Tomasello 1999, 2006).

2. *Andrew Whiten: The evolution of deep social mind.* Primate Machiavellian cog-nition reflects the fact that reproductive success is likely to be secured by harassment and deception as much as by cooperation. In humans, strikingly different cognitive developments reflect novel strategies of cooperation whose roots lie in "counterdominance"—resistance to being physically dominated by others. Within increasingly stable coalitions, status began to be earned in novel ways, social rewards accruing to those perceived by their peers as especially cooperative and self-aware. Selection pressures favored such psychological inno-vations as imaginative empathy, joint attention, moral judgment, project-oriented collaboration and the ability to evaluate one's own behavior from the stand-point of others. Underpinning enhanced probabilities of cultural transmission and cumulative cultural evolution, these developments led to the establishment of hunter-gatherer-style egalitarianism in association with "deeply social" minds (Whiten 1999).

3. *Christopher Boehm: From counterdominance to reverse dominance.* During the later stages of human evolution, counterdominance tipped over into "reverse dominance". Humans became so resistant to being intimidated or dominated

that they remained constantly on guard, ready at any moment to band together in countering perceived threats. As coalitions organized in this way regularly defeated all opposition, they established themselves collectively as the dominant force. Society became "moral" when everyone was embraced within the same coalition, evaluating the behavior of its individual members from this new collective standpoint (Boehm 2001).

4. *Robin Dunbar: Social brain, gossip, and grooming.* Seeking safety in numbers, evolving humans formed larger groups. Among primates, larger group sizes lead to greater internal competition, raising levels of harassment and associated stress. Negotiating larger groups also selects for a larger neocortex, placing females in particular under more reproductive stress. Increase of Machiavellian intelligence is a specifically female problem, in terms of meeting reproductive costs. Dunbar proposes a strategy for cutting costs of time budgets—the vocal grooming and gossip model, which offers a precursor to language. Subordinates buffer themselves by forming defensive alliances, maintaining friendships through manual grooming. But as such alliances became progressively larger, pressure mounted to find a cheaper, more efficient way of maintaining social bonds. The solution was to switch to vocal grooming. By using sounds instead of fingers, evolving humans could service multiple allies at once while leaving their hands free for practical tasks. Vocal "gossip" had its origins here (Dunbar 1996).

14.8 Dominance and Reverse Dominance

Psychologist *Michael Tomasello* studies the cognitive interface between humans and other primates. The special thing about humans, in his account, is *cooperation in pursuit of a goal held jointly in mind.* An element of contractual understanding is involved, since commitment would collapse without confidence that future gains will be shared. Resource sharing is in this way bound up with an orientation toward the future. There has to be a dream or vision, those sharing it committing themselves to whatever forms of collaboration are needed to secure its practical implementation.

So how and why did *Homo sapiens* begin collaborating in this special way? The fact that wild-living apes don't even point things out to one another shifts attention from cognitive mechanisms to competitive and cooperative strategies. Declarative pointing presupposes individuals so trusting and cooperative that they are willing to decide collaboratively on the perspective to be adopted toward the world. Humans during the course of evolution established such "we"-intentionality. Linguistic rules and symbols—complex elaborations on the simple theme of pointing—are in Tomasello's view culturally inherited patterns that evolved and became transmitted from the moment when this development occurred. As to why it occurred, Tomasello offers no evolutionary explanation, remarking with refreshing candor "I really have no idea" (Tomasello 2003, pp. 108–109).

Andrew Whiten offers at least the beginnings of an idea. The struggle to resist being dominated has an inherent tendency to bring together unrelated individuals who might not previously have been allies. In Whiten's model, humans retain their primate heritage of "Machiavellian" strategic intelligence, initially without undergoing any psychological rupture or break. But as they developed increasingly effective strategies of resistance, the benefits of imposing dominance on others became matched by the associated costs. Eventually a stalemate was reached: instead of everyone competing to find someone else to dominate, the winning strategy was "don't mess with me"—a generalized refusal to be dominated. As this strategy became evolutionarily stable, it altered the trajectory of cognitive and cultural evolution, leading to the emergence of distinctively modern human psychology.

Whiten avoids the conundrums and paradoxes associated with the topic of symbolism. Boehm does little better, barely mentioning ritual, religion, or language. Yet Boehm takes one notable step in the necessary direction. Tomasello, as we have seen, links the evolution of symbolism with collaboration in pursuit of a shared vision or goal. Boehm in this context offers a concrete proposal. The vision that mattered was a political one. The aim was to take hold of primate-style dominance and turn it upside down. No longer should physical violence or threat be allowed to determine access to resources or status within the group. Humanity's first moral community was committed to the ideal of an egalitarian order, turning dominance on its head.

According to Boehm, the strategy of resisting dominance leads eventually to full-scale revolution. But how exactly did this happen? Boehm asks us to envisage a coalition expanding until eventually it includes everyone. This is a demanding concept, since a coalition by definition presupposes a boundary between insiders and outsiders. Given that primate dominance is always in some sense sexual, it would follow that a model of counterdominance culminating in reverse dominance should take account of this. Could male-versus-female conflict and cooperation lead to a coalition embracing everyone? Boehm (2001, pp. 167–169) does consider distinctively female strategies, but curiously only when dealing with chimpanzees. His arguments about the evolution of human hunter-gatherer egalitarianism are surprisingly unisex.

If we are to consider counterdominance and reverse dominance in human evolution, the most critical issue becomes reproductive counterdominance. How do these models deal with the question of reproductive skew among males? Bowles points to reproductive leveling among predominantly monogamous hunter-gatherers as critical to egalitarianism (Bowles 2006). To explore the evolutionary establishment of egalitarianism on this reproductive level, we must bring into consideration the energetic requirements of females.

According to the Social Brain hypothesis (Dunbar 1996, 2003), the factor driving increase in neocortex size in hominin ancestors was increasing group size. In the case of early *Homo*, as climate dried towards the end of the Pliocene, groups needed to be bigger for protection in more open environments. In the case of later *Homo*, during the Pleistocene, the main danger of predation was likely to have been from other human groups. Under these pressures for increasing group size, *Homo* was selected for increased Machiavellian intelligence to negotiate increasing social complexity. Pawlowski et al. show that as neocortex size increases in primates, the correlation of male rank with mating success is progressively under-mined (Pawlowski et al., 1998). Selection for increased social intelligence therefore goes hand in hand with greater reproductive leveling.

But whatever the specific selection pressures were, these larger brain sizes in later *Homo*, along with their larger bodies, led to increased costs of reproduction for females. It is now time to consider how the extra energetic requirements of mothers of large-brained offspring were being met. We turn to models for sexual strategies and investment.

14.9 Female Coalitionary Strategies

1. *Sarah Hrdy: The origins of mutual understanding.* Ape mothers are insuffi-ciently trusting to allow others to hold their babies. *Homo erectus* mothers, facing increasingly heavy childcare burdens, enhanced their fitness by relinquishing young offspring to trustworthy allocarers. However, this was only possible if female kin were living close together (see Hawkes below). Distinctively human cognition evolved in this context, as mothers probed potential allocarers for their cooperative intentions. Infants monitoring the intentions and feelings of mothers and others became adept at perspective-taking and integrating multiple perspec-tives. Offspring more skilled in reading the intentions of others and eliciting their help were better nourished and more likely to survive. Female strategies of cooperative childcare can explain how and why humans became cognitively and emotionally "modern" (Hrdy 2009).
2. *Kristen Hawkes: Grandmothering and show-off hunting in human evolution.* Together with her colleagues James O'Connell and Nick Blurton Jones, Hawkes offer two key arguments for investment in offspring at different stages of human evolution. The "grandmother" hypothesis (Hawkes et al., 1998; O'Connell et al., 1999) argues for the beginnings of humanlike life history in early *H. erec-tus*. Burdened with increasingly heavy childcare costs, evolving *Homo* mothers sought help from the most reliable source: female kin and especially their own mothers. Post-reproductive life-spans extended as older females came under selection pressure to invest in the offspring of their daughters. With drying of climate in the Early Pleistocene, and scarcity of accessible foods for weanlings, older females stepped in, providing gathered foods such as tubers to these young offspring. In terms of life history this selected for relatively early weaning (hence short interbirth intervals) along with longer childhood dependency on adult pro-visioning, and delay in sexual maturity, along with longer life-spans. Males were intermittently or unreliably involved in supporting offspring at this stage, but during the Middle to Late Pleistocene (associating to *H. heidelbergensis*), hunt-ing strategies become more effective and reliable. Males were motivated to hunt big game as "show offs". Rather than hunt small-to-medium game for their own offspring alone, they demonstrated quality by generously providing big game for the whole camp (Hawkes and Bliege Bird 2002). Thus females gained male investment via mating effort rather than specifically paternal strategies.

3. *Camilla Power: Female cosmetic coalitions*. The evolution of concealed ovulation, extended receptivity, and increased reproductive synchrony in the human female forced males to spend more time in female company. Potential philanderers were deprived of the information they needed to successfully rove from one female to the next, picking and choosing between females on the basis of current fertility cues. However, one signal—menstruation—was left salient, giving away this kind of information to philanderers. As an indicator of imminent fertility, menstruation will trigger conflict both between males, who may compete for the cycling female, and between females, who may compete for male investment. In the absence of countermeasures, mothers who are pregnant or lactating may be at risk of losing male investment to the cycling female. The rapid increase in neocortex size characteristic of human evolution over the last half million years meant mothers could no longer tolerate such risks; it was in their individual fitness interest to prioritize future economic security over short-term sexual favor-seeking. Counterdominant female coalitions on this basis responded by "painting up" with false signals, representing all members of the coalition as uniformly "fertile". Investor males—whose offspring might have better chances of survival—had a fitness interest in colluding with the corresponding fictions. The evolutionary stability of female strategies of cosmetic bonding and adornment culminated in the transition to symbolic ritual, religion, and language (Power and Aiello 1997; Power 1999, 2009).

14.10 On Cooperative Breeding

Sarah Hrdy effectively combines the "grandmother" model with Tomasello's arguments for intersubjectivity as the basis for human culture and cognition. Pregnancy and postnatal childcare in *Homo* were such heavy burdens that they offer the most convincing context for the development of novel cooperative strategies. Alone of the great apes, we became cooperative breeders. Hrdy's arguments about the effects of alloparenting on human cognitive evolution are persuasive. Her focus on changing female strategies and on consequences for infant psychology are necessary and welcome. Demographically flexible cooperative breeding networks could act as a safety net compensating for extreme variability of male commitment to investment.

Neither Hrdy nor *Kristen Hawkes*, whose model she acknowledges as the initial steps into cooperative breeding, aim to deal with symbolic culture. Both models also keep males as investors in the margins, with female kin getting on with the job, not expecting regular investment from males. Males enter the picture only late, becoming more reliable hunters as female sexual choice drives them to intensified mating effort. There is no clear argument from Hawkes as to what causes the shift in male behavior and productivity between *H. erectus* and subsequent encephalized humans. In fact, in her life history models she does not take much account of increasing brain size even though this is critical in adding to female costs. Among Hadza bow-and-arrow hunters to this day, males are only intermittently successful, an observation which led Hawkes to doubt the validity of the model of "man the hunter" provisioning his own offspring.

Camilla Power concurs with Hrdy's and Hawkes's initial position of female kin-related social structures among *H. erectus*. Because female fertility is altered by the grandmother strategy, since mothers with allocare support would tend to have shorter interbirth intervals and be fertile more often, this must affect male behavior. More dominant males might attempt to target fertile females opportunistically, moving from one to another, while less dominant males could pursue a strategy of hanging around more reliably, offering provisioning and protective support to a particular female and her kin. As interbirth intervals shortened, investor males who waited around, rather than competed for other mates, should get more reproductive benefits. Such a picture of variability in male commitment fits Hrdy's observations of stark differences among modern human fathers.

Power argues that while such variability may have been tolerable for less encephalized early *H. erectus*, as brains rapidly expanded during the Late Middle Pleistocene (from ca. 500,000 to 150,000 ky), female fitness was increasingly affected by male investment. In these conditions among *H. heidelbergensis*, sporadically in Eurasia, and increasingly regularly in Africa, females resorted to the cosmetic strategy from ca. 300,000 ky. This had the effect of rejecting male philanderers who were not prepared to work and invest, while promoting the rewards to male investors—in the form of Hawkes's big game hunting show offs.

An advantage of Power and colleagues' model (Knight et al., 1995) is that the emergence of symbolism is intrinsic to the strategy. Symbols are socially accepted fakes, and in Power's model that means cosmetics. But were pigments necessarily used by women alone? Evolving human males had little Darwinian reason to alter or transform their biologically perceptible identity. With females, matters had always been more complex. The evolving human female had good reason to conceal external signs of ovulation, given that philanderer males might use such information to their advantage. The use of blood-red cosmetics to scramble menstrual signals was in that sense nothing new. Power's model does not exclude males from using cosmetics; but there is no good Darwinian reason why males should "fake" with cosmetics first. At present, the Female Cosmetic Coalitions (FCC) model is the only Darwinian explanation as to why the ochre is so prominent at Blombos and other Middle Stone Age sites.

The FCC model posits counterdominance leading to reverse dominance. In this case, however, both the initial dominance and its subsequent reversal are gendered. The model applies a standard behavioral ecological approach (one distinguishing sexual strategies and male and female tradeoffs) to the suggestions of Whiten and Boehm. Females concealing ovulation and extending sexual receptivity are already promoting "counterdominance" on a sexual level, since the strategy discriminates against dominant males in favor of subordinates more likely to invest time and energy. When the scrambling of reproductive signals is extended to menstruation, the effect is to tip "counterdominance" into "reverse dominance". When a female begins to menstruate, her senior female kin have every interest in surrounding her, identifying with her attractions and "painting up" to spread those attractions around. But they also have every interest in barring male access to her except on their terms (cf. Knight 1991).

Hawkes's model of male hunting as a "show-off" strategy needs to be placed in this wider sexual and political context. After all, there are many different ways in which males might show off, not all of them conducive to symbolic culture. Males could resort to violence and threat, "showing off" in terms of aggression and fighting skills. The Female Cosmetic Coalitions model can explain how they were successfully corralled into showing off productively rather than destructively.

14.11 Sex and Symbolism

Whereas Chase argues that symbolic culture emerges in order to enforce cooperation between strangers, Power sets out from selfish-gene theory and stays with it throughout. "There is no reason to believe that symbolic culture was ever essential to survival", writes Chase (1994, pp. 626–628). But in that case, why invest so much energy in the necessary rituals? Chase has contributed to the conceptual definition of symbolic culture, but in the absence of any evolutionary theory he lacks specific predictions about exactly what taboos, what laws, what rules would be collectively enforced. By contrast, Power and colleagues offer an array of specific predictions testable against the archaeological, fossil, and ethnographic records (Power 2009, Table 14.2, p. 273; for detailed ethnographic tests see especially Watts 2005).

But how exactly does the model generate such detailed predictions? In pursuing their direct reproductive interests, women "gang up" on anyone in their own ranks threatening to prove a weak link in the chain. A female who has begun cycling comes potentially into that category: in view of her special attractions she might be tempted to break ranks. Abandoning his current partner, any would-be philanderer will be on the look-out for a new partner signaling that she is of the same species as himself, of the opposite sex, and currently available to be impregnated. This immediately gives us the predicted signature of "reverse dominance". The defiant, cosmetically adorned coalition must bond tightly with the target of philanderer attention. Reversing her perceived biological identity, they signal collectively: *Wrong species, wrong sex, wrong time!*

Note that we now have a coalition that might in principle extend to embrace everybody, as Boehm's argument demands. On the one hand, the entire female community has an interest in joining, irrespective of kinship or previous friendship or familiarity—all should benefit over the long term by making philandering an unplayable game. But the coalition of females should also expect much male support. Brothers and sons might be expected to defend the interests of their female kin. Meanwhile, investor males should have an interest in ganging up against potential philanderers seeking to impregnate their long-term mates. On all these grounds, we might expect the "reverse dominance/reverse reality" coalition to succeed in imposing its message.

There is cognitive difficulty in believing in counter-reality. It is not easy to accept that biological reality can be so completely reversed—that the categories of human versus animal, female versus male, menstrual blood versus hunting blood can be switched around in this way. But such tricks—the stuff of mythology the world over—are not arbitrary cultural inventions. Reverse dominance will generate them by conceptual necessity. The message that results is patently false. The biological female undergoing her "initiatory" ordeal is not a male, not an animal, and not mortally wounded. But if everyone accepts the reversal, it is an institutional fact. And not just any institutional fact. If the argument is accepted, reverse sexual dominance conjures up Rappaport's Ultimate Sacred Postulate—the symbolic truth underpinning all others.

14.12 Conclusion

In this chapter, I have tried to show how the problem of the emergence of symbolic culture might be solved. In revisiting a set of currently prominent models—all of which offer insights—I have asked how they might be parsimoniously fitted together.

My aim has not been to set up Female Cosmetic Coalitions in opposition to the other models considered here. Chase is correct to view symbolic culture as a means of enforcing cooperation between strangers. But we require more than a statement: we need a Darwinian explanation. Rappaport and Sosis are surely correct about the importance of ritual, but to construct a testable theory we need to specify which rituals, when, where, and by whom. Donald is persuasive in his arguments about mimesis. But mimesis is "faking it": if everyone is just acting, why should anyone believe? Similar theoretical difficulties afflict Sperber: how, when, where, and why did patent falsehoods become trusted by evolving humans as valid intellectual currency? Whiten's model is persuasive but unfortunately avoids the topic of sex, as does Boehm's. What political purposes might have been sufficiently constant and unifying to produce "deep social mind"? Tomasello posits commitment to shared goals as a condition of language's evolutionary emergence. Can we specify whose goals? Hrdy reminds us that half the human population is female, and that novel strategies of social cognition and cooperation are most likely to have been driven by females and infants. But why stop there, given increasing reproductive costs associated with encephalization after *H. erectus*? Why not posit the emergence of symbolism as a continuation of the previous logic of female allocare strategies? Hawkes brings male mating effort back into the picture, but without explaining why symbolism had anything to do with it.

Lewis comes into a rather different category. Instead of proposing yet another cultural origins theory, his purpose is to persuade scholars researching modern human origins of the relevance of hunter-gatherer ethnography. The Mbendjele forest people who inspire Lewis's vision challenge the conceptual distinctions central to so much western evolutionary psychology and social science. Language, play, and ritual are cut from the same cloth. Religion is not a different thing from childhood pretend-play: it is pretend-play taken seriously and enjoyed also by adults. Hunting is not necessarily a different thing from speaking or listening: from a Mbendjele perspective, it is a matter of talking to and listening to the forest. Lewis argues persuasively that such interconnections need to be borne in mind by those of us struggling to explain the evolutionary emergence of human symbolic culture. It may be that everything is simpler than we thought.

References

Alcorta CS, Sosis R (2005) Ritual, emotion, and sacred symbols. The evolution of religion as an adaptive complex. Human Nature 16:323–359

Boehm C (2001) *Hierarchy in the Forest. The Evolution of Egalitarian Behavior.* Harvard University Press, Cambridge, MA

Bowles S (2006) Group competition, reproductive leveling, and the evolution of human altruism. Science 314:1569–1572

Burling R (2005) The Talking Ape. How language evolved. Oxford: Oxford University Press

Chase PG (1994) On symbols and the palaeolithic. Current Anthropology 35:627–629

Chase PG (1999) Symbolism as reference and symbolism as culture. In: Dunbar RIM, Knight C, Power C (eds) *The Evolution of Culture: An Interdisciplinary View.* Edinburgh University Press, Edinburgh

Chomsky N (2000) *New Horizons in the Study of Language and Mind.* Cambridge University Press, Cambridge, MA

Chomsky N (2005) Three factors in language design. Linguistic Inquiry 36(1):1–22

Christiansen MH, Kirby S (2003) Language evolution: the hardest problem in science? In: Christiansen MH, Kirby S (eds) *Language Evolution.* Oxford University Press, Oxford

d'Errico F (2003) The invisible frontier: a multiple species model for the origin of behavioral modernity. Evolutionary Anthropology 12:188–202

d'Errico F, Lawson G, Vanhaeren M, van Niekerk K (2004) Nassarius kraussianus Shell Beads from Blombos Cave: evidence for symbolic behaviour in the Middle Stone Age. Journal of Human Evolution 48:3–24

d'Errico F, Vanhaeren M (2009) Earliest personal ornaments and their significance for the origin of language debate. In: Botha R, Knight C (eds) *The Cradle of Language.* Oxford University Press, Oxford

de Saussure F (1983 [1915]) *Course in General Linguistics.* Transl Harris R. Duckworth, London

Donald M (1991) *Origins of the Modern Mind. Three Stages in the Evolution of Culture and Cognition.* Harvard University Press, Cambridge, MA

Donald M (2001) *A Mind So Rare.* Norton, New York, NY

Dunbar RIM (1996) *Grooming, Gossip and the Evolution of Language.* Faber and Faber, London

Dunbar RIM (2003) The social brain: mind, language and society in evolutionary perspective. Annual Review of Anthropology 32:163–181

Fitch WT, Hauser MD, Chomsky N (2005) The evolution of the language faculty: clarifications and implications. Cognition 97(2):179–210

Grafen A (1990) Biological signals as handicaps. Journal of Theoretical Biology 144:517–546

Guilford T, Dawkins MS (1991) Receiver psychology and the evolution of animal signals. Animal Behaviour 42:1–14

Hauser MD, Chomsky N, Fitch WT (2002) The faculty of language: what is it, who has it, and how did it evolve? Science 298:1569–1579

Hawkes K, Bliege Bird R (2002) Showing off, handicap signalling, and the evolution of men's work. Evolutionary Anthropology 11:58–67

Hawkes K, O'Connell JF, Blurton Jones NG, Alvarez H, Charnov EL (1998) Grandmothering, menopause, and the evolution of human life histories. Proceedings of the National Academy of Sciences of the USA 95:1336–1339

Henshilwood CS, d'Errico F, Vanhaeren M, van Niekerk K, Jacobs Z (2004) Middle Stone Age shell beads from South Africa. Science 304:404

Henshilwood CS, d'Errico F, Watts I (in press) Engraved Ochre from the Middle Stone Age levels, South Africa. Journal of Human Evolution

Henshilwood CS, d'Errico F, Yates R, Jacobs Z, Tribolo C, Duller GAT, Mercier N, Sealy JC, Valladas H, Watts I, Wintle AG (2002) Emergence of modern human behavior: middle stone age engravings from South Africa. Science 295:1278–1280

Henshilwood CS, Dubreuil B (2009) Reading the artifacts: gleaning language skills from the Middle Stone Age in southern Africa. In: Botha R, Knight C (eds) *The Cradle of Language*. Oxford University Press, Oxford

Hrdy SB (2009) *Mothers and Others: The Evolutionary Origins of Mutual Understanding*. Belknap Press of Harvard University Press, Cambridge, MA

Klein RG (1999) *The Human Career: Human Biological and Cultural Origins*. University of Chicago Press, Chicago, IL

Klein RG, Edgar B (2002) *The Dawn of Human Culture*. Wiley, New York, NY

Knight C (1991) *Blood Relations: Menstruation and the Origins of Culture*. Yale University Press, London

Knight C (2004) Decoding Chomsky. European Review 12:581–603

Knight C (2009) Language, ochre and the rule of law. In: Botha R, Knight C (eds) *The Cradle of Language*. Oxford University Press, Oxford

Knight C, Power C, Watts I (1995) The human symbolic revolution: a Darwinian account. Cambridge Archaeological Journal 5(1):75–114

Krebs JR, Dawkins R (1984) Animal signals: mind-reading and manipulation. In: Krebs JR, Davies NB (eds) *Behavioural Ecology: An Evolutionary Approach*. Blackwell, Oxford, pp 380–402

Lewis J (2009) As well as words: Congo pygmy hunting, mimicry, and play. In: Botha R, Knight C (eds) *The Cradle of Language*. Oxford University Press, Oxford

Marean CW, Bar-Matthews M, Bernatchez J, Fisher E, Goldberg P, Herries AIR, Jacobs Z, Jerardino A, Karkanas P, Minichillo T, Nilssen PJ, Thompson E, Watts I, Williams HM (2007) Early human use of marine resources and pigment in South Africa during the Middle Pleistocene. Nature 449:905–908

Maynard Smith J, Harper D (2003) *Animal Signals*. Oxford University Press, Oxford

McBrearty S (2007) Down with the revolution. In: Mellars P, Boyle K, Bar-Yosef O, Stringer C (eds) *Rethinking the Human Revolution: New Behavioural and Biological Perspectives on the Origin and Dispersal of Modern Humans*. McDonald Institute for Archaeological Research, Cambridge, MA

McBrearty S, Brooks A (2000) The revolution that wasn't: a new interpretation of the origin of modern human behavior. Journal of Human Evolution 39:453–563

Mellars PA, Stringer C (eds) (1989) *The Human Revolution. Behavioural and Biological Perspectives in the Origins of Modern Humans*. Edinburgh University Press, Edinburgh

O'Connell JG, Hawkes K, Blurton Jones NG (1999) Grandmothering and the evolution of *Homo erectus*. Journal of Human Evolution 36:461–485

Pawlowski B, Lowen CL, Dunbar RIM (1998) Neocortex size, social skills and mating success in primates. Behaviour 135:357–368

Pinker S (1994) *The Language Instinct*. Penguin, London

Power C (1999) Beauty magic: the origins of art. In: Dunbar RIM, Knight C, Power C (eds) *The Evolution of Culture*. Edinburgh University Press, Edinburgh

Power C (2009) Sexual selection models for the emergence of symbolic communication: why they should be reversed. In: Botha R, Knight C (eds) *The Cradle of Language*. Oxford University Press, Oxford

Power C, Aiello LC (1997) Female proto-symbolic strategies. In: Hager LD (ed) *Women in Human Evolution*. Routledge, New York, NY

Rappaport RA (1999) *Ritual and Religion in the Making of Humanity*. Cambridge University Press, Cambridge, MA

Searle JR (1996) *The Construction of Social Reality*. Penguin, London

Sosis R (2003) Why aren't we all Hutterites? Costly signalling theory and religious behavior. Human Nature 14:91–127

Sperber D (1975) *Rethinking Symbolism*. Cambridge University Press, Cambridge, MA

Sperber D (2005) A pragmatic perspective on the evolution of mindreading, communication and language. Paper delivered to the Morris Symposium on the Evolution of Language. Stony Brook, New York, NY

Sperber D, Wilson D (1986) *Relevance: Communication and Cognition*. Blackwell, Oxford

Tooby J, Cosmides L (1992) The psychological foundations of culture. In: Barkow J, Cosmides L, Tooby J (eds) *The Adapted Mind: Evolutionary Psychology and the Generation of Culture*. New York: Oxford University Press, pp. 19–136

Tooby J, Cosmides L (1995) Foreword. In: Baron-Cohen S (ed) *Mindblindness. An essay on autism and theory of mind*. Cambridge, MA: MIT Press, pp. xi–xviii

Tomasello M (1999) *The Cultural Origins of Human Cognition*. Harvard University Press, Cambridge, MA

Tomasello M (2003) Different origins of symbols and grammar. In: Christiansen MH, Kirby S (eds) *Language Evolution*. Oxford University Press, Oxford

Tomasello M (2006) Why don't apes point?. In: Enfield NJ, Levinson SC (eds) *Roots of Human Sociality: Culture, Cognition and Interaction*. Berg, Oxford

Watts I (1999) The origin of symbolic culture. In: Dunbar RIM, Knight C, Power C (eds) *The Evolution of Culture*. Edinburgh University Press, Edinburgh

Watts I (2005) 'Time, too, grows on the moon': some evidence for Knight's theory of a Human Universal. In: James W, Mills D (eds) *The Qualities of Time: Anthropological Approaches*. Berg, New York

Watts I (2009) Red ochre, body painting, and language: interpreting the Blombos ochre. In: Botha R, Knight C (eds) *The Cradle of Language*. Oxford University Press, Oxford

Whiten A (1999) The evolution of deep docial mind in humans. In: Corballis M, Lea SEG (eds) *The Descent of Mind: Psychological Perspectives on Hominid Evolution*. Oxford University Press, Oxford

Zahavi A (1975) Mate selection: A selection for a handicap. Journal of Theoretical Biology 53:205–214

Zahavi A (1993) The fallacy of conventional signalling. Philosophical Transactions of the Royal Society of London 340:227–230

Zahavi A, Zahavi A (1997) *The Handicap Principle: A Missing Piece in Darwin's Puzzle*. Oxford University Press, New York, NY

Chapter 15
Belief in Melanesia

Wulf Schiefenhövel

Abstract The Eipo, a Papuan group in the highlands of West-New Guinea, had a neolithic tool kit and an animistic religion when they were first studied in the 1970s. They believed that there were close links between one's health (and disease) and one's behavior in society, particularly, that wrongdoing causes mishap, sickness and death. Thereby, Eipo religion was not only linked to wellbeing and health, but there was, furthermore, a juncture between the extrahuman powers *(isa,* i.e. souls of the dead, spirits of creator "gods", of nature etc., who were thought to interfere in the lives of people) and the primordial juridicial system of the Eipo protecting the social system with its canon of conduct. Persons who committed a breach of sacrosanct norms were, so the unshakable belief, punished by one of the *isa* powers whose main function was the safeguarding of social rules and the perpetuation of society and cosmos. The Eipo were also convinced that disease and death had no natural cause (except for people dying of old age) but were always the consequence of interference by *isa.* Healing was, thus, a religious domain and faith a very powerful companion of all therapeutic settings. This has repercussions for our own society, where modern medical and psychological science only slowly grasps the full impact of placebo and halo effects which are important adjuvants of medical intervention. In a similar vein, religion can contribute to stress reduction – a classic function of belief systems in pre-scientific cultures like that of the Eipo, but also relevant for industrialised societies. The Eipo accepted Christianity around 1980, a consequence of more political than transcendental concern. It is quite striking, how smooth this change from animism to monotheism has been so far and that the new religion has helped them, in many ways, to cope with the abrupt process of acculturation.

W. Schiefenhövel (✉)
Max-Planck-Institut für Verhaltensphysiologie, 82346 Andechs, Germany
e-mail: schiefen@orn.mpg.de

U.J. Frey et al. (eds.), *Homo Novus – A Human Without Illusions,*
The Frontiers Collection, DOI 10.1007/978-3-642-12142-5_15,
© Springer-Verlag Berlin Heidelberg 2010

15.1 Case History: Death of an Eipo Man

Ebna[1] was an unmarried man from the village of Munggona, about 23 years old, unusually tall and very fit like most people in this physically demanding mountain region. One night, some of his family woke me up and urged me to see him because "he is dying". The patient was rolling on the floor in pain and had vomited three times; there was no blood in the material he had brought up. He complained of severe abdominal pain. His blood pressure was 110/80 RR, his pulse 58 per minute, the temperature 36.7°C, the abdomen soft. I gave him a pain killer and a mild vasopressor. The symptoms resided somewhat. A few hours later his breathing rate was very high, about 70 per minute and the breathing very shallow. E. now complained of pain in both sides of the chest. The following day his temperature rose to 38.1°C and his lungs showed signs of beginning pneumonia, I therefore gave an intramuscular injection of depot-penicillin, very effective medication in those early days of contact. On the second day his temperature had gone down to 37.3°C, breathing frequency was about 40 per minute, the condition of the lungs improved. One of the traditional healers (*ninye kwetenenang*), who had treated E. with the usual religious ritual invoking helpful spirits, said to me: "You and I, both of us have cured him". So, there was optimism.

On the third day, E., walking slowly with a stick, came to my house at the fringe of the village. He smiled a little and made the impression of a person who had overcome a serious disease. Two days later I was called again: "E. is really dying now!". E. lay on the floor, moaning. When I punctured his vein to give him a drip I observed that the viscosity of his blood was very high. He had eaten and drunk very little in the last days, perhaps also in the time before the onset of serious symptoms. His blood pressure was 100/60 RR, his pulse 72, his belly soft as all the days before. E. had a fresh skin wound, the size of a man's palm, in the upper abdomen. He had pressed a hot stone from the fireplace against this part of his body, where he said he had the worst pain. In all my decades of treating patients in Melanesia before and after this incident I have never seen a similar case of self-aggression—except suicide. E. died, while the drip was still running, bemoaned by his family with the typical, emotionally moving songs of grief (Schiefenhövel 1985; Eibl-Eibesfeldt et al. 1989). No autopsy was performed, but my assessment is that E.'s death was a case of thanatomania or psychogenic death, as the clinically observed symptoms did not seem to explain the fatal outcome. He might have given himself up and stopped drinking, which would have been important in his condition of raised temperature. A thus provoked breakdown in the electrolyte-fluid system might have contributed to his death, even though there were only a few days with severe symptoms.

In the days before, when I talked to him about the cause of his being sick, he told me that he was certain that the spirits (*isa*) of wild animals (arboreal and nocturnal

---

[1]I give the real names of the persons described in this chapter. I have followed this good ethnographic tradition since my first publications and do not see any possible harm for the persons mentioned.

marsupials of the genus *Phalanger*), which he had hunted, had caused his body to be affected by their angry revenge—in Eipo pathogenic theory one of the typical causes of becoming sick. These beliefs are very powerful, they can produce the conviction that one is guilty and has become an outsider in the community; they may thus be the basis for thanatomania or psychogenic death (Cannon 1942; Levi-Strauss 1967; Lester 1972; Stumpfe 1973, 1974, 1976), which could bring death via a stress-induced collapse of the adrenal gland or another mechanism.

Psychogenic death does not seem at all absent in our own secular world: the fact that prisoners in concentration camps and even in normal prisons have died without medical reason and that some widowers or widows die very soon after the death of their partner are indications of this (Stumpfe 1973, 1974). A volume (Carr et al. 2005) on the adjustments of widowed elderly persons in the United States found that quite a large number of the bereaved manage their loss quite well, but still there are cases where such lonely lives end with psychogenic death.

E.'s elder brother Babesikna,[1] one of the impressive personalities of Munggona and candidate to become a "big man", *sisinang* in the Eipo language, was visibly shocked by E.'s death and remained, for hours, in a kind of stupor, holding the corpse on his lap. Then he took care, in the very decisive and active way typical for him, of the mortuary rites (Schiefenhövel 1985). He first had a tree prepared in the usual way for placing the corpse in its crown by cleaning off smaller branches, twigs, and leaves. Yet, after some days, he decided that this was not a worthy enough abode for his brother and placed him in a tall Casuarina tree, which was equally cleared in the crown and given a roof and simple protective bark walls around the dead body so that it was not exposed to the almost daily rainfall (about 6,000 mm/year). The corpse was fixed in such a way that his face pointed in the direction of the central cordillera, where the ancestors of his clan were believed to have come from. This is always done to facilitate an easy, straight journey of the soul (*isa*) to the ancestral abode, so that it does not linger around, filled with longing to continue staying with the living, as that is likely, so the belief, to cause problems.

B., squatting in the high tree beside his dead brother, sang mourning songs for several months. Every day. His grief was enormous. Soon, he reinterpreted E.'s self-diagnosis, namely that he had become the victim of angry spirits of the wild animals hunted by men. B. claimed that his beloved younger brother had not died because of animal spirits but because someone from the neighboring village of D. had killed him with "black magic" (*kire*). B. actually shot the accused man from this village with an arrow, who later died of the wound. It seemed to me, the ethnographic observer, that the very marked aggression that was intertwined with his deep grief needed an outlet and that this was the reason for him to reinterpret E.'s death cause. If someone dies of suspected interference from animal spirits, nothing much can be done, no revenge is possible against the invisible enemy. But when a person, woman or man, is accused of committing *kire*, a monstrous crime in the eyes of the Eipo, then a stage for action is prepared: revenge with bow and arrow—which we witnessed in one other case, where a woman was accused of having killed a child through *kire*. From the time the suspected "sorcerer" died, B.'s mood changed: He smiled and felt joy again, despite the fact that he was, any minute of the day, in danger of being killed in revenge by the relatives of the suspected sorcerer. He had fulfilled what he must have perceived as his duty towards his brother and, at the same time, ambitious as he was, placed himself in the center of a politically mean- ingful chain of events, which made him, through his killing the "sorcerer", visible and a topic of admiration.

15.2 Health and Wellbeing as a Goal: The Juncture Between Extra-Human Powers and an Incipient Juridicial System

The above described case exemplifies five major beliefs held by the Eipo in the early days of our research project, in which their culture functioned along age old traditions without interference from outside (Schiefenhövel 1976, 1991). Very sim- ilar religious concepts were found, in equally unacculturated early contact situations, among the Dani (Heider 1970, 1991), Yali or Jalé (Koch 1974; Zöllner 1977) in the west of the Mek cultures and the Nalum (Hylkema 1974), who are part of the Ok cultures, in the east:

1. disease is always the consequence of interference by some extra-human power; these powers can be represented by
2. nature spirits or creator spirits or spirits of the dead or
3. human-made "black magic", which involves a supernatural power source, as well as
4. the conviction that the souls of the dead, whether recently or long deceased, can interfere in the lives of the living. The Eipo society, as countless others characterized by an animistic belief system, had a
5. tight juncture between the religious realm of extra-human powers (all classed as *isa*) and the, as it were, juridical realm (see below). People who committed a breach of the sacrosanct norms were punished by one of the *isa* powers, whose main function was the safeguarding of social rules so that order could prevail in the group and in the cosmos.

During our fieldwork on Kaileuna, one of the Trobriand Islands, Papua New Guinea, a girl of about 3 years was brought to me by her father for medical treat- ment; the mother was on a visit in a neighboring village. I could not detect any symptoms, the temperature was normal and the child did not appear sick. In the night, I was called by the grandmother of the girl who was now "dangerously sick". Indeed, the little patient was unconscious, had very high temperature and convul- sions. Obviously a case of cerebral *Malaria tropica*, which runs in daily cycles, had not produced fever when the child had been brought in the morning and can kill within hours. I tried everything at my disposal, including intravenous quinine, as *ultima ratio*, but could not save the girl's life.

At the onset of severe symptoms people in the family and the village started to speculate about the cause of the disease and came to the univocal conclusion that the father was guilty. He had, undisputed by him, stolen some betelnuts (*Areca catechu*), a mild, very much sought after hallucinogenic stimulant, from a tree that was protected by a taboo (*tabu* in Kilivila, the Trobriand language). In pre-Christian days, these taboo signs were usually a ring of leaves, sometimes twisted to a wreath, wound around the tree or other valuable object to be protected. If somebody crossed this sign by climbing over it with his body, so the belief, the sanctioning power of the spirits would punish the wrongdoer. In this case, the taboo sign was a small wooden plank with the written inscription *tabu pela tapwororo* (taboo by the church). The idea behind this Christianized form of sanction is that God is angry with people because they crucified his son. This anthropomorphically conceived divine anger and revenge motivation is "tapped" and will cause disease, accident, or mishap to the wrongdoer. In this case, the very much beloved daughter of the couple, the only girl in a row of boys, was the victim. At one stage of fighting death, the local priest had been called to ask God to take the spell away. He brought the bible, read a chapter, and prayed, but it was of no avail.

Religious syncretism is the norm in today's Melanesian societies; the Trobriand example demonstrates well how animistic and Christian concepts are intertwined. This does not cause any feeling of uneasiness, not even in the native priests (who are more evangelists as they have very little formal training in theology).

Among the other peoples of Melanesia, whether Papuan by origin (arrival probably before 50,000 years b.p.) or Austronesian (arrival in the westernmost part of New Guinea approximately 4,000 b.p.), the belief that wrongdoing causes sickness was ubiquitous (see the section on religion in the chapters on a large number of ethnic groups in Melanesia in Hays 1991, and the detailed descriptions of religious beliefs and practices in Lawrence and Meggit 1965, as well as in Swain and Trompf 1995), at least as long as data were collected before the impact of missionaries who have succeeded in transforming mainland and island New Guinea into a large predominantly Christian region, despite the fact that the Indonesian Republic, to which the easternmost Province Papua belongs, is the largest Muslim nation in the world.

Punishment in the form of damaging one's health, gardening success, or other endeavors are very tangible sanctions, which could also affect a close relative, e.g. a brother or a sister or one's child, as in the case described. Punishment was thought to happen *hic et nunc*, that is without much time delay. Religions, like the Christian belief, that postpone punishment for sinful or award for good behavior to the day of judgment have taken a revolutionary step. To discuss the consequences of this strategic turn is, however, outside the scope of this chapter.

As disease, accident, or other mishap was frequent, and as the sanction not necessarily needed to be addressed to the actual wrongdoer, the primordial belief-system could not easily be falsified: Always, somebody does something wrong and always somebody is struck by disease or bad luck. It is important, in this context, to state that these "egalitarian" societies had no institutions that functioned to keep law and order and to make sure that socio-religious rules were kept. Social conduct was controlled by gossip, dispute, or verbal or physical aggression. When a conflict had cost a person's life, the law of revenge sent aggression up an escalating spiral—until the parties involved decided that enough blood had been lost (Schiefenhövel 2001). In other words, no third party was established tha could act as mediator, ombudsman, judge, or the like. This fact has been well described by Koch in his account on war and peace in Jalémo, the culture and language group neighboring the Mek in the west (Koch 1974).

It seems, therefore, that in these socio-politically simple societies, which functioned along a strictly meritocratic principle and had no heritable chiefs, the threat directed to health and wellbeing of oneself or one's kin and executed by extra-human powers was an important safeguard of norm control (for the similar role of "black magic" see below). Religion of this type was servicing the social and political coherence of the group and, at the same time, protecting individuals from being harmed or killed as they would try to stay in the range of accepted behaviors and actions. Most probably this is another ancient trait that might have been at work among early members of our species.

That disease is punishment for wrongdoing still is, surprisingly perhaps, a belief present in some regions of Germany and Austria (probably other countries as well), even though most patients, including those with life threatening diseases such as cancer, attribute their condition to other, secular factors, for example "too much stress in life" (Fischer 1993). I remember that the priest of our catholic diaspora community in a small Siegerland town (in southern Westphalia, where a fundamentalist, evangelical version of Protestantism is common) menaced in his sermon: "God will punish you at the organ you have sinned with!"—It was quite clear which part of the body he was referring to. So it seems that the juncture between extra-human powers and the juridical, punishing system, with human wellbeing and health at its center, is still present even in our modern societies.

15.3 "Black Magic" a Possible Second System of Ensuring Norm-Oriented Behavior

In a somewhat different light, but perhaps in a similar framework, one could interpret the possible function of "black magic", also a very widespread, probably universal trait of human societies. It is quite surprising how common forms of "black magic" still are, even in the middle of Europe. One only has to think about the Italian fear of "malocchio" (eeil eye) and other forms of "sorcery" and the many counterstrategies that people take refuge in.

Here, in contrast to the belief in vengeful spirits described above, human actors are involved: A, who wants to harm somebody (X), and B, who performs the allegedly harmful ritual. It seems at first sight that maintaining social order and harmony is not the main motive of both actors, rather A's hatred of X and B's greed, as he or she is often paid for this service, or perhaps her or his agreeing with A's judgment that B is a bad person and should be punished. On the other hand one could view "black magic" as a stepped-up and religiously ritualized form of envy, one main function of which is to improve or regulate one's own position in the group (Schoeck 1987; Eibl-Eibesfeldt 1989). In this perspective, "black magic" could also be understood as, *cum grano salis*, a mechanism of norm control and egalitarianism. People, aware of the threat of ever looming "black magic" could be induced to stay in the middle level of society and lead a life on the submissive side of options, which would be accepted by others, show softness, generosity, and altruism rather than tough egocentrism. In this way, harmful "magic" would function in a rather similar way as the typical extra-human powers safeguarding social harmony.

Protestant Sweden is the classic country where envy (*avundsjuka*, envy-sickness) is seen as a negative and socially unacceptable emotion—the so-called *jantelag*. Modern Swedish students learn, by way of leaflets listing undesirable actions and attitudes, to behave in a modest way that would not arouse any envy-sickness in others. And, actually, most people in Sweden try to avoid enticing feelings of envy (Jordan 2002). It is, therefore, no surprise that this country is one of the most egalitarian and conformist nations in the developed world.

I believe that both the punishing power of spirits and the threat of becoming a victim of envy-driven "black magic" are functional mechanisms that protect the survival and reproductive success of individuals as well as the survival and wellbeing of the group. Given the belief in these two punishing agencies, which are perceived to be a more or less constant threat, it is really surprising that some individuals do *not* stay in the safe middle level of the rank hierarchy but move, driven by ambition, vitality, and intelligence, to leading positions, where they are much more vulnerable than the less ambitious ones. Usually, these people (in the Melanesian context almost exclusively men, the so-called "big men" with probably rather high testosterone levels) are not only driven by burning ambition but are also socially very skillful and able to gain respect and to buffer their own success against the envy of others.

These are preliminary ideas. I could not find, in the literature, an interpretation of the possible function of "black magic" similar to that sketched out here, but it seems a worthwhile exercise to analyze the very widespread practice of harmful "magic" in a functional way and not just see it as some spill-over of basically malevolent human nature.

In a number of Melanesian cultures, and probably beyond, the people believed to be able to conduct effective "black magic" are, at the same time, the healers, i.e., the ones who conduct "white", healing, helpful magic. It is, conceptually, understandable that societies attribute both "powers" to the same person. What, on the other hand, happens in the mind of such a double-headed healer-sorcerer, how does he handle his enormous power and how does he or she avoid being killed when the precarious balance can't be kept . . . as happens from time to time? Questions which, similarly, have not been dealt with sufficiently in the literature of ethnological and religious studies.

One disturbing effect of the Christian faith having been accepted by the Eipo (see below) around the year 1979 has to do with the fact that belief in spirits is very much reduced now. Yet, spirit action was considered the major cause of disease and death in the times before. Today, loss of one's relative is not blamed on these extra-human powers any more but on *kire*, "black magic", the formerly alternative explanatory principle. During my last stay there in 2008 I was told: "When you lived with us long ago there was little *kire*, now there is lots of it". Quite understandable, if the spirits can't be the cause of death, then it must be *kire*, performed by humans. This has obviously led to a kind of witch hunt. Some people, women and men, have been singled out as being a "sorcerer" or "sorceress", verbally attacked in public and symbolically, as my informants said, hit or touched. One woman, who was held responsible for a large number of deaths and allegedly admitted to have carried out *kire*, committed suicide, a male "sorcerer" died through disease, which could perhaps have been the effect of social stress, but there is no proof of that.

The Eipo were, when they told me about these events, quite proud that nobody was physically killed any more as in the old days—actually all killing has stopped, even the intergroup war with the people of the Famek valley. And they were happy that they had found the "evil" people who they believe had killed their relatives. They don't seem to understand that the massive psychological pressure against accused members of their village communities is inhumane and also un-Christian. I don't know how they will solve this problem. It seems, from what I have experienced in other parts of Melanesia, extremely unlikely that the concept of death being brought about by "black magic" will disappear soon. When somebody of the family of a professor for anthropology at the University of Papua New Guinea died, this colleague told me, deeply convinced, that "poison", a new word for and similar concept to "black magic", had been the cause. One will have a hard time finding any local person, no matter how educated she or he is, who won't attribute death to such a concept. A long way lies ahead for the native people of Melanesia to understand disease and dying as a biological phenomenon and it becomes clear that a traditional belief system that has grown over time is not so easily replaced, and if some piece of it is broken out, cracks may happen and new injustice be created.

15.4 Healing, a Religious Domain

During my first field stay in New Guinea, in 1966, while collecting ethno-medical data in the Gulf of Papua, I became witness of a blind healer, named Kaiei,[1] attending a sick elderly woman who complained of pain in the back and lack of strength. The healer was led to her house by one of his daughters, climbed on the platform, where the patient was seated, and started to bite her back in a relatively gentle way, at least no blood appeared. In the very unspectacular way he conducted he whole ritual he stood up, drew his *laplap*, the typical long waistcloth men wore in this region, aside and began to pull a longish thing out of his scrotum. It turned out to be a 12 cm long piece of carefully twisted copper wire. What a therapy! I, a medical student in the midst of university education, was quite stunned. And all performed very calmly, almost without words and definitely without grand gestures or the like.

The idea was, as I learned, that this foreign body had entered the sick woman's body and was responsible for her bad condition. The job of the healer (*meamea tauna* in the Motu vernacular spoken in this area, *vada tauna* is the term for a man who performs "black magic") was to extract this foreign element from the patient's body, thereby freeing her from the disease. The patient went to the garden the next day—not so surprisingly. She probably had suffered (I did not carry out an examination) from hookworms or malaria or perhaps both or another chronic disease for which the psychological push of this psychosomatic curing ritual was a sufficient therapeutic momentum.

Kaiei, I was told, always pulled longish, lancet-type things from his scrotum, pieces of wood, arrowheads, etc. What, I couldn't help thinking, goes through the mind of such healer? Obviously, he carefully hides the object in his scrotum before he sees the patient, where he produces it as cause of the sickness. He must know what he is doing: tricking his patients and the onlookers. Furthermore: In this case, did he produce this unusual receptacle by an operation which he or somebody carried out? Was it the effect of an accidental injury? Did he learn this technique from somebody who utilized the same method?—I was too shy to ask this gentle frail old man all these intimate questions. As a matter of fact, I never asked any of the healers whom I met in all those decades to tell me the tricks of their trade. Possibly, these people, who are usually much respected in their groups, conduct something like *pia fraus*, pious fraud, believing that the effect their tricks have, namely curing people, is the important part and the trick is just a kind of vehicle to bring that about.

I have documented other extraction rituals in New Guinea. One was performed by Mosubiya,[1] the most powerful healer on the Trobriand Island Kaileuna. One of my neighbors had presented his child, a boy of about $2\frac{1}{2}$ years of age who had recently been weaned. He had a mild upper respiratory tract infection, which I treated with a mentholated ointment—a therapy much liked by most parents, as it does not cause pain, smells strong and nice, and involves massage. The father, however, was not happy with my medical performance and called the healer, who came from his village at the southern tip of the island, several hours away. I was very surprised that he seated himself right in my house, the lion's den, so to speak, the practice of his competitor. He said: "I know you have treated this boy already. Did you remove the four stones from his chest? They are the cause of his sickness." I replied that I had not done this and also did not know how to perform such therapy. "Okay, then I will demonstrate to you, how I do that."

He handled the little patient, who was crying as soon as one looked into his eyes, extremely well by passing his arm behind the father's back on whose lap the child was sitting. Thereby, he made contact in a nonthreatening way. The boy calmed down. The healer then sent one of the girls, who had come with him, to fetch some plants. He took them in his right hand, with which he calmly and inconspicuously made some movements inside the typical men's handbag that was on his lap. He then turned to the child, rhythmically hitting his back with the bunch of plants and reciting, quite loudly and impressively, a sacred curing chant. Not long and there was a loud bang: the first stone had fallen out, right against the metal box on which father and child were sitting. Quite a spectacular performance. Stones number two and three appeared in the same stunning way, the fourth fell, silently, through a crevice in the floor to the sand below. The healer looked at me and said: "Number four, did you see?" The people around my house were clearly pleased and proud about this convincing performance. The healer behaved with dignity and colleagiality. The boy got better. This was attributed to the treatment by the healer.

In a similar case a girl with a dangerously large abscess behind her eyeball, which could have led to life-threatening encephalitis, was given a depot penicillin injection by me. The father of the girl observed that, in contrast to the normal course of events after such injection, no improvement occurred—a very intelligent judgment. It turned out that the penicillin was not effective anymore; it was damaged by having been kept in the tropical climate too long. The same healer, M., was called in and performed the same kind of extraction ritual, "removing" some plant mass and small stones from the girl's head. Our little son Fridtjof, 5 years old, who stayed with me during this field trip, watched the scene with great interest. When we were walking home to get a different kind of antibiotic he said "Daddy, that was a big lie". "Why do you think so?" "Daddy, I watched very carefully. With all those things taken out from behind the eye, it should have moved inwards a lot, but it didn't do that at all!" Here was the little European kid, using a scientific approach and convinced of charlatanry, lost for the esoteric world of miracle healing. The local people don't worry about such minor flaws. They firmly believe in the efficacy of their curing rituals. I then went back to the patient and injected, in the dim light of a flickering kerosine lantern, megacillin into her tiny vein—high noon in the middle of the night in a Trobriand village. The abscess disappeared rather well, a strabismus remained but everybody was very happy that the girl survived. Again, the healer was given the bigger share of credit for that than the white doctor.

I was never told that this very respected and powerful healer also conducted harmful rituals. It is possible. I viewed him as a knowledgeable colleague. When his wife was sick, he called me. The respect it seemed was mutual.

Traditional midwives and healers are, I am sure, the first specialized professionals in the history of our species. Not only did they relieve pain and bring about, sometimes at least, wellbeing and health but they were, together with other specialists for the extra-human sphere, providing meaning, a framework to explain and understand what happens with people in difficult and dangerous life events and why they occur at all. It is safe to say that most of these personalities are quite impressive and up to their job (Schiefenhövel 1986), however limited their therapeutic inventory may be.

The pathogenic concept that disease is brought about by foreign bodies that have found their way into the body of people is very widespread. I wouldn't be surprised if this was another universal. Possibly the observation that splinters, arrows, and other objects entering skin and tissue cause infections and other problems have led to this generalized idea. Consequently, one has to try and get these bodies out. This is done in quite different ways. So-called "psychic surgery", which was, some decades ago, very popular among white patients who traveled all the way from Europe or North America to Bagio in the Philippines, is one of these forms (Schiefenhövel 1986), hand-tailored to the needs and expectations of a white clientele for whom "surgery" is the epitome of medical professionalism. In these kinds of rituals the healer puts, it seems, his hands deep into the abdominal cavity, which is miraculously "opened", blood is flowing copiously, an object (often a piece of tissue of animal origin supposed to be the "sick" body part) is pulled out, then the hands of the "surgeon" are taken off… and the abdominal wall is closed again. "Transmaterialization" and other esoteric terms have been used to describe and explain this indeed impressive sleight of hand, as modern western man, disenchanted by the materialist ways of his own medicine and society, wants to believe that the Philippino specialists, who act in a religiously spiced up scenery and are perceived to have a kind of holy aura around them, can perform miracles, therapeutic feats inaccessible to even the most famous western surgeon. The blood that so impressively drips from the belly to the floor is from a chicken or other animal... not from the patient, as was found out by simple blood tests. On the other hand, could not the psychosomatic effect of such exotic and thereby impressive, religiously meaningful curing rituals have beneficial effects? Should one try to prevent terminally ill persons or patients with chronic complaints for which they cannot find relief within modern medicine from spending the money on such a long and physically demanding trip to a tropical country? This question may not be relevant any more, at least not with regard to the "psychic surgeons" from Bagio and other places in the Philippines—their time seems to be over.

Faith is a very powerful companion of all therapeutic settings, and only slowly is modern medical and psychological science grasping the full impact of placebo and halo effects, which are so important for a good result of medical intervention. In the west we have ousted the personality cult by which former doctors were surrounded. Now, the medic has become the partner of the patient, has been stripped of his demi-god like status of yesteryear. Modern therapy has gained, no doubt, from this shift, but, at the same time, has lost the important religious elements of curing rituals. Perhaps this is why so many patients, disappointed by modern medicine for one reason or another, seek healing and meaning in the manifold esoteric methods and theories offered by ever more market-oriented providers of alternative medicine.

15.5 Religion as Stress Reduction

In the book *The Biological Evolution of Religious Mind and Behavior* edited with Eckart Voland (Voland and Schiefenhövel 2009) I have argued ("Explaining the Inexplicable", Schiefenhövel 2009) that one important building block of evolved religiosity is the human need to satisfy the causality urge so typical for our species. The often very thoughtful cosmologies around the world and in the one thousand or so cultures of Melanesia are a good proof of this hypothesis. Humans need to have answers to nagging questions like: Where did the earth and all the things on it come from? What is the essence and destination of humans? How can harmony and balance, health, wealth, fertility of humans and nature be safeguarded? What must be done in times of individual and group crisis and when massive disasters threaten to take the lives of many, if not all? This was the case when the two strong earthquakes of 1976 affected the Mek region. During this time, the old religion was still intact; the members of the German Research Team (funded by the Deutsche Forschungsgemeinschaft) did not, of course, carry out any missionary work in the Eipomek valley. On the contrary, their interest in the traditions of the local people might have helped to corroborate trust in the old way of life. At least at that point in time.

Consequently, the old concept of how earthquakes are brought about was still in place. The idea was that a giant (*Memnye*, i.e., the tabooed one) normally sleeps soundly deep down in the earth, but every now and then moves in his sleep, like humans do, and thereby causes the mountains to shake and tremble and massive landslides to gush down into the valleys where the houses of the people are, covering large areas of fertile garden land and causing other terrible destruction. It does not matter that the Wegnerian science of plate tectonics (generally accepted only in the 1960s!) provides a more plausible explanation of earthquakes than the theory of the sleeping giant. The important point is that causality concepts are at work in the human mind, which reduce stress and anxiety. That should have been a rather powerful adaptive mechanism in our past.

Malinowski, the founder of modern functionalism and thereby a kind of distant cousin of human ethologists and the like, argued along similar lines. In the Trobriand Islands where he did his influential and groundbreaking fieldwork, learning the local language, Kilivila, and employing an emic approach, he found that magic was used for various purposes, e.g., to kill enemies and to prevent being killed, to ease the birth of a child, to protect fishermen, and to ensure harvest and success of fishing. He hypothesized (Malinowski 1973) that magic works in those cases where people need explanations but do not have access to science. He saw that one important function of religious rituals is to exert control over events and conditions that are outside the human sphere of influence and are thereby possibly dangerous, like the untamed elements of nature. Religion (or "magic" as he phrased it) thereby reduces anxiety. And, I would add, anxiety is a very potent mechanism damaging, if continuously activated, human health and wellbeing via pathways connecting psyche, immune, endocrine, and neurobiological systems.

Almost always, religious concepts of the kind I have described above portray deeply anthropomorphic images. That is also the case with Christian religion: the benign or vengeful father God, Mother Mary, and all the innumerable saints who dwell in their celestial abode, smuggling a rather polytheistic tinge into the monotheistic faith. Kunz, Richert and Smith, Brüne, and Frey (all in Voland and Schiefenhövel 2009) as well as the classic author on this issue, Feuerbach (1854) and other authors before and after him have discussed the very human face all religions reflect.

15.6 Religion as a Mechanism to Manage Drastic Cultural Change

In a recent book (Voland and Schiefenhövel 2009), I have given an account of the rather surprising way in which the Eipo in the highlands of west New Guinea have so far coped with the advent of Christian religion and the other set of revolutionary changes that have happened since the mid 1970s (Schiefenhövel 2009). It seems to me that this is a very promising development, but it remains to be seen, of course, whether it will continue without major setbacks.

In the early days of missionary work in Melanesia, which started around the middle of the nineteenth century, quite a large number of white priests were killed. The best known case is probably that of James Chalmers, who was generally much liked, even venerated, by the people of the south coast of the then British Crown Colony of Papua, who gave him the name Tamate, father. This did not, however, prevent his being killed by headhunters of Goaribari Island in the Gulf of Papua who just had completed a new men's house and needed a trophy skull for its inauguration (Lovett 1902). But, in most regions the gospel and the new way of life was met with quite open minds, not so much because the Old and New Testaments were seen as much more attractive than their old beliefs but because the white people obviously came from a culture so much superior in material aspects. The local people thought that taking over the new religion would give them access to these goods. This is the basic core of the well-known and quite widespread "cargo cult movements" (Lawrence 1964). They implied a rather typical chain of events.

First, as mentioned, there was eagerness to be baptized or become, in other forms, associated with the new religion. When the hopes of gaining access to "cargo" were not fulfilled, the positive attitude turned into a negative one, often including aggression against the protagonists of the new belief. Sometimes the local people suspected that the missionaries had not given them the final, decisive secret words because they wanted to keep the goods for themselves. Often, the old religion was revived (*ribabal*, as it is called in Neomelanesian Pidgin/Tok Pisin) and sometimes very drastic steps were taken, such as destroying gardens and other vital assets in a desperate attempt to force the person's own ancestors to appear and deliver goods. It is, for the scientific mind, quite strange to accept that waves of these kinds of millenarian prophecies swept over the country many times. They still have not stopped. How could people believe that a new prophet would be more trustworthy than the old ones who had dramatically failed? But, we must not forget, the same happens in our own midst: The end of the world has been forecast over and over again, and still there are many people who believe the heralding of Jesus' imminent second arrival.

Melvin Lasky has described the world's big millenarian movements in the book *Utopia and Revolution* (Lasky 1976). His assessment, in my view, very well explains the universal and constant human longing for a better world, for better times. The Christian faith is one of the big millenarian beliefs and many Christians live in the expectation of the coming of Christ. Interestingly, this element of the new religion does not seem to have played a major role in the development of the utopian movements in Melanesia. I think it is more likely that such concepts appear in people's brains without necessary historic precursors or models. It is probably just one of the big dreams we all share: Things will get better.

It is quite possible that also the Eipo won't be immune to the obviously rather powerful lure of "cargo cult" ideas. Recently, 1 year, a new millenarian wave went eastward from the meanwhile quite developed Dani area in the Balim valley and Wamen as the central highland town—the massive influx of goods and much money around making it a classic starting point for such a utopian movement. It also reached the Eipomek valley (Heeschen, private communication). The direct western neighbors of the Eipo, the inhabitants of the Famek valley, connected a newly erected men's house by a long wire to one of the rocks in the vicinity of the village. The idea was that by performing the right ritual energy would be transmitted to the rock, which would then split open and give access to an enormous amount of money, which would spill out from this secret treasure chamber.

In Eipomek, those of our informants who had been the most intelligent in the early days of our contact in the 1970s prevented the "cargo cult" movement from playing any significant role in their villages. During my last visit there in 2008 I gained the firm impression that the Eipo have understood that success in life won't come through religiously induced miracles but through becoming educated and through hard work. It will be very interesting to monitor the situation and see whether this competitive, performance-oriented attitude will prevail or whether the local people will, at some point in time, be seduced by the promises of an imminent utopia that has caught countless Melanesian minds in the past and is still hanging over those small communities who have lost their religious roots and are trying to grow new ones.

References

Cannon WB (1942) Voodoo death. American Anthropologist 44:160–181

Carr D, Nesse R, Wortman C (2005) *Spousal Bereavement in Late Life.* Springer, New York, NY

Eibl-Eibesfeldt I (1989) *Human Ethology.* Aldine de Gruyter, Hawthorne, NY

Eibl-Eibesfeldt I, Schiefenhövel W, Heeschen V (1989) *Kommunikation bei den Eipo. Eine humanethologische Bestandsaufnahme.* Reimers, Berlin

Feuerbach LA (1854) *The Essence of Christianity.* Chapman, London

Fischer E (1993) *Warum ist das gerade mir passiert? Wie wir Krankheit deuten und bewältigen.* Herder, Freiburg

Hays T (1991) *Encyclopedia of World Cultures, Vol. II, Oceania.* Hall, Boston, MA

Heider K (1970) *The Dugum Dani: A Papuan Culture in the Highlands of West New Guinea.* Aldine, Chicago, IL

Heider KG (1991) Dani. In: Hays TE (ed) *Encyclopedia of World Cultures. Vol. II, Oceania.* Hall, Boston, MA

Hylkema S (1974) Mannen in het dragnet. Mens- en wereldbeeld van de Nalum (Sterrengebergte). Verhandelingen van het Koninklijk Instituut voor Taal-, Land- en Volkenkunde, no. 67

Jordan S (2002) *Neid. Ein interkultureller Vergleich einer vernachlässigten Emotion.* Diplomarbeit, Naturwissenschaftliche Fakultät der Universität Innsbruck

Koch KF (1974) *War and Peace in Jalémo.* Harvard University Press, Cambridge, MA

Lasky MJ (1976) *Utopia and Revolution.* University of Chicago Press, Chicago, IL

Lawrence P (1964) *Road Belong Cargo. A Study of the Cargo Movement in the Southern Madang District, New Guinea.* Melbourne University Press, Carlton, VIC

Lawrence P, Meggit MJ (1965) *Gods, Ghosts and Men in Melanesia. Some Religions of Australian New Guinea and the New Hebrides.* Oxford University Press, Melbourne

Lester D (1972) Voodoo death—some thoughts on an old phenomenon. American Anthropologist 74:386–390

Levi-Strauss C (1967) *Strukturale Anthropologie.* Suhrkamp, Frankfurt

Lovett R (1902) *James Chalmers. His Autobiography and Letters.* The Religious Tract Society, Oxford

Malinowski B (1973) *Magie, Wissenschaft und Religion.* Suhrkamp, Frankfurt

Schiefenhövel W (1976) Die Eipo-Leute des Berglands von Indonesisch-Neuguinea: Kurzer Überblick über den Lebensraum und seine Menschen. Einführung zu den Eipo-Filmen des Humanethologischen Filmarchivs der Max-Planck-Gesellschaft. Homo 26(4):263–275

Schiefenhövel W (1985) Sterben und Tod bei den Eipo im Hochland von West-Neuguinea. In: Sich D, Figge HH, Hinderling P (eds) *Sterben und Tod—eine kulturvergleichende Analyse.* Vieweg, Braunschweig

Schiefenhövel W (1986) Extraktionszauber. Domäne der Heilkundigen. In: Schiefenhövel W, Schuler J, Pöschl R (eds) *Traditionelle Heilkundige—Ärztliche Persönlichkeiten im Vergleich der Kulturen und medizinischen Systeme.* curare, Journal of Medical Anthropology and Transcultural Psychiatry, Special Volume 5, Vieweg, Braunschweig

Schiefenhövel W (1991) Eipo. In: Hays TE (ed) *Encyclopedia of World Cultures. Vol. II, Oceania.* Hall, Boston, MA

Schiefenhövel W (2001) Kampf, Krieg und Versöhnung bei den Eipo im Bergland von West-Neuguinea. Zur Evolutionsbiologie und Kulturanthropologie aggressiven Verhaltens. In: Fikentscher W (ed) *Begegnung und Konflikt—eine kulturanthropologische Bestandsaufnahme.* Bayerische Akademie der Wissenschaften, Philosophisch-Historische Klasse, Abhandlungen, Neue Folge, Heft 120. C.H. Beck, Munich

Schiefenhövel W (2009) Explaining the Inexplicable: Traditional and Syncretistic Religiosity in Melanesia. In: Voland E, Schiefenhövel W (eds) *The Biological Evolution of Religious Mind and Behavior.* Springer, Heidelberg

Schoeck H (1969, 2nd ed 1987) *Envy: A Theory of Social Behavior.* Harcourt, Brace & World, New York, NY

Stumpfe K-D (1973) *Der psychogene Tod.* Hippokrates, Stuttgart

Stumpfe K-D (1974) Der psychogene Tod in der Kriegsgefangenschaft und Maßnahmen zu seiner Verhütung und Therapie. Wehrmedizinische Monatszeitschrift 18:46–51

Stumpfe K-D (1976) Der psychogene Tod des Menschen als Folge eines Todeszaubers. Anthropos 71:525–532

Swain T, Trompf G (1995) *The Religions of Oceania.* Library of Religious Beliefs and Practices. Routledge, London

Voland E, Schiefenhövel W (2009) *The Biological Evolution of Religious Mind and Behavior.* Springer, Heidelberg

Zöllner S (1977) *Lebensbaum und Schweinekult. Die Religion der Jalî im Bergland von Irian Jaya (West-Neu-Guinea).* Theologischer Verlag R. Brockhaus, Witten

Illusion Number 6
We Are Free in What We Want

Chapter 16
Free Will

Insights from Neurobiology

Gerhard Roth

Abstract The traditional concept of free will is based on the assumption that humans are able to decide and to act by "immaterial" mental causation. Accordingly, they can decide and act differently under otherwise identical physical, physiological, and psychological conditions (alternativism). This alternativistic concept of free will constitutes—among others—the basis of the criminal justice system of the Western World and the justification of legal punishment. However, modern neuroscience and psychology can demonstrate that our intentions to act and to execute voluntary actions are guided by the interaction of conscious, preconscious, and fully unconscious motives, deriving from cognitive, executive, and emotional-limbic brain centers. During the entire process of the preparation and execution of voluntary actions, there is no "causal gap", in which an immaterial force could become determinative. Furthermore, the sensation of executing "freely willed" actions can be explained by experiments based on recording brain activity as well as electrical stimulation of certain cortical areas. Humans feel free, if—under the absence of external or internal constrictions—they "can do what they want" based on their conscious or unconscious motives, but they cannot form their will by a second-order will, as the philosophers Hume and Schopenhauer formulated long ago. This has strong implications, among others, for the criminal justice system.

16.1 Introduction

Since antiquity, there has been a dispute among theologians, philosophers, criminal law experts, and scientists about the question of whether human beings have freedom to will and to act. However, until recently, most experts in these fields considered and still consider this question as either obsolete or uninteresting, although for a variety of reasons.

G. Roth (✉)
Institut für Hirnforschung, Universität Bremen, 28359 Bremen, Germany
e-mail: Gerhard.roth@uni-bremen.de

U.J. Frey et al. (eds.), *Homo Novus – A Human Without Illusions*,
The Frontiers Collection, DOI 10.1007/978-3-642-12142-5_16,
© Springer-Verlag Berlin Heidelberg 2010

Among idealistic philosophers, there is the conviction, most prominently formulated by Immanuel Kant, that freedom of will is a phenomenon that transcends the "kingdom of nature" and is therefore located beyond the reach of empirical investigations. Inside the "kingdom of nature" only deterministic processes occur, in the sense of an endless chain of cause and effect. Freedom of will represents an exception by having its cause *only in itself*—a principle called "self-determination", "agent's autonomy", or "mental causation"—and not in preceding causes, be those physical or psychological such as pre-experience, character, or motivation. Accordingly, Kant considered it useless to aim for an empirical proof or disproof of freedom of will, that being a moral postulate and not a fact (Kant 1904, 1907).

In criminal law of the Western World, the concept of freedom of will is—as explicitly declared, among others, by the German Bundesgerichtshof (BGH) and the American Supreme Court—the theoretical basis of the concept of culpability, blame, and justification for criminal punishment. These institutions hold that individuals are personally responsible for their actions and have the ability to choose between good and evil. Accordingly, any mentally healthy person is able to decide between two alternatives, i.e., to commit the crime or to refrain from it—a principle called "alternativism". Within this context it appears irrelevant whether a free will *actually exists*. "The law treats man's conduct as autonomous and willed, not because it is, but because it is desirable to proceed as if it were" (Jones 2003). Accordingly, free will is considered a *necessary* or *desirable social fiction*, and without this social fiction criminal sanctions cannot be justified. Consequently, following the lines of Kant, any attempt to empirically test the existence of a free will seems irrelevant.

Another argument against empirically investigating the "real" existence of free will is popular among physicists and goes as follows. If we are not impressed by the above arguments and try to empirically study the psychological and neurobiological conditions of voluntary actions, we are confronted with an irreducible difficulty resulting from the mentioned principle of "alternativism", i.e., the ability to decide or act otherwise under *identical* physical, physiological and psychological conditions. In order to prove the existence or nonexistence of such mental causation, we would need to test a person repeatedly under identical physical, physiological, and psychological conditions. This—physicists argue—is impossible ("you cannot enter the same river twice!"), and therefore any empirical test in favor or against the existence of free will in the classical, alternativistic sense is impossible from first principles.

A very popular argument among philosophers and social scientists in favor of free will points to the strong subjective feeling of being free in most of my decisions. Indeed, in many cases of decision making I have the undeniable feeling that (1) I, as a conscious and reflective being, am the only instance controlling my willful actions; (2) I could decide and act *otherwise*, if I only wanted it. This strong feeling—so goes the argument of many philosophers—cannot be an illusion, and any empirical test of the veridicality of that feeling is unnecessary and superfluous.

Let us now briefly discuss these—admittedly heterogeneous—arguments against the possibility of empirically testing the existence of a free will. The first argument seems irrefutable, because the phenomenon of free will appears to escape empirical proof *by definition*. However, it includes an error in reasoning. Even an idealistic defender of freedom of will must admit that the *consequences* of an *immaterial* "free-willed" decision must become visible at one point or another of the sequence of the *material* processes in our body or brain. Whenever an individual threatens a potential victim with a revolver to kill him, then certain muscles of his or her index finger must be contracted in order to pull the trigger of the revolver. This muscle contraction must be caused by the activity of specific motor neurons in a segment of the spinal cord of the murderer. The activity of spinal motor centers must be caused by a number of brain centers located in the cerebral cortex, the basal ganglia, and the cerebellum, which interact in a characteristic manner as will be described further below. These higher motor activities, in turn, are induced by activities at still higher motor and executive levels involving the activity of the prefrontal and parietal cortex, which themselves interact with emotional and motivational centers in the cortical and subcortical limbic system. Thus, with much experimental effort and some luck, we will be able to trace the finger movement back to brain regions where motives, intentions, and desires arise as the ultimate causes of behavior.

The decisive question now is whether, in the detailed analysis of such processes, we are confronted with a closed chain of neuronal-psychological events, or whether we discover a *causal gap*, in the sense that there is a process that *in no way* can be linked causally or at least with high statistical probability to preceding processes. If such a causal gap exists, it could be identical with the site and moment of intrusion of mental causation acting upon brain processes, and this event would then give the entire process of action preparation a direction not explainable in neurobiological terms. However, if we are *unable* to discover such a gap and instead could reconstruct a chain of neurobiological processes that eventually lead to the index finger movement with certainty or at least with high probability, then neither free will nor mental causation exist, or if they do, they are irrelevant to the explanation of the observed action.

We thus come to the conclusion that although free will as a possibly nonnatural or metaphysical entity might be unobservable per se, its impact on our actions must be observable and thus testable at one point from the first intentions to act to the finger movement, otherwise it would play no role in our material life.

The second argument pretending that freedom of will is a pure normative legal construct and therefore immune to any empirical disproof, is a weak argument. In a famous statement from 1952, the German BGH did *not* say that human beings *should be treated* as being free and responsible and having moral self-determination—this would be the correct formulation of a normative construct—but that they *are* free by their very nature. This is a claim of fact, not a norm. In court, an affirmation of fact needs to be empirically proven. In a criminal case the court is obliged to demonstrate with sufficient probability that the offender has indeed committed the crime, otherwise the court must declare him or her not guilty. In the same sense, reasons for the exclusion or reduction of guilt as listed in paragraphs 20 and 21 of the German Criminal Law (StGB) have to be empirically proven by experts (e.g., forensic psychiatrists), and the court cannot overrule such expert statements without necessity.

The third argument dealing with the principle of nonrepeatability of once-happened situations is, of course, irrefutable, but at the same time easy to circumvent. To prove the existence or nonexistence of freedom of will, it is unnecessary to make tests under purely identical conditions. Rather, it is sufficient to demonstrate that under similar physical, physiological, or psychological conditions people decide and act similarly; the rest might be due to the complexity of processes investigated, noise, or irreducible variability of the methods applied. Eighty percent prediction validity would suffice. Weather forecasts are often much worse in their predictive power, but nobody doubts that weather is a deterministic-chaotic phenomenon that can be described in terms of natural science.

Let us now turn to the fourth argument, which states that our strong subjective feeling of being able to decide freely (at least in many cases) cannot be an illusion. It is the weakest of all arguments. From the sensation of how things are or happen in a certain way it does *not* follow that this is *really* the case. This holds for the impression that the sun "rises" in the morning as well as for many perceptual illusions, some of which are so strong that we would bet our bottom dollar on it. Another important fact is that in most cases of daily decisions, viz. all automated ones, we do not execute any act of will, but just do them, while we nevertheless attribute them to us. Finally, there are many reports that individuals are forced by hypnosis or brain stimulation to execute certain movements that they afterward consider to be "willed" (Wegner 2002). I will come back to these phenomena again.

In summary, all four arguments against the possibility of empirically proving the existence of freedom of will can be refuted or circumvented. Regardless of the "true" nature of freedom of will, such a capacity must be manifest in our material world, and this can be tested. Let us now have a look at the present state of knowledge of neuroscience and psychology regarding the control and subjective experience of so-called free-willed actions.

16.2 The Experimental Proof of Freedom of Will

16.2.1 The Neurobiology of Voluntary Actions

In order to understand the various ways of empirically and experimentally testing the existence or nonexistence of freedom of will, we have to briefly consider the processes that take place in our brain during preparation and execution of *voluntary actions*, i.e., of those movements which—as opposed to pure reflexes—can be executed in a flexible way and which we perceive as being caused *by ourselves* (Passingham 1993).

According to present knowledge all such voluntary actions involve the activity of the following cortical areas: (1) primary motor cortex (MC) responsible for detailed control of (mostly distal) muscles, (2) lateral premotor cortex (PMC) and medial supplementary motor cortex (SMA), which have to do with more global guidance of bodily actions, and (3) pre-supplementary motor cortex (pre-SMA), which is active during the preparation and planning of movements and even during imagined movements. Pre-SMA also contributes to the *awareness* of being the author of one's own deeds (Haggard 2005, 2009; Haggard et al. 2002). Accordingly, it was believed by some philosopher-neuroscientists to be the seat of "free will" (Eccles 1982). Accordingly, pre-SMA is active mostly or only for self-paced actions, while PMC and SMA are also active for externally induced or triggered actions; the MC, however, is active even during unconscious and highly automated movements. These cortical motor, premotor, and supplementary motor areas interact with the posterior parietal cortex, which is involved in planning and preparing intentional movements (e.g., goal-directed eye and arm–hand–finger movements) and with the prefrontal, including frontopolar, cortex, which is responsible for the deliberation of intentions to act, their alternatives and consequences, as well as the maintenance or "shielding" of such intentions during the execution of actions (Jeannerod 1997, 2003).

These motor and executive cortical areas send massive fiber pathways either directly (via the cortico-spinal or pyramidal tract) or indirectly (via relay nuclei in the brainstem) to those motor segments that are responsible for the final activation of specific muscles. Thus, conscious action planning arises in the prefrontal-frontopolar and posterior parietal cortices (origin of desires and intentions to act as well as the sequence of single motor acts). This activity then stimulates the pre-SMA, then SMA and PMC and MC, before activity runs "down" via pyramidal and extrapyramidal pathways to the motor segments of the spinal cord, which eventually start the intended movement.

So far, we have not mentioned a brain center indispensable for the execution of willed actions, viz. the *basal ganglia*, consisting of the putamen, nucleus caudatus and globus pallidus, together forming the striato-pallidum of the telencephalon, the nucleus subthalamicus situated in the diencephalon, and the substantia nigra situated in the midbrain tegmentum (Nieuwenhuys et al. 1988; Graybiel et al. 1994). These parts are closely connected, partly via thalamic relay nuclei, with the prefrontal, premotor, and parietal cortices as well as with the entire limbic system, predominantly the amygdala and the mesolimbic system (ventral tegmental area, nucleus accumbens). The basal ganglia are the site of our procedural or action memory in the sense that all voluntary movements that have been repeatedly and successfully executed are stored here (McHaffie et al. 2005; Grillner et al. 2005). Before a new voluntary movement is executed, it must be sent by the cortex to the basal ganglia for compliance with this memory. Likewise, the basal ganglia are the seat of all stereotyped actions and of habits, which typically can be executed without conscious control. The basal ganglia are under strong control of the amygdala and the mesolimbic system and in this manner by the *unconscious emotional memory*.

The entire process of activation mediated by the basal ganglia in the context of planning and executing actions is characterized by a complex interaction between excitatory and inhibitory centers and pathways, in which, however, inhibition predominates. In order to execute a certain movement, it is essential to dis-inhibit one specific sequence of muscle activity stored in the basal ganglia and to strongly suppress all alternatives. Otherwise, these alternatives would severely disrupt the execution of the intended action, as happens in certain motor diseases such as

Huntington's chorea. The selective dis-inhibition of just one motor sequence is bound to the release of the neuromodulator *dopamine*. Dopamine is produced by neurons located in the substantia nigra pars compacta and transported via a fiber tract to the corpus striatum, where it exerts both an excitatory and inhibitory effect on neurons projecting to the globus pallidus and the substantia nigra pars reticulata, depending on the nature of striatal dopamine receptors (D1, D2). This eventually leads to the *suppression* of unwanted and the *release* of wanted movements, and the excitatory activity related to the wanted movement is then sent, via thalamic relay nuclei (ventrolateral and ventral posterior nuclei), to the cortical motor, premotor, and executive areas mentioned above. The volleys of excitation from the thalamic relay nuclei lead, in combination with activity from the prefrontal and parietal cortex, to the formation of the so-called readiness potential (RP, "Bereitschaftspotenzial"; Kornhuber and Deecke 1965; Brunia and van Boxtel 2000), which will be explained further below. Without the dopaminergic release–suppression mechanism the cortex by itself cannot release or guide voluntary movements.

The question now is, *who* or *what* controls the release of dopamine via the neuronal pathway from the substantia nigra pars compacta to the striatum? The answer is that this control is exerted primarily by three important centers of the limbic system already mentioned, viz., the amygdala, the mesolimbic system and the hippocampus. The *amygdala* is the most important center for inborn affects and emotions as well as emotional conditioning and recognition of emotional communicative signals such as glance, facial expressions, gestures, and voices. It identifies and stores the positive or negative consequences of events in the environment and of the organism's own actions (LeDoux 2000). The *mesolimbic system* is related to the registration of reward and to reward expectation (Schultz 1998). Amygdala and mesolimbic system together form the emotional memory. Finally, the *hippocampus* is relevant for the formation and organization of spatial and context memory and in humans of declarative-episodic (including autobiographic) memory. The hippocampus contributes context information to the affective-emotional memories stored by the amygdala and mesolimbic system (Schacter 1996). If now certain actions are consciously planned by the prefrontal and parietal cortex, these intentions run from these areas to the emotional memory system of the amygdala, mesolimbic system, and hippocampus; the pre-experience available regarding the intended act is checked. If such pre-experience is positive, then the emotional memory plus context system sends a "go" signal to the substantia nigra pars compacta to release dopamine to the striatum, leading to a selective dis-inhibition of the appropriate sequence of movements.

This final "check" occurs a few seconds before the onset of an intended movement and consists of several "loops" of activity between cortex, basal ganglia, limbic system, and thalamic relay nuclei. During that time the readiness potential starts building up. The RP is formed by an increase in synchronous activity of specific neuronal assemblies, first of cortical pre-SMA, then of SMA proper and PMC, and finally of MC motor neurons, until the assembly passes a level of activity necessary for the release of the movement. This activity is then sent via the pyramidal and extrapyramidal pathways to the spinal cord and from there to the appropriate muscles, and the intended action is executed (Brunia and van Boxtel 2000).

Importantly, the RP is a combination of activity arriving from the prefrontal and posterior parietal cortex, representing the intention to act on the one hand, and from the basal ganglia (via thalamus), representing the "vote" of the limbic system and of the basal ganglia on the other. This interaction, however, is asymmetrical. The basal ganglia plus motor cortex and cerebellum can start highly automated actions without the contribution of the prefrontal and posterior parietal cortex (we just do things), but the opposite is not the case, i.e., without the contribution of the basal ganglia, willed action cannot be started by the mere activity of prefrontal and posterior parietal cortex, as can be seen in recordings of RPs from brains of patients suffering from Parkinson's disease (Brunia and van Boxtel 2000). Here, the activity coming from the prefrontal and posterior parietal cortex are present, but the component arriving from the basal ganglia is missing, because these patients suffer from an insufficient production and release of dopamine from the substantia nigra to the striatum, and therefore no "go" signal for the striato-pallidum occurs. As a consequence, the purely prefrontal-parietal BP cannot pass the critical level necessary for starting the intended movement, even though the patient strongly wanted it.

The RP begins to build up 1–2 s before the onset of movement and occurs first symmetrically, i.e., in both hemispheres of the cortex in the pre-SMA, SMA, and parietal regions (called *symmetrical* RP), and then shifts to the hemisphere contralateral to the side of the body parts to be moved (called *lateralized* RP), because it is (almost always) the contralateral premotor and motor cortex that controls body and limb movements. As soon as the lateralized RP occurs, an intended movement cannot be stopped any more voluntarily. This means that at least at the moment of the lateralized RP, the intended movement is fully determined.

16.2.2 Libet's Experiment and Its Consequences

The RP was discovered in the 1960s by the two German-Austrian neurologists Kornhuber and Deecke (1965). In the early 1980s, the American neurobiologist Benjamin Libet, in a series of ingenious experiments, made use of the fact that the RP (at least the lateralized part) reliably predicts the execution of an intended movement. Libet's idea was the following: Assuming that the RP precedes the onset of a willed action by 1–2 s and reflects brain processes necessary for the release of the intended act, then we should be able to determine whether the act of will *precedes* (as an idealist/dualist would expect) or *follows* (as a materialist/physicalist would expect) the onset of the RP. If the former turned out to be the case, then—in combination with a visible "causal gap" as defined above—this would strongly indicate that the act of will caused the RP and with this the intended action in an idealistic sense. If, however, the act of will *followed* the onset of the RP and no "causal gap" was detectable, then the role of the act of will as a "mental cause" of the RP and resulting action is unlikely.

The RP can be filtered out from the ongoing electroencephalogram (EEG). What is needed, however, is a way to determine the moment of the act of will with sufficient precision. Libet solved this problem in the way that during recording the RP subjects watched a clock hand rotating on a screen with a period of roughly 3 s. Subjects were asked to decide to move one hand at a freely chosen moment and to remember the position of the rotating clock hand exactly at the moment of decision. This remembered and reported moment was then compared with the moment of RP onset (Libet et al. 1983; Libet 1990).

Libet, an adherent to the concept of free will, was very surprised by the results of his own experiments, which showed that the moment of "free" decision on average came about half a second *after* the onset of RP. This would mean that—at least under these experimental conditions—in most cases the act of will could not have caused the onset of the RP, and Libet admitted that in these cases the brain had already and unconsciously decided to start the hand movement before the subject came to know that.

Because of several insufficiencies of Libet's experimental design, the British-German psychologists Patrick Haggard and Martin Eimer repeated these experiments under improved conditions (Haggard and Eimer 1999). They concentrated on the onset of the lateralized RP, which, compared to the symmetrical RP, more precisely predicts a given movement, and they introduced a "choice" version, i.e., subjects could not only decide on the moment of hand movement (here pressing a key), but also whether this was done with the left or right hand. By applying both versions, they confirmed the findings of Libet. In the meantime, a number of other "Libet-like" experiments have been carried out with further improvements of RP registration method or with functional magnetic resonance imaging (fMRI). Recently, Haynes from Humboldt University and colleagues (Soon et al. 2008) were able to demonstrate that unconscious activity in the prefrontal-frontopolar cortex and in the posterior cingulate cortex predicts with high probability a certain decision or intention to act, and this prediction was possible up to 10 s before the onset of the movement.

Both kinds of experiments—those based on EEG-RP onset and the fMRI experiment on frontopolar and cingulate activity—confirm the view that willed actions are preceded and prepared by conscious as well as unconscious brain processes as a combination of prefrontal and posterior parietal activity and of activity coming from the basal ganglia via thalamus plus activity from limbic centers, which together give rise to the RP, as described above. During this sequence of brain processes no "causal gap" is visible; rather—based on the registration of prefrontal-parietal plus basal ganglia activity—neuroscientists are able to predict with the high reliability of 70–80% a certain decision or act of will. This probability is high considering the complexity of brain events, and in simpler cases (e.g., face recognition) predictions with up to 100% validity based on unconscious neuronal activity preceding the conscious recognition statement can be reached (JD Haynes, personal communication). If instead of neural activity an immaterial act of will were decisive, then the predictive power of the knowledge of neuronal activity would be close to chance (50% for a dual decision).

However, the statement often ascribed to neuroscientists and psychologists that "the brain has already decided before our conscious self becomes aware of that decision" needs some clarification, because in a narrow sense it is true only in cases of short-range and pre-programmed movements. In cases of longer-range decisions with extended periods of deliberation the conscious self, of course, participates in the process of decision, and the process may take a completely different direction depending on the outcomes of deliberation, e.g., when we suddenly become aware of highly unwanted, but hitherto unreflected consequences of the intended action. At the same time, unconsciously or pre-consciously working brain centers (amygdala, mesolimbic system, hippocampus) determine *which* desires, intentions, and thoughts will "pop up" during the period of deliberation, and sometimes we bitterly regret that some negative consequences of our actions just did not come to our mind before our decision. Thus, conscious acts of will contribute *only* to our decisions as long as they are "permitted" or supported by our unconscious or preconscious brain. Most importantly, conscious acts of will are instantiated by neuronal processes in the same way as unconscious and preconscious preparations of acts—there is absolutely no sign of a "purely mental" causation of actions.

16.3 Determinism of Motives

Interestingly, the traditional concept of freedom of will, as is at the heart of present criminal law, can be disproven without any help from neuroscience. In order to do so, we start with the assumption put forward by defenders of the concept of free will that our free-willed decisions must be "reasonable", i.e. founded on moral or logical rules or acceptable motives. In contrast, actions for which neither the actor nor a psychological or psychiatric expert can find any such motives, will be regarded as "random" or "irrational" rather than "free". The philosopher Kant accepted only the universally given and inborn "moral law" as the basis of free-willed decisions, which is independent of any other instance—even God's will. For Kant, nobody can do anything truly "good", even if guided by any motives, be they very positive such as love or friendship. Rather, the good must be done for the sake of itself, because otherwise moral would be based on egoistic motives (at least in the long run), which for Kant would be a contradiction in itself.

Today's psychologists and anthropologists would strongly object to such a concept of pure morality, because in their (and most other present-day scientists') eyes moral rules and conscience do not come from nowhere or are "inborn" (as Kant believed), but depend on culture, education, and personal experience. Many ways of behaving represent absolute moral norms at some times and/or in some countries, while at other times and/or in other countries they appear morally inacceptable or at least neutral. Accordingly, any given norm knows some exceptions (even killing one's own father or mother!).

We thus arrive at the concept of *motive determinism*: with the exception of random events in our minds, brains, and bodies (the strength and effect of which on human behavior is unknown) our decisions and actions are guided by *motives*. These may be genetically fixed; result from prenatal, early postnatal, or later personal experience; from trial-and-error learning, education, insight, reasoning, imitation, etc.; and can influence our behavior consciously, preconsciously, or unconsciously. They are stored in our declarative, emotional, and procedural memories and are recalled in a specific way in certain circumstances that require a decision. Usually, some motives promote each other, while others compete. What we will eventually do depends on the outcome of the sometimes simple and sometimes extremely complex interaction of positive or negative motives, and in most cases this interaction will result in *one dominant motive* at a given moment. This dominant motive then forms our will, and a special "volitional" mechanism will shield this motive against competing motives, until an intended action has started and eventually executed.

Freedom of will has no place in such a concept. If actions are guided by motives on closer inspection, then they cannot be "free" in Kant's sense of pure morality. If they are not guided by motives, then they must appear "un-motivated" and random events. This holds for normal people as well as for criminals.

16.4 How Does the Feeling of "Freely" Deciding Arise?

The dominant condition for feeling "free" is the *absence of external or internal constrictions*. An external constriction can be a revolver in front of my chest; an internal force can consist in the compulsion to wash my hands every few minutes or an intolerable fear of something. If no such external or internal forces exist, I am physically, physiologically, or psychologically free to do or not do certain things or to choose between alternatives. Let us assume that right now my desire or will is to drink a cup of coffee, and this desire will urge me to reach for the cup of coffee in front of me. As long as I am not extremely thirsty or coffee-addicted, I could refrain from doing so, e.g., if another person likewise wants to drink that cup of coffee. The same holds for the situation that both a cup of coffee and a cup of tea stand in front of me. I have to decide which one I will drink, and will take the tea, because at that time in the afternoon I am used to drinking tea instead of coffee. If somebody asks me whether I consider my decisions as being "free", I would answer "of course!", because I could have acted otherwise, of only if *I had wanted otherwise*.

The main reason for my feeling free is that I had *really existing alternatives*, i.e., to drink or not to drink, if there was only one kind of beverage in front of me, or to drink coffee or tea, if both were available. It is the existence of physical (a beverage really exists in front of me), physiological (my arms are not too tired), and psychological (I have the will to drink) alternatives that counts for the feeling of freedom of choice. I could act otherwise, if these conditions, especially my desires, were different, and in the same way I can *imagine* afterward having acted otherwise, if the conditions had been different. This situation is fully compatible with the fact that my choice is determined by my motives. People feel free, if they have a choice, even if this choice is highly theoretical. I can, for example, *imagine* giving up my present job, which I love, while everybody, including me, knows (based on the knowledge of my character) that at least for the next few years this will be highly unlikely. What is important is that my motives are *not so strong* that in my self-awareness they overrule any alternative.

I have just described the conditions for feeling free, i.e., the existence of "true" alternatives. Which mechanisms underlie the feeling of being the author of one's own deeds? Recent studies demonstrate that the sensation of having *willed* a certain action or *attributing it to myself* does depend on a prediction of the specific pattern of sensory re-afferents that will appear after a movement has been executed as planned. This prediction or "forward model", which is probably identical with the *efference copy* concept of Helmholtz, includes a *prediction* about the information from the muscles, skin, joints, and possibly the visual aspect of the intended movement (Blakemore et al. 2002; Haggard et al. 2002). During and after the execution of the movement, it is compared with the actual sensory information. If both sets of information are more or less identical, then we have the feeling that "it was *me* who determined that action". If the two sets, however, do *not* coincide, then the feeling arises that an external force was responsible for that discrepancy, even if we do not sense the presence of such an external force. If in rare cases the sensory re-afferent pathways are lesioned by a disease, a cut, or by anesthetizing the re-afferent nerves, then the subjects or patients report that an "alien force" was moving their hand, or that their hand did not belong to them anymore.

This sensation is easy to understand, because our actual body scheme depends on the continuous comparison of information of the intended with that of the actually executed movements and body posture. Every part of the body that "obeys" the conscious or unconscious motor commands of our nervous system is regarded as belonging "to us" and will be *self-attributed*, and this is even the case with prostheses and a blind person's stick (which after long training is felt like a part of the body); every part of the body that does not "behave" according to the forward model is excluded from the body scheme and considered "alien". This sensation is strikingly immune to knowledge: a patient will continue to claim that this is not his hand or that it is moved by an "alien force" even if we carefully explain the causative lesion or neurological deficit. This is not different from the fact that optical illusions do not disappear even when we understand how they arise.

The idea that self-attribution of a certain body movement is strictly connected to the formation of a sensorial forward model or "efference copy" arising in the posterior parietal cortex (PPC) is confirmed by older as well as very recent brain stimulation experiments, where the skull had been opened (cf. Wegner 2002). By stimulation of certain areas limb movements are elicited, of which the patients sometimes report they had felt an "urge" or "will" to execute such a movement. Recently, a French group (Desmurget et al. 2009) published results from such direct stimulation of the cortex in patients prior to tumor surgery. Interestingly, the authors found remarkable differences in the consequences of stimulation of the SMA/pre-SMA on the one hand and the PPC on the other. When SMA/pre-SMA was electrically stimulated with increasing strength, the urge or will of the patients to execute a certain movement (e.g., raise the left arm) became likewise stronger and stronger, until they actually moved the limb. In contrast, when the PPC was stimulated with increasing strength, the urge or will to move the limb likewise increased, but the actual movement never occurred; rather, at very high stimulation strengths the patients had the illusionary impression of having executed the movement. For example they said "I just said something—what was it?" If instead of SMA/pre-SMA or PPC the primary or secondary motor cortex was stimulated, then the patients reported that something had *forced* them to move the limb.

Recently, Patrick Haggard (Haggard 2009) interpreted these findings of the French group as well as earlier findings of the same kind in the way that the (natural or electrically induced) activity of pre-SMA represents the immediate "act of will", while the activity of PPC corresponds to the intention to act and the formation of the "efference copy". This would explain why a very strong stimulation of PPC leads to the illusion of actually having executed the movement—the strong prediction of a movement replaces the actual execution. In contrast, strong stimulation of pre-SMA leads to the actual execution, because pre-SMA is strongly linked to SMA, premotor cortex, and motor cortex, from where the pyramidal tract descends to the brainstem and spinal cord. In the case that only the primary motor cortex was stimulated, neither pre-SMA nor PPC activity preceded this activity, and the patient's brain, very logically, concludes from this fact that the movement was not "willed".

All this indicates that the sensation "it is me who executes the movement" more strongly depends on the intensity of the *intention* to act rather than on the actual execution. Accordingly, in our daily life we sometimes ascribe actions to ourselves that we had strongly intended but forgotten to execute, because our attention was distracted, and on the other hand we can do things that are classical "voluntary actions" without an explicit act of will but with self-attribution, provided these are highly automated. Here, the efference copy is probably highly automated, too, leading to an "automatic" self-attribution.

16.5 What Follows from All This for the Concept of Free Will?

A skeptical philosopher could argue that all these empirical findings are very interesting but completely irrelevant for our daily life. In order to have the sensation of feeling free in our decisions and actions, it is sufficient for us to have—besides an intact brain and nervous system—the absence of external and internal forces and the "possibility to do what we want". The will that drives my actions is *my* will, i.e., results from *my* motives, and as long as I follow my own will in the presence of real or imagined alternatives, I feel free. This shows that motive determinism and the sensation of freedom of deciding and of acting do not contradict each other. As the eminent British philosopher David Hume said, all this is possible only because we subjectively do not have insight into the conditions of our will; we simply take our will as given. Only if we start with reflections based on an idealistic-dualistic concept of freedom of will in philosophy or in the theory of criminal law do the above discussed findings become relevant.

Then we learn from them that neither a self-determined will nor a hierarchy of conscious wills is logically and empirically possible. Consciously and subjectively, our will is something *immediately given*, while objectively (from the perspective of a psychologist or neurobiologist) it is *caused by motives*. In terms of criminal law this means that a delinquent, as well as any of us, is determined by his or her motives, be they conscious, preconscious, or unconscious, or more precisely by that specific motive that becomes dominant at the very moment of the criminal act. In all cases where the defendant admits that he or she was fully aware of his or her own actions (e.g. by carefully preparing or hiding them), and that he or she would have had alternatives to act or not to act, to shoot or not to shoot, it can be said that he or she was *only subjectively free* in the sense of true or imagined alternatives. There might have been a chance that during the last minute before committing the crime, the defendant could have become aware of the risks or consequences of the crime, or some other obstacle might have occurred, and he or she could suddenly have refrained from killing the victim. But whether such ideas or obstacles pop up is not under the conscious control of the defendant.

Thus, if we stick to the idea that humans are able to willingly influence their own will as a basis in classical criminal law, we are asking the delinquent to do something logically and empirically impossible. The only solution to this dilemma lies in the abandonment of the entire concept of "immaterial" free will. It has been demonstrated by a number of experts of criminal law, including Claus Roxin from Munich University, that this is possible without a complete reorganization of the criminal justice system (Roxin 2006). Grischa Merkel from the University of Rostock and I have expanded these ideas in the sense that society defines which actions are considered tolerable and which are punishable (Merkel and Roth 2008). This is done in the framework of defending the criminal law system by sanctioning delinquents and thus deterring potential wrongdoers from breaking the law, but with a much stronger emphasis on the rehabilitation of the criminals.

Abandoning the concept of moral guilt removes an important and fatal aspect of punishment, viz. the desire of individuals and society for retribution and revenge. In Kant's eyes, the idea of retribution (the *talion principle*: an eye for an eye) was essential, and punishment had to be painful. However, as every psychologist and every expert of the penal system knows, punishment is a rather ineffective means of meliorating a criminal character, because it mostly evokes a strong desire for revenge in the punished individual. It cannot be denied that deterrence has a certain effect on the prevention of norm breaking, but this effect is the weaker the more severe the criminal act (cf. the discussion about the death penalty). Even more regrettably, limiting punishment to the role of revenge prevents the criminal law system from concentrating on therapeutic effects: the criminal is being punished only because he or she "deserves" it. This contradicts not only human rights, but in the long run also economic reasoning because it is an extremely expensive way in contrast to means of re-education and therapy.

It cannot be denied that sanctions are necessary to enforce the law, but to have any effect in the future there needs to be a minimum of cooperation of the delinquent. Importantly, however, cooperation in the form of therapy cannot be extorted, therefore the delinquent must play an active role in the process of protecting the norm by choosing between therapy and traditional sanctions—after careful consideration of safety interests.

Acknowledgments I thank Dr. Grischa Merkel, Rostock, and Dr. Mattias von der Tann, London, for criticism and helpful comments.

References

Blakemore S-J, Wolpert DM, Frith CD (2002) Abnormalities in the awareness of action. Trends in Cognitive Sciences 6:237–242

Brunia CHM, van Boxtel GJM (2000) Motor preparation. In: Cacioppo JT, Tassinary LG, Berntson GG (eds) *Handbook of Psychophysiology*, 2nd ed. Cambridge University Press, Cambridge

Desmurget M, Reilly KT, Richard N, Szathmari A, Mottolese C, Sirigu A (2009) Movement intention after parietal cortex stimulation in humans. Science 324:1–813

Eccles JC (1982) The initiation of voluntary movements by the supplementary motor area. European Archives of Psychiatry and Clinical Neuroscience 231:423–441

Graybiel AM, Aosaki T, Flaherty AW, Kimura M (1994) The basal ganglia and adaptive motor control. Science 265:1826–1831

Grillner S, Hellgren J, Ménard A, Saitoh K, Wikström MA (2005) Mechanisms for selection of basic motor programs—roles for the striatum and pallidum. Trends in Neurosciences 28: 364–370

Haggard P (2005) Conscious intention and motor cognition. Trends in Cognitive Sciences 9: 290–295

Haggard P (2009) The sources of human volition. Science 324:731–733

Haggard P, Eimer M (1999) On the relation between brain potentials and the awareness of voluntary movements. Experimental Brain Research 126:128–133

Haggard P, Clark S, Kalogeras J (2002) Voluntary action and conscious awareness. Nature Neuroscience 4:382–385

Jeannerod M (1997) *The Cognitive Neuroscience of Action*. Blackwell, Oxford

Jeannerod M (2003) Self-generated actions. In: Maasen S, Prinz W, Roth G (eds) *Voluntary Action*. Oxford University Press, New York, NY

Jones M (2003) Overcoming the myth of free will in criminal law: the true impact of the genetic revolution. Duke Law Journal 52:1031–1053

Kant I (1904) *Kritik der reinen Vernunft*. Akademie Ausgabe III, Reimer, Berlin

Kant I (1907) *Metaphysik der Sitten/Rechtslehre*. Akademie Ausgabe VI, Reimer, Berlin

Kornhuber HH, Deecke L (1965) Hirnpotentialänderungen bei Willkürbewegungen und passiven Bewegungen des Menschen: Bereitschaftspotential und reafferente Potentiale. Pflügers Archiv European Journal of Physiology 284:1–17

LeDoux J (2000) Emotion circuits in the brain. Annual Review of Neuroscience 23:55–184

Libet B (1990) Cerebral processes that distinguish conscious experience from unconscious mental functions. In: Eccles JC, Creutzfeldt OD (eds) *The Principles of Design and Operation of the Brain*. Pontificae Academiae Scientiarum Scripta Varia 78:185–202

Libet B, Gleason CA, Wright EW, Pearl DK (1983) Time of conscious intention to act in relation to onset of cerebral activity (readiness-potential). Brain 106:623–642

McHaffie JG, Stanford TR, Stein BE, Véronique Coizet V, Redgrave P (2005) Subcortical loops through the basal ganglia. Trends in Neurosciences 28:401–407

Merkel G, Roth G (2008) Freiheitsgefühl, Schuld und Strafe. In: Grün KJ, Friedman M, Roth G (eds) *Entmoralisierung des Rechts*. Vandenhoeck & Ruprecht, Göttingen

Nieuwenhuys R, Voogd J, van Huijzen C (1988) *The Human Central Nervous System*. Springer, Heidelberg

Passingham R (1993) *The Frontal Lobes and Voluntary Action*. Oxford University Press, Oxford

Roxin C (2006) *Strafrecht Allgemeiner Teil, Band I: Grundlagen Aufbau der Verbrechenslehre*, 4th ed. C.H. Beck, Munich

Schacter DL (1996) *Searching for Memory. The Brain, the Mind, and the Past.* Basic Books, New York, NY

Schultz W (1998) Predictive reward signals of dopamine neurons. Journal of Neurophysiology 80:1–27

Soon CS, Brass M, Heinze H-J, Haynes J-D (2008) Unconscious determinants of free decisions in the human brain. Nature Neurosciences 11:543–545

Wegner D (2002) *The Illusion of Conscious Will.* Bradford Books, MIT Press, Cambridge, MA

Chapter 17
Could I Have Done Otherwise?

Naturalism, Free Will, and Responsibility

Gerhard Vollmer

Abstract Naturalism is a universal philosophical position between natural philosophy, epistemology, and anthropology. Its main characteristics are its claim to universality and the limitation of means for the description and explanation of the world. Naturalists try to expose and to dispose of illusions. They have confidence in the findings of science while avoiding dogmatism. Since free will in the traditional libertarian sense (I, you, he, she could have done otherwise) contradicts both classical and modern physics—and for other good reasons—naturalists will deny free will in this very sense. How then should we understand our talk about free will and about responsibility, and how should we understand our praxis of punishment? If we want to keep these concepts, we must explicate them with more sophistication. Our concept of free will must be compatibilist, i.e., compatible with determinism. From responsibility we must subtract the element of free will altogether or—if we keep it—say how we use it. From the functions of punishment we must remove the elements of guilt and reproach, of expiation and retaliation. If we carefully do so we may indeed keep our praxis of talking about free will, of attributing responsibility, and of imposing punishment.

17.1 What Do We Want to Mean by Naturalism?

Naturalism is a universal philosophical position between natural philosophy, epistemology, and anthropology. Its main characteristics are its *claim to universality* and the *limitation of means* for the description and explanation of the world.

The *claim to universality* is essential. Even Immanuel Kant would call himself a naturalist, but "of a special kind". He demands that everything in science should be formulated and explained *naturally*—and not in theological language (Kant 1788, A

G. Vollmer (✉)
Seminar für Philosophie, Technische Universität zu Braunschweig,
38106 Braunschweig, Germany
e-mail: g.vollmer@tu-bs.de

U.J. Frey et al. (eds.), *Homo Novus – A Human Without Illusions*,
The Frontiers Collection, DOI 10.1007/978-3-642-12142-5_17,
© Springer-Verlag Berlin Heidelberg 2010

126f). This sounds quite naturalistic. But at the same time he sets a *limit* to science, mainly to the explanation by causes: In order to explain organized systems, especially to explain the *expediency* of organismic structures, *teleological* explanations must be called upon; a Newton of the grass blade is simply impossible. Kant is a naturalist only with regard to physics, not with regard to biology, much less with regard to psychology, epistemology, or ethics.

In this respect modern naturalism is more ambitious: It dispenses with teleology altogether. The undeniable expediency of organismic structures is explained by natural selection, hence ultimately by a *causally* effective principle. Thus a grass-blade Newton is possible; whether Charles Darwin is already the complete Newton or whether Gregor Mendel, Ronald Fisher, Julian Huxley, Ernst Mayr, and Manfred Eigen should be added is a question of the history of science and therefore—with regard to our problem—a minor one. What is crucial is that with Darwin's selection theory all living beings and therefore biology as a whole are included in the framework of naturalistic explanations, such that teleological explanations become dispensable and Thomas Aquinas's or William Paley's teleological proofs for the existence of God lose their persuasive power.

The renunciation of teleology is itself a typical example of the second characteristic of naturalism—the programmatic *limitation of means*. Not that certain means of description and explanation were prohibited from the very beginning; it is rather a *principle of economy* that makes us prefer, under competing but otherwise equivalent hypotheses, theories, models, and thought systems, the more economical, the simpler, the more fundamental ones. What is crucial is that we use it as a principle for theory selection and as an argument. Naturalism's claim to universality mentioned above is certainly compatible with this principle, even if it does not automatically follow from it.

Our characterization of naturalism by its universality and by its limitation of means is not very precise yet. I shall try to give it more profile by presenting its *program*. (For a more detailed characterization see Vollmer 2007.) This program consists of five parts:

- It demands and draws a *cosmic overall picture*, a "Weltbild" (world conception).
- It assigns *man* a particular (and in the end rather modest) place in the universe.
- It includes *all* human faculties, in particular language, cognition, scientific research, moral behavior, aesthetic judgment.
- Modern naturalism is *evolutionary*: All complex systems consist of and develop from simpler subsystems. Evolution in general leads from the simple to the complex. (There are exceptions.)
- On this basis naturalism demands and develops

  – a naturalistic anthropology,
  – a naturalistic epistemology,
  – a naturalistic methodology of research,
  – a naturalistic ethics,
  – a naturalistic aesthetics.

Naturalism follows and promotes the program of *enlightenment*. It dispenses with many illusions. Teleology is one of them. Eternal life is another. Supernatural beings—gods, angels, devils, ghosts—are still others. The contents of the present volume give more examples.

Another illusion, if we follow modern neuroscience, is *free will*. If free will doesn't exist what are we to do with concepts such as guilt and responsibility? If free will doesn't exist, may we still reproach or even punish people for doing wrong? Do responsibility and punishment presuppose free will? To discuss these questions in the context of naturalism is the task of this contribution.

17.2 Why Responsibility?

Anyone who does something, will make mistakes. Thus, it cannot be our aim to make no mistakes at all; for then we could not do anything—and just that could be a big mistake. But we may try hard to recognize and to eliminate our mistakes as soon as possible and to improve. Most important, then, is not that we act flawlessly but that we *act to the best of our knowledge and conscience.*

Knowledge alone is not enough. We are able to make very many things happen, perhaps too many. Indeed: we can do more than we are entitled to. "Knowledge without conscience is unguarded explosive." (Hans Derendinger) But conscience alone is not enough either. It is not sufficient to aim at the good; we should also be able to do it. "There is nothing good unless you do it." ("Es gibt nichts Gutes außer man tut es." Erich Kästner) And it is evident that knowledge enhances our abilities. "Knowledge is power." (Francis Bacon) What is decisive then is the right combination of knowledge and conscience. We can say it even more succinctly: Our goal must be to act *responsibly*. It's as simple as that.

Is it possible to be a bit more precise? What then is responsibility? Evidently a difficult concept. "I take responsibility," says the politician or the mountain guide and thus claims competence and a leading role. (And sometimes it might only mean: "Let me do my own thing, don't interfere!") But what happens if still something does happen? May we reproach him? Does he have to resign? Must he pay the damages? All damages or only the foreseeable ones? Responsibility—is it attributed, may we take it of our own free will, or do we have it inevitably, like it or not?

May an animal be responsible? Of course it may *cause* something. The dog biting the child has shown a specific behavior. But is he also responsible, is he guilty, did he *act* at all? We are ready to see the guilt with the persons involved: with the owner, who trained the dog to bite humans; with the child, who teased the dog; with the parents, who didn't watch the child closely enough. Yet, more often than not, the dog is blamed, even punished, and ethologists affirm that dogs may even have a bad conscience. Nevertheless, we don't ascribe responsibility to them. What is missing for the dog to be responsible? (An allusion to Herder's question: "What is the most manlike animal, the ape, lacking to become a human?" In answering his question Herder spots the difference: It is language. Likewise, language is necessary for the ability to take responsibility; but it is by no means sufficient, as we can

see with any 4-year-old child, who has language but is not ascribed responsibility.
If it is not language—what is it then?)

17.3 On the Special Character of Explications

Before we venture an answer we should—following a practice often used by
philosophers and sometimes smiled at—reconsider how this question might be
meant. "What is" questions suggest that there is a definite answer to be found. There
might be people who know this answer, and many others who don't. Thus in asking
an astronomer "What is the polar star?" or a chemist "What is gold?" we take it that
besides many false answers there is one correct answer that we hope to get. Perhaps
the expert can show us the polar star or an object made of gold and tell us distinctive
traits that help us to recognize such things, such materials, or such properties.

With abstract concepts such as law, justice, or responsibility, this is quite different and markedly more difficult. We cannot point at an object and say "This is
responsibility." Nor is it sufficient to point at a person and call him responsible (or
irresponsible). Plato's dialogues take ample profit from this difficulty when their
protagonist Socrates exposes the answers of his interlocutors as hasty. What we can
do, however, is to sketch actions and decisions and investigate whether and why they
are responsible (or irresponsible) actions and decisions.

However, the decision will not always be easy. Another star, a satellite, or a flying
plane may be mistaken for the polar star. And how often might people have taken
brass or fool's gold for real gold? All the more abstract concepts like *responsibility*
offer a latitude of vagueness, leading to unclearness and misunderstandings. We will
immediately see this if we try to *define* this concept. (Why don't you try? You will
not be blamed, or held responsible!)

Even if we boast of a reliable judgment when acting or when a decision has to be
taken as to whether something is responsible, we will still have trouble giving a useful, explicit *definition*. And if we come upon a useful definition we sometimes still
doubt whether it is adequate, that is, whether—intentionally—it really says what we
mean, and whether—extensionally—it is neither too narrow nor too broad.

Isn't that strange? On the one hand we are by no means free in our definition, for
the term "responsibility" already has a meaning that we want to hit and are afraid to
miss. Hence we want to *state* descriptively what we and others normally mean by
"responsibility". And with such statements we may easily err. (What we are after
here is, classically speaking, a *real definition*.)

On the other hand, we cannot enter a laboratory and find out by observation,
measurement, or experiment what's the matter with responsibility. And due to the
vagueness mentioned we are not totally constrained. Inside this latitude of meaning we can still *stipulate* or determine prescriptively what we *want* to mean by
responsibility. (This is a typical trait of a *nominal definition*.)

Such a definition combining descriptive and stipulative elements we call concept
sharpening or *explication*. An *explication* in this sense answers the question "What
does it mean?" Hence it is not an *explanation*, which answers "why" questions.

Looking for explications we do better not to ask "What *is* responsibility?" but rather "What do we *want* to mean by responsibility?" In saying this we express that we not only state, but also fix the meaning, that there are other possibilities, and that if need be we are ready to think over our proposal. Beyond that, by using "we" instead of "I", we make manifest that we are not interested in a private language, but that we want to be understood by others and to understand them.

17.4 What Do We Want to Mean by Responsibility?

As in other cases we can incorporate in our explication more traits or fewer. In the following we restrict ourselves to two extreme cases: a maximum definition with many traits (full list) and a minimum definition with one single trait (in italics). What then do we want to mean by "responsibility"? We say:

Responsibility is the social charge and the ability

- consciously (knowing and considering the alternatives),
- of one's own free will (freedom of the will is presupposed here!),
- with reasons (not arbitrarily or at random),
- and following the personal order of values (the person must have preferred her decision to the alternatives known to her),
- to decide for (or against) an act (there must be at least two possibilities of deciding or acting),
- *to let the act (or omission) be attributed (imputed) to oneself* (that is, to regard oneself as the originator of the act and of its consequences, and to accept praise and criticism, reward and punishment for the act),
- and to answer for the foreseeable consequences (this point being especially difficult and therefore controversial).

According to the minimal definition (here in italics) responsibility is essentially *attributability* (*imputability*; in German *Zurechenbarkeit*). In fact, both concepts are nearly synonymous; at least it will be difficult to name cases where someone is responsible for an act that would not be attributed to him, or, vice versa, where we attribute an act to someone without holding him responsible for it.

Some of our starting questions might be answered with this minimal definition: Responsibility can be attributed and can be taken consciously. But since the definition calls upon a *social* charge, it does not refer to the laws of nature. Problems arise, however, if someone is assigned responsibility he is unable or unwilling to take. In such cases he will not be ready to be attributed the consequences of his act. It is extremely impressive how Max Frisch (1911–1991) in his stage-play *Andorra* makes the persons involved in the murdering of a supposed Jew step to the foreground between the scenes and stress with good arguments "I'm not guilty!" They don't deny their actions, but are unwilling for the consequences to be attributed to them, especially to be reproached therewith, i.e., to take responsibility. They mollify their conscience by declaring that they did not foresee or ever want what happened.

The maximum explication names more elements as connected with responsibility frequently if not always. Since they transcend the minimal claim, most of them are debated. In everyday cases and in law it is especially problematic how far we have to compensate for the *damage* we have caused. As a rule we act under uncertainty and under risk and cannot foresee all consequences. And how will we answer for our deeds if the consequence is death? An eye for an eye, a tooth for a tooth, a life for a life? And if there are two victims?

We could continue with further deep considerations. However, we shall turn to another aspect. In our maximum definition mention is made of free will. In fact, the maximum definition *presupposes* free will. But do we have free will? The spontaneous answer is normally a decisive yes. In general we *feel* free. Although not all our wishes are fulfilled, we still feel that we may choose between several alternatives. Proponents of free will we call *libertarians*. Intuitively, we are libertarians to begin with.

On the other hand, we know that the problem of free will is a much discussed philosophical riddle. How come that philosophers are fiercely discussing a problem that seems, to our intuition, to be so easily solved? Evidently, there are reasons to doubt the freedom of our will (or of our decisions or of our acting). What reasons can there be? Before we turn to these reasons we should ask ourselves what we *mean* by free will. And before we delve into terminology we reflect why the problem of free will might be of interest to us even if we are not philosophers.

17.5 Why Is the Problem of Free Will Relevant?

After all, this problem could be relevant only to philosophers. Of course, nobody can be forced to find a special question interesting, but we can at least try to show which *further* questions are connected with the problem of the existence of free will, not all of them being purely philosophical.

- First of all, it is a factual question capable—if only in principle—of a factual answer. It is at least conceivable that empirical sciences such as psychology, neurobiology, or psycho- and neuropathology provide us with a convincing answer. So far, this has not been possible; hence we could consider the problem of free will as an open problem of the empirical sciences. However, it is also conceivable that this question cannot be decided empirically at all. Even then the empirical sciences can contribute to a meaningful discussion: They could find out, e.g., how the freedom of our will, of our decisions, or of our actions is restricted.
- Considerations on free will influence both our world picture and our view of man? In particular our concept of man is not complete as long as we have no conclusive answer to this question. The problem of free will then belongs to biological as well as to philosophical anthropology.
- Such an answer has also a bearing on our self-esteem. Traditionally we are prone to ascribe to us freedom of our will; we are even proud of it and are not easily dissuaded from having a free will. How empirical results may interfere with

our self-esteem is exemplified by the three famous "offences" by Copernicus, Darwin, and Freud. (The term "offence"—in German "Kränkung"—and the counting up to three go back to Freud himself. It is easy, however, to specify more than three such offences. For this see Vollmer 1995.)

- That our will is free is also—some prominent exceptions notwithstanding—the doctrine of Christian *theology*. It plays a major role in the questions of sin and redemption, of heaven and hell, of God's vindication for the suffering in the world made by him. (This divine vindication was called *theodicy* (in French *theodicée*) by Gottfried Wilhelm Leibniz (1646–1716), whence we also talk about the problem of theodicy.) However, determinism or even predestination is advocated by Augustine, Luther, Zwingli, and Calvin.

- Quite independent of religious and theological questions, freedom of will is also a premise of many *ethical systems*. The concepts of "good" and "evil", of "guilt" and "expiation", of "repentance" and "retaliation" are barely meaningful without this premise. And very often the concept of "responsibility" is used in such a manner that it *presupposes* freedom of the will. (That is why we incorporated "of one's own free will" in our maximum definition.) Anyway, the determinist denying free will must check carefully which of these concepts may still be used by him and how they then are to be explicated.

- Our *criminal law* is based on the concept of guilt. What is more, not the penal code, but the (German) Federal Supreme Court explicitly declares: "Punishment presupposes guilt. Guilt is reproachability. By the disapproval implied in the term guilt the delinquent is reproached for not having behaved lawfully, for having opted for injustice although he could have behaved lawfully, could have opted for the law. The inner reason for the guilt reproach lies in the fact that man is designed for free, responsible, moral self-determination and therefore able to opt for the law and against injustice." (Federal Supreme Court, cited by Engisch 1963, pp. 1, 2). Whoever denies free will, must therefore ponder upon the meaning and purpose of punishment.

- How we *educate* our and other people's children also depends on whether or not we believe in free will. It makes a difference whether we reproach a child not only for having behaved wrongly, but *badly*, whether we point to the child's behavior being forbidden as such or to the unwanted consequences that should be avoided, whether we put forward God, the Bible, the State, or the social community as the norm-giving authority. From bad marks to extra work to expulsion from the school: Even an education completely oriented to the child will point to the negative consequences. Thus, the problem of free will is meaningful for education and educational science, and here for both theory and practice.

- Finally there is a close connection to the *mind–body problem*: Whoever votes for free will in the usual sense, must be a *dualist*, starting from two levels of being; for on the purely physical level the gaps in the causal net claimed or postulated by the libertarian do not exist according to all we know.

Possibly this network of problems gives us some motivation to dig deeper into the problems of free will.

17.6 What Do We Want to Mean by Free Will?

Traditionally we say with Edward Moore: A person has *acted freely* if and only if in this same situation—hence under the same laws of nature and under the same boundary and initial conditions—she *could have acted otherwise*. It is decisive that the conditions are in fact *identical* or the same; for under slightly *other* conditions it is no miracle if we act differently, since our behavior could be influenced by these very conditions. This applies to all what could be different: other natural laws, other external conditions, but also other inner states, be it other motives, other wishes, other characters, another education, or new experiences.

We must admit that this formulation reproduces our intuitive concept of free will quite well. Not only that: In general we are deeply convinced that we have in fact this kind of free will, that we could have done otherwise.

Evidently, this is also the conviction of the (German) Federal Supreme Court, which binds, in the above formulation, punishment to the condition that the delinquent "could have opted for the law". Quite similarly, the argument related to the reform of the criminal law in 1960 (cited by Engisch 1963, p. 2) reads: "Reproachability presupposes that the actor when acting was able to evade the specific formation of the will leading to the act." The Supreme Court and the creators of the reform would not have chosen these formulations had they not been convinced that humans, including, for that matter, also delinquents, have free will.

But do they have it? Do we possess free will? And how can we find out?

The explication given, natural as it might seem to be, comes with a characteristic difficulty: We can never exactly reconstruct the same situation. The supposed repetition occurs either at another place or at another time, in most cases even both. And the person may have changed since. But then the conditions are not the same. If, however, the situation is only *similar*, that is, still *different* in some way, then we cannot exclude any more that just this difference is decisive for our deviant behavior. Thus we *cannot test empirically* whether the *same* person in the *same* situation does indeed behave differently at least sometimes (which we then would explain by her having free will). Whether the person has indeed acted freely cannot be empirically established.

It doesn't help to replace freedom of will by freedom of decision and to ask whether our decision was free. For the libertarian it is clear that the person would have acted differently if only she had wanted to, hence if she had decided differently. Hence, we could say as well: A person has *decided freely* if and only if she *could have decided otherwise*. Nevertheless: Whether she could have decided otherwise is *not testable* either.

Thus, we cannot find out by observation whether, in the sense of the explication given, we—or anybody else—have free will. This is the main reason why we have here—have had for a long time and still have now—a *philosophical* problem. That does not mean, however, that we can't say anything reasonable about it. On the contrary: Just *because* we cannot hope for a decision by empirical science, by psychology, say, or by neurobiology, *arguments* are in special demand—and *arguments* are a main concern of philosophy.

17.7 Do We Have This Kind of Free Will?

Since intuitively we are quite convinced to have free will, we first investigate what could count against it.

The main argument against free will is *determinism of the natural sciences.* Unfortunately, the term "determinism" has two different meanings. Determinism in general is the thesis of natural philosophy that all events—with the possible exception of the origin of the universe—are effects, hence have causes, are causally determined, hence are inevitable, happen by natural necessity. However, they need not be predetermined, predestined by someone, by a divine being for example, and they need not be predictable.

The opposite of this determinism of natural philosophy we call *indeterminism.* Indeterminism implies nonpredictability, but not vice versa. There are more obstacles to predictability than chance: lack of knowledge about the laws of nature, lack of data about initial conditions, lack of better memory, lack of computing power, slow calculation, deterministic chaos, self-reference, etc. Even in a perfectly deterministic world not all events can be predicted by us.

As far as free will is concerned, *determinism* is the concept according to which all our decisions and acts are causally determined by previous factors. There is no nice term for the opposite position. Leaning on the term "libertarian", we might name it *libertarianism.*

According to general determinism, under identical conditions exactly the same must happen. There can then be no freedom of the will in the traditional sense formulated above. Determinism in natural philosophy includes determinism for the question of free will, hence excludes free will.

Our world, however, doesn't seem to be deterministic. In particular quantum theory (in its usual interpretation) teaches us that in the micro-area there is absolute chance, that there are undetermined events: nuclear decay, decay of unstable particles, of neutrons, etc. Therefore the strict determinism of natural philosophy is false, indeterminism is true.

Is free will saved by this indeterminism of physics? No! It is true that chance events make it possible that under the same conditions something different happens, therefore it is indeed possible that we decide or act otherwise. However, the fact that a decision comes about in part by chance can by no means satisfy the libertarian; what is happening by chance we can neither produce nor predict. Thus, for chance events we cannot be held more responsible than for causally necessary events. The physicist Pascual Jordan (1902–1980) and the philosopher Hans Jonas (1903–1993) have tried to use the indeterminism of physics to explain free will. Somehow microscopic chance events should amplify up to the macroscopic level. But then our decisions or acts would be accidental as well. Therefore physical indeterminism cannot save freedom of will.

We could try to locate the will and hence free will on a level totally different from the level covered by the natural sciences. With regard to the body–mind problem, we would, as was mentioned, have to turn dualists. Many respected thinkers have been dualists: René Descartes (1596–1659), Immanuel Kant (1724–1804), Karl Raimund

Popper (1902–1994), John Eccles (1903–1997). But by that move our problem is not solved either. For, if free will, wherever it might be located, comes to decisions and thus causes acts, that is, has *effects* on the level of matter and energy, then it is evident that, at least, some *information* is transmitted, and, at least in our physical world, any information transfer is connected with an energy transfer. Since this energy does not come from the physical world its appearance must violate the law of energy conservation on the physical level.

The resolute libertarian must accept then that on the physical level energy conservation is continuously violated. For Descartes, the arch-dualist, this was no problem since he didn't know the law of energy conservation. Popper and Eccles (1977, pp. 541, 543, 564), however, explicitly concede that energy conservation must be violated. This conception is by no means satisfying: It is very implausible that just on the level of the brain continuous violations of energy conservation occur whereas physicists do not find the slightest trace of such violations on the microscopic level.

The naturalist cannot accept such an anomaly. For him, there is no good reason to turn dualist and he will therefore stick to the simpler solution, monism, mostly in the form of an identity theory.

There are more arguments against free will. Pothast (1980) gives a sensible collection. Amusing and at the same time instructive is Smullyan (1981), who presents a dialogue between God and a mortal who wishes to give up his supposedly free will. For a more extensive discussion see Walter (2001), or Honderich (1993). All of them come to the conclusion that free will in the traditional sense does *not* exist. What shall we do if there is no free will in the traditional sense?

17.8 What Should We Do If There Is No Free Will?

That there should be no free will is in plain contradiction to our intuition. Do we then have to correct our intuition? It is also in contradiction to our linguistic usage. Do we then have to correct our linguistic use as well? There is still another possibility: We can also try to change our definition. Is it possible to explicate free will such that it is compatible with the determinism of natural philosophy? In fact this way out is widely used. Such positions are called compatibilist theories. Some striking formulations are listed here:

- Freedom is insight into necessity (Engels 1878, following Hegel).
- Free will ultimately arises from and consists in the nonpredictability of our (determined) decisions and acts (Wittgenstein 1922, 5.1362; Planck 1936).
- Free will is the ability to do what we most eagerly want to do, which is to follow our strongest impulse (Jeans 1943, Chap. VII).
- Free will is absence of external compulsion. "If the act springs from his own character [. . .], then we say that he acted freely" (Carnap 1966, p. 219).
- Free will is dominance of the rational control system over the instinctive one (Büchel 1981, p. 256).

These explications are not equivalent. Hegel's remark is tricky, but not very helpful. The trick is that a sovereign, possibly a dictator, can give his people the illusion of being free if only they understand *why* they are not and cannot be free. Wittgenstein's and Planck's position is called *epistemic indeterminism*. This is a misnomer because it is a fully *deterministic* position. We only have the impression, or *illusion*, of being free, because, for fundamental reasons, we are unable to disentangle the processes in our brain and to predict our own decisions and acts. The problem with Jeans's explication is that an addicted person might most strongly want drugs and get them, but still feel compelled by his needs. This also applies to Carnap's proposal: Is being addicted part of my own character? Why not?

The most useful explication seems to be the last one. Like all other definitions it presupposes determinism, but it distinguishes between different hierarchically connected control systems in our brain, and calls a decision free if it is determined by the uppermost system, the rational system. Since rationality is nearly exclusively human, Büchel's explication allows free will to be located with humans, thus satisfying our intuition.

Let us be quite clear what can be achieved with such an explication. All positions cited presuppose determinism! For them free will in the traditional sense does not exist. It would be possible then, and for the sake of clarity perhaps even preferable, not to talk of free will any more. It is the strong intuition that we own "something like" free will that motivates people to look for an alternative explication. They prefer a solution that allows them to follow intuition and language use, but if it is needed they must be ready to give a more refined explication that reveals the deterministic background.

It is quite similar with concepts such as "punishment" and "responsibility".

17.9 Punishment and Responsibility in a Deterministic World

What is the role of *punishment* if our acts are determined? At first sight it seems to make no sense to punish someone who could not do otherwise. But this first sight is deceptive. When inquiring as to the meaning of punishment we are inquiring as to its *function*. Now we may distinguish several functions of punishment:

- expiation or retribution (the offended law shall be "repaired");
- reform or deterrence of the delinquent (special prevention);
- deterrence of others (general prevention);
- protection of the community (e.g. by imprisonment).

We shall not delve into the problem of whether and how these aims may be achieved. What we are after is the question: What functions can be fulfilled by punishment in a deterministic world, that is, in a world without free will in the traditional sense. The idea of expiation or retribution is indeed lost. But all other functions are kept! By the threat of punishment or by punishment itself potential actors may be influenced, even *determined*, to refrain from a deed they are inclined to commit. (For a detailed discussion see Honderich 2005.)

We may even say that in a deterministic world such measures are most effective. And don't we continually try to push our fellow humans to certain acts? To attain this we are even especially inventive: We use arguments, threats, promises, prizes, bribes, extortion. Admittedly, in doing so we are not always successful. But does this count against determinism?

How about responsibility? With this question we return to the considerations above that motivated us to think about free will. Our minimal explication "Responsibility is attributability" remains, of course. In our maximum explication, however, the element of free will is lost. Our explication will then be slightly shorter: "Responsibility is the social charge and the ability consciously, with reasons, and following the personal order of values, to decide for (or against) an act, to let the act be attributed (imputed) to oneself, and to answer for the foreseeable consequences."

Someone reading this explication unprepared will not even miss the element of *free* decision or act. There are two reasons for that: On the one hand, the explication is still extensive enough and difficult to grasp totally. On the other hand, when hearing the expression "to decide" we can barely help hearing "to decide freely". Whoever *decides* something has several alternatives, of which he prefers and finally chooses one. And still the decision may be perfectly determined! A businessman who can keep his shop only by raising a loan may *decide* to do so, but what else *can* he do in view of his strong wish to keep the basis of his existence? At the time he proves to be *responsible* regardless of being assigned this responsibility or having taken it.

This consideration shows that it is perfectly possible to talk about *responsibility* if we are under mental or physical pressure and even if we lived in an utterly determined world. In a determined world there can be free will, punishment, and responsibility. It is true, we must be more careful in defining and using these concepts, but in doing so and just by doing so we may make ourselves perfectly understood.

References

Büchel W (1981) *Die Macht des Fortschritts*. Langen-Müller/Herbig, Munich

Carnap R (1966) *Philosophical Foundations of Physics*. Basic Books, New York, NY

Engels F (1878) Herrn Eugen Dührings Umwälzung der Wissenschaft, genannt "Anti-Dühring"

Engisch K (1963) *Die Lehre von der Willensfreiheit in der strafrechtsphilosophischen Doktrin der Gegenwart*. de Gruyter, Berlin

Honderich T (1993) *How Free Are You? The Determinism Problem*. Oxford University Press, Oxford (revised 2002)

Honderich T (2005) *Punishment, the Supposed Justifications Revisited*. Pluto, Ann Arbor, MI

Jeans J (1943) *Physics and Philosophy*. Cambridge University Press, Cambridge

Kant I (1788) Über den Gebrauch teleologischer Prinzipien in der Philosophie. Teutsche Merkur 36–52; 123–136 [In: *Kants Werke. Abhandlungen nach 1781*. Akademie-Textausgabe, Bd. VIII, de Gruyter, Berlin, 1968, pp 157–184]

Planck M (1936) Vom Wesen der Willensfreiheit. Lecture 1936. In: Planck M (ed) (1949) *Vorträge und Erinnerungen*. Hirzel, Stuttgart, pp 301–317

Popper KR, Eccles JC (1977) *The Self and Its Brain*. Springer, Heidelberg

Pothast U (1980) *Die Unzulänglichkeit der Freiheitsbeweise*. Suhrkamp, Frankfurt

Smullyan RM (1981) Is God a taoist? In: Hofstadter DR, Dennett DC (eds) *The Mind's I*. Harvester, Brighton, pp 321–341

Vollmer G (1995) Die vierte bis siebte Kränkung des Menschen. In: Vollmer G (ed) *Auf der Suche nach der Ordnung*. Hirzel, Stuttgart, pp 43–59

Vollmer G (2007) Can everything be rationally explained everywhere in the world? In: Gasser G (ed) *How Successful Is Naturalism?* Ontos, Frankfurt, pp 25–48 (German 1994)

Walter H (2001) *Neurophilosophy of Free Will: From Libertarian Illusions to a Concept of Natural Autonomy*. MIT Press, Cambridge, MA (paper 2009; German 1998)

Wittgenstein L (1922) *Tractatus Logico-Philosophicus*. Routledge, London

Epilogue

Chapter 18
Modern Illusions of Humankind

Ulrich J. Frey

Abstract Some scientific discoveries contradict core beliefs of our perception of ourselves and the world. Core beliefs include the illusion of free will or being largely independent of our species' biological roots. Nevertheless, these "illusions" persist, which is not exclusively due to historical or ideological reasons. It is argued that conceptual revolutions imply restructuring one's beliefs. This is costly in various ways, e.g. in the form of time, resources, and social networks, and is uncertain in its outcome. In an evolutionary sense, these costs are considerable and should lead to initial resistance against their adoption via psychological mechanisms. In addition, adopting new beliefs could always be a manipulation for the ends of somebody else, which should be avoided, too. This article presents these illusions, together with the scientific evidence exposing them as illusions, and explores why there is so much resistance to giving them up.

18.1 Destroyed Illusions

18.1.1 Intuitions Versus Science

Science is—among other things—notable for contradicting intuitions. Throughout history, many facts have been crystal clear to our intuition and could not have been otherwise—except that scientific discoveries put an end to these illusions. Illusions are understood to be misinterpretations of reality, usually shared by many people. They are not quite such a fixed belief as a delusion, which is held up even when faced with massive contradictory evidence, but they are more fixed than merely misapprehensions.

Some core beliefs include being superior to all other animals, being in charge of most decisions concerning one's own life, being largely independent of our species'

U.J. Frey (✉)
Zentrum für Philosophie und Grundlagen der Wissenschaft, Universität Gießen,
35394 Gießen, Germany
e-mail: ulrich.frey@phil.uni-giessen.de

U.J. Frey et al. (eds.), *Homo Novus – A Human Without Illusions*,
The Frontiers Collection, DOI 10.1007/978-3-642-12142-5_18,
© Springer-Verlag Berlin Heidelberg 2010

biological roots (e.g. cultural achievements such as religion have severed us from the necessities of evolution), and having a free will. These beliefs constitute the very perception of ourselves and the world. Unfortunately, they rest upon illusions and certain peculiarities of our cognitive abilities.

This chapter is an analysis of such illusions, why we believe in them in the first place, why they still persist, and why we stubbornly resist giving them up. Resistance in itself is hardly a surprise as our intuitions maintain quite forcibly every day that they are right, even when modern science dispels them quite convincingly. But there are more and interesting reasons for such a resistance (see below). This chapter describes four illusions, gives evidence why they are illusions, and explains the resistance to giving them up. The analysis of the described phenomena throughout this chapter distinguishes between the proximate level of explanation, which is mainly psychological here, and the ultimate level of explanation, which is the evolutionary reason for the existence of such phenomena. Thus, for example, the psychological resistance to a new concept or idea is subjective and accessible for each individual, but the evolutionary reason for it remains hidden.

18.1.2 Illusions and Insults

As scientific advances shatter illusions, a good point to start with are scientific "revolutions" and the alleged resistance to them. Freud labeled the three greatest as *insults* to humanity (Freud 1969): The Copernican revolution (the earth is not the center of the solar system but is just one planet among others), the Darwinian insight that humans are descended from apes and not created by God, and Freud's own discovery that humans are not as rational as they think, but largely governed by uncontrollable unconscious forces and emotions. Once disillusionment takes place, pride and the perception of ourselves and the world suffer so as to feel "hurt", as Freud implies.

This needs to—and has been—qualified in several respects. First, this article contends that we do not feel hurt but reject these ideas because of their negative consequences. Second, in a historical sense: neither the Copernican nor the Freudian "insult" were initially regarded as an insult (Brague 1994; Kuhn 1957/1981; Kraiker 1992). The Darwinian controversy, however, still rages today. Third, whether a discovery is regarded as an insult depends critically on the idea of humankind and the world-view prevalent at a certain time (Vollmer 1995). Fourth, because an insult is at first negated by the majority (i.e., the consequences are not taken seriously), insults often do not acquire their rank until later on. In some cases this does not happen before the scientific breakthrough has become conventional wisdom. Fifth, still more scientific breakthroughs since Freud could be regarded as insults (see Table 18.1 below). Sixth, these and other insults are at present in different phases of their "potential to disturb", which usually increases at first, reaches a peak, and then declines. Seventh, an insult is directed against persons and therefore subjective, the disillusionment by scientific revolutions is, however, not directed against individuals.

Table 18.1 Illusions, Insults and Consequences

Illusion in this book	Illusion destroyed	Consequences for perception of ourselves and the world	Traditional schema (see Vollmer 1995)
Illusion 1: Humans are exceptional	Unique position in the universe Superior position among creatures Superior brain and thinking abilities	(A) My life is unimportant (B) I am a creature among countless others (C) I am nothing special	Copernicus (A, C) (heliocentrism) Darwin (B, C) (evolutionary theory) Sociobiology (B, C) (behavior is adaption-driven) Artificial intelligence (C) (thinking computers)
Illusions 2, 3, 4: We are independent of our biological roots Illusions 5, 6: See below	Control over own life (decisions, actions)	(A) I am substantially extrinsically determined by biological factors (B) I have little control over my decisions and actions	Sociobiology (A, B) (behavior is adaption-driven) Neurobiology (B) (no free will) Freud (B) (the unconscious)
Illusion 5: Cultural achievements are a purely social construction	Unique position in the universe Superior position among creatures	(A) My life is meaningless (B) My life has no purpose (C) There is no plan of overriding importance	Sociobiology (A, B, C) (behavior is adaption-driven)
Illusion 6: We have a free will	Control over own life (decisions, actions)	(A) I have little control over my decisions and actions	Sociobiology (A) (behavior is adaption-driven) Neurobiology (A) (no free will)

Take as an example the Copernican revolution, which has long been accepted by almost everybody. Still, some consequences *derived* from this world view are problematic: Take the feeling of utter loneliness and insignificance of living on a planet that is just one among billions of others and humankind as being alone in this infinite universe, which bothers quite a few people.

Why would we call revolutions insults? And why is there so much resistance to them? A preliminary answer anticipates the argument of this chapter: Insults destroy illusions and illusions are strongly shielded against criticism or corrections because their function is to protect a coherent, positive self-perception. This in turn is essential to our well-being and capable of acting or guarding against manipulations from others. Some critics, however, doubt that this is true:

> Should the naturalistic research program be successful, thus the allegation, it follows that
> we have to structurally revise our self-concept (concept of humans), because central tenets

about ourselves would then emerge as illusionary. It is exactly this proposition that shall be refuted in this book (Pauen 2007, p. 24, translated).

Conceptual revolutions, however, include high costs for change in several ways (see below for a detailed explanation). Particularly strong resistance is to be expected whenever a discovery destroys more than one illusion or if it is blatantly against our intuitions.

Table 18.1 is an overview of old and new insults, to be found under the heading "traditional schema". Instead of using this usual classification I propose a new schema focusing on the underlying conceptual shocks (under the heading "consequences"). The schema is deliberately formulated in personal form so as to load the consequences emotionally and to underline the distinction between a first person perspective being "harassed" by scientific insults in a third person perspective. An example for that is, although I feel quite important and special myself, a third person perspective tells me that this is not the case.

As is apparent, three new insults have been added: artificial intelligence, threatening to construct an intelligence on par with or superior to humans; sociobiology, with its diverse claims, most notably that humans do show phenotypic altruism, but which can be explained by genotypic egoism. This in turn appears to devaluate morality, emotions such as love or cultural achievements such as religion as parts of reproductive strategies; finally neurobiology, with its claim that there is no free will.

To understand why there is so much resistance to adapting a new conceptual model as suggested by the first column, the third column lists the consequences. (Trying to imagine all of them as true should make you feel uneasy.) The focus on consequences makes us understand which revolution is or will be the most dramatic—sociobiology. With its diverse implications it is the only insult present in all four rows. In contrast, true artificial intelligence—if it is ever invented—will not be a severe shock. Its implication—we are creatures among others—has been hotly debated since Darwin. Plus, it is ourselves who will design AI, therefore there is the argument that it is eventually our intelligence expressed here.

Note that all insults are facets of a particular way to explain the world— naturalism. Most naturalists argue in regard to *methodology* "Use findings in the natural sciences to the greatest extent possible", in regard to *ontology* "Take materialism as foundation", and in regard to *epistemology* "Work with hypothetical realism" (Sukopp 2006).

18.2 Illusion 1: Humans are Exceptionally Exceptional

18.2.1 Description of Illusion

This first illusion of humans being exceptional is very old and very pervasive. The following short sketch can only highlight a few historical milestones concentrating on our surmised exceptional position in the universe (Copernican revolution) and on earth among creatures (Darwinian revolution). Christianity's creation myth probably has had the greatest impact. It characterizes humans to be like God, assigns to us the task of subduing the earth, to "Rule over the fish of the sea and the birds of the air and over every living creature that moves on the ground." (Genesis 1, 26–29). It is also evident in the Hindu and Buddhist karmic system of being promoted and demoted from animal to human and vice versa by reincarnation.

A similar fascination for this illusion is apparent in both Plato "human's likeness with God" (Protagoras 322a) and Aristotle (Politics, 1.2; Nicomacean Ethics 10.7). The latter systemized the world taxonomically—the scala naturae—taking for granted that humans belong at the top of this hierarchy below the fixed stars and the unmoved mover. Throughout the Middle Ages, the Renaissance and the early modern period there is no notable exception to this conception of humans (Flasch 1987). This interspecies comparison does not come as a particular surprise, since, until very recently, humans had very little information about other species. It seemed sufficient to point out that humans have exclusively developed religion, global cooperation, trade, writing, a code of law, and other advances of civilization.

Switching from interspecies to intra-species comparison, the same tendency is apparent. Within humanity it is rather common (!) to see one group as exceptional, too. For example, the oldest monotheistic religion, Judaism, sees itself as God's chosen people. Various other religions and ethnic groups have claimed to be the new chosen people, too. Indeed, all five monotheistic world religions insist on their claim to exclusivity, *only they* being exceptional and different from the rest.

This can even be generalized to the widespread upward reevaluation of one's own group, nation or religion as *exceptional*. The frequent distinction of many civilizations from strangers as *barbarians* is evidence for this. In ancient and modern times, many nations have nurtured the belief that they are exceptional: China (Sinocentrism), Japan, and the United States of America are prominent examples for the widespread concept of *supremacism*. "God's own country" is a phrase used by at least six countries around the world (New Zealand, Australia, USA, Rhodesia, Kerala, and the county of Yorkshire).

The same illusion can be found not only for nations and religions, but also individuals. Godlike, exceptional figures have been present throughout the ages. Prominent examples are Egyptian kings or Japanese emperors, and there are lists of people *all* claiming to be the messiah. An exceptional status/rank is cited rather commonly to legitimate claims to political (e.g., Hitler, Kim Jong-il) or intellectual leadership (e.g., Nietzsche 1883/1999, the chapter "Das Zeichen" for himself, or, more generally, the chapter "Vom höheren Menschen" for the "Übermensch").

18.2.2 Why Is It an Illusion?

It was Darwin in *The Origin of Species* and *The Descent of Man* who told us that humans are qualitatively not different from animals in general and descended from apes in particular. Modern biology adds that dolphins are able to hand down cultural techniques as well. Soft sponges are used as protection of the snout

(Krützen et al. 2005). Meerkats teach their pups using a sophisticated "curriculum" (Thornton and McAuliffe 2006). Chimpanzees have a concept of fairness (de Waal 1989; Brosnan and de Waal 2003) and that young chimpanzees have a better working memory than humans (Inoue and Matsuzawa 2007). Other primates are able to master most (abstract) structures of language (linked with the names Koko, Washoe, and Kanzi, but see the chapter by Fischer in this volume for limitations). Rooks use tools to solve meta-problems insightfully at the first attempt (Bird and Emery 2009) and so on. It is even safe to say that there is no "ascending ladder" of cognitive abilities, but only special skills for different ecological niches (Güntürkün 1996).

Still, it is argued that humans are far more sophisticated in all these "techniques" and that they uniquely *combine* all these singular phenomena known from other species. However, to choose culture as the criterion for exceptionality is rather arbitrary. Other criteria for success might be the sheer number of members or species (31,000 fish species make up 50% of all vertebrata; 3,00,000–3,50,000 species of beetles make up around 25% or more of all life forms, Losito 2002), the expansion in all habitats (bacteria), or social complexity (bees, ants, and termites)—but there is no valid measure to rank creatures in any way (Voland 2007a). But *even if there were* a valid criterion, it would not do, since evolution is fundamentally a-teleological. There is no direction and therefore no goal in evolution. Therefore, humans as the "crown of creation" is no more than a myth, an illusion.

There remains the allegedly exceptional position in the universe. However, since Copernicus we know that the earth is only one planet among others, since Galilei we know that there are millions of planets. Today, we are aware that we are at the rim of one spiral arm of our galaxy, one of over 100 billion galaxies, each with up to a trillion (10^{12}) stars. This clearly strips us of our exceptionality. Then again, it does not. If the earth were indeed the only planet with life on it, this would *enhance* our exceptional position. Although conservative estimates (using the Drake equation) arrive at about 10 higher civilizations sharing with us the universe, this is still guesswork (Cirkovic 2003). We simply don't know.

It is more important to consider how we came to be: if by natural selection, then this process consists of "chance and necessity", making us the product of a blind process. Again, this process would not reward us with an exceptional position. Furthermore, it is plausible that this process did not occur only once, given the staggeringly huge number of planets. And, if we could rewind the evolutionary clock, most probably nothing like humans would evolve a second time.

18.2.3 Why Do We Have This Illusion?

Three main factors contribute to this illusion: One is historical, one sociobiological and one psychological. First, historically, biology simply did not know (until some 40 years ago, see Sommer 1989) that animals do have culture, tradition, etc. Consequently, differences between humans and animals were more conspicuous than their similarities ever since antiquity. All humans are descended from one common ancestor, and evolutionary psychologists speak of "human nature" in the singular.

Second, humans have a tendency to differentiate quite markedly between their own and foreign groups. Sociobiology readily explains this tendency as an ingroup–outgroup behavior stemming from our Pleistocene past spent in small groups with up to 150 persons (Williams 1968; Dunbar 1998). Inclusive fitness theory (Hamilton 1964a, b) and strong reciprocity models (Gintis et al. 2003) explain within-group solidarity. Most group members are related to each other in one way or another. Other groups were possibly the most dangerous competitors for resources (Alexander 1987; Choi and Bowles 2007), hence the out-group hostility. Foreigners are unknown (epistemic distance and risk factor) and not related (genealogical distance). Probably one of the primary problems to be solved was how to uphold cooperation, trust, and solidarity within the group. One way of doing this is to make the own group special in some way, e.g. via paintings, behaviors, customs, and more, making it exceptional (Söling 2002; Sosis and Bressler 2003).

Third, the simple psychological fact that we perceive the world from a first-person perspective means we know more about ourselves than we do about others. This holds true for every degree of familiarity: we know more about friends than about strangers, more about our culture than about foreign cultures, more about humans than animals, and so on. However, the less we know about someone, the less important this person, group, or species usually seems to us. Consequently, other people or groups cannot be as important as we are. Again, this makes us exceptional. Quite a number of attributional biases contribute to that ill-formed perception, e.g., humans tend to ascribe positive outcomes to themselves, negative to others (for an overview, see Plous 1993).

18.2.4 Resistance

The modern tale of Copernicus is that of a revolutionary fight against the Church. But, this picture is wrong (see Kuhn 1957/1981 and Koestler 1959). His main work, *De revolutionibus*, appeared in 1543, the year Copernicus died. It was preceded by the *Commentariolus* by some 30 years, where the basic tenets of the heliocentric system were outlined and circulated among astronomers. The theory had been known by the Pope since 1533; moreover, Copernicus was actually urged by the Archbishop von Schönberg to publish it. From the beginning, there was a small but growing number of followers. Most of them, however, used the mathematical formulas and dismissed the heliocentric concept. Resistance built up around 1600, as Luther, Melanchthon, and others insisted on the literal truth of the bible (Brague 1994; Kuhn 1957/1981; relevant are Joshua 10:12–13; Chronicles 16:30; Psalms 104:5; Ecclesiastes 1:5). One reason may have been that in 1613 Galilei found important empirical evidence for the model, which was up to then purely theoretical.

It is true (see Brague 2006) that the heliocentric concept "promotes" the earth and does not "demote" it in the hierarchy of planets, as is often purported. But this could not belittle the resistance, as this theory created many new theological problems: What is the role of humans, when they can spatially no longer be found between devils and angels? What about the importance of the fall of man if there were many planets? How could living beings on other planets know that Jesus existed and be saved? These and a host of other equally perturbing questions fuelled resistance, manifesting themselves in 1616 in the official banning of Copernicus's teachings as heresy (see Kuhn 1957/1981).

18.2.5 Reasons for Resistance

The Darwinian revolution met with heavy resistance, which was mainly directed against two ideas: that humans were descended from apes and that morality loses its legitimacy through evolutionary theory (Desmond and Moore 1994). In the case of Copernicus (but of course also Darwin) Christian teachings and the literal truth of the Bible were discredited. Such an analysis would, however, only describe the surface phenomena.

The next section focuses on two fundamental underlying causes for the historical, social, and psychological reasons for resisting the idea that humans are not exceptional.

Every true conceptual revolution means that the perception of ourselves and the world both change. This change is costly in two very basic evolutionary senses. First, creatures should be, and in fact are, very conservative, concerning

- *behavior* ("better safe than sorry", which expresses risk bias; i.e. the very heavy cost of missed real alarms vs. the low cost of reacting to false alarms, see e.g., Nesse 2001),
- *strategies* (e.g., the strategy "win–stay, lose–shift" outperforms "tit-for-tat", see Nowak and Sigmund 1993), or
- *decisions* (e.g. the recognition heuristic, see Goldstein and Gigerenzer 1999), as change often implies unknown and unpredictable situations, which are potentially dangerous and risky (e.g., conservatism bias, see Frey 2007 for details).

Second, to restructure fundamental patterns—mental activities, behavior, social position, etc.—requires time, and cognitive and other resources. It is also in a major sense open to failure. Both basic principles entail significant implications.

If external events (here the belief in the new heliocentric system) trigger a massive reconceptualization of the perception of the world, then the gravest danger—in an evolutionary sense—is not the cost of restructuring, although this can be formidable indeed, but the very real possibility of being manipulated by others *for their benefit and to one's own harm*. This can be described in the terms of the extended phenotype theory (Dawkins 1990) or, more generally, by gene-centered evolutionary theories. Therefore, a deep mistrust against possible manipulation is an essential evolutionary safeguard for all creatures (Dawkins and Krebs 1978). However, there is the opposite phenomenon, too. Particularly humans may follow a leader blindly when they have ascertained that he or she represents their interests with all the known consequences.

In a psychological sense a reconstruction of self-perception is costly, too. There is robust evidence that humans are consistently biased towards optimism, affecting e.g., their own accomplishments and future prospects (Taylor 1989/1995). A recent poll of more than 1,50,000 adults in 140 countries representing 95% of the world's population concludes that 98% "expect the next 5 years to be as good or better than their current life" and that optimism is "a universal phenomenon" (Gallagher 2009). To which extent this reflects positive economic tendencies is another matter.

In combination with a necessarily ego-centered outlook on the world this leads to an overestimation of the own person and a too optimistic view of opportunities for oneself. A conceptual revolution with the hidden message at its core that humans are insignificant and meaningless in a vast universe runs contrary to the deeply entrenched biases of leading a life that matters. It also denies the justification in an optimistic belief of being able to achieve own's goals and to be at least of some sort of importance. Feeling insignificant in a huge universe "deaf for one's cries", can be overwhelming indeed, as testified e.g. by Nietzsche, Camus, or Monod. However, not all persons share it. Thus, this feeling of absurdity is by no means universal.

In addition, it is a necessity for humans to have consistent and coherent models of the world (Frey 2007). Otherwise, they would not be able to act at all, but remain paralyzed by contradictory models and choices. Therefore, logically inconsistent or contradictory strategies and heuristics have to be discarded (Gigerenzer and Todd 1999).

The quite significant resistance of humans shown above can therefore be explained by basic evolutionary mechanisms. These consider costs and benefits, risk, and social competition. It should not be surprising that even high-level decisions are influenced by these general cost–benefit considerations.

18.3 Illusions 2, 3, 4: We Are Independent of Our Sociobiological Roots

18.3.1 Description of Illusions

The illusions numbered in this volume as 2, 3, and 4 have in common that they postulate an independence of biological roots (see Table 18.1), which is not the case. On the contrary, scientific evidence points to the fact that humans are overwhelmingly determined in their decisions and subsequently their lives, although rational reasons for decisions are readily at hand.

These three illusions are perhaps the most interesting, because they are *not yet perceived* by most people to be illusions. It is common knowledge that humans are alone in the universe and share common ancestors with animals. However, it is not yet canonical in science and much less usual for the general public to view human behavior from a sociobiological perspective, which holds the belief in independence of biological roots to be an illusion. Thus, the multiple and far-reaching *implications* have not been realized in philosophy, politics, and other areas.

Historically, the biological nature of humans has not played a role at all in most humanities. Take philosophy as an example. From Plato and Aristotle to the Christian philosophers in the middle ages—empirical biological findings were thought to be unimportant and ranged rather low in esteem and low in the pyramid of human knowledge. The philosophical idea of humans was a-biological. Even many empiricists (Bacon, Locke, and Hume) did not rate the *specific* biological nature of humans as decisive for their insights. Surprisingly, this did not change with Darwin's *The Descent of Man* (see especially the third and fourth chapter). Philosophical anthropology after Darwin (e.g., Pleßner, Scheler) acknowledged that humans were biological beings but failed to integrate this new body of knowledge into philosophical theories. Instead, what *distinguished* humans from their animal roots became essential:

> There is absolutely no way from 'homo naturalis' and his hypothetically constructed prehistory to the 'humans' of cultural history, [...]. (Scheler 1955, p. 174, translated)

It was not until the 1960s (see Vollmer 1975/2002, pp. 277ff for a brief history) that evolutionary epistemology emerged as a fully fledged discipline. In addition, it took 20 more years for philosophers of science to recognize the evolutionary roots of human cognitive abilities and their role in science (Tweney 2001; Dunbar 1997). There is no evolutionary-cognitive philosophy of science up to now (Frey 2007).

Summing up, on the one hand empirical facts about the evolutionary basis of "higher" cultural activities were simply not known until very recently. On the other hand, the introspection of our reasoning abilities is opaque—we are not able to intuitively perceive our genetic and cultural imprinting, our peculiarities or errors. Hence, the illusion that "higher" cultural activities are quite independent of biological roots.

Especially *morality* and *rationality* have always been regarded as uniquely human, separating us from animals. The ability to reason, proceed logically, deducing and inducing laws, and in consequence being capable of developing an ethical system and to adhere to its rules, has always been seen as principally *freeing* us humans from our biological roots, demonstrating our independence from them. Although there have always been attacks on rationality, e.g., skepticism, empiricism, the Freudian revolution or the heuristics-and-biases challenges by Nobel prize winners Kahneman and Tversky, in most scientific disciplines the dominant conception of humans has been rational to the core (e.g., homo oeconomicus in economics).

Eminent philosophers such as Plato and Kant (and most others) insisted that to act morally is the ultimate goal in life. The precondition for morality is free will. That is, we are moral of our own volition—we freely and rationally choose to be moral (Kant 1788/1995, A 1, 171–186, A 289).

It is not only moral philosophy, but also religion that focuses on ethical teachings. Virtues such as charity and love *for others* (altruism) are not arbitrary, but essential for morality.

18.3.2 Why Are They Illusions?

As is shown in more detail below, sociobiology dispels the above mentioned illusions:

- "Higher" cultural abilities (e.g., the ability to reason and its organized form, science) are independent of our biological roots.
- Behavior and decisions are rational, intentional, and controlled considerations.

- Morals and emotions (like love and care) are of our own volition and an expression of altruism.

Not quite so, say evolutionary epistemology (1), life history theory (2), and the theory of gene-centered evolution (3), respectively.

Evolutionary epistemology (Vollmer 1975/2002) and cognitive philosophy of science (Frey 2007) make it quite clear that it is necessary to take human evolution into account when human rationality is analyzed. Human minds are adapted to a certain environment with specific demands for hunters and gatherers, but not to modern science. Nevertheless, we are able to transfer problem-solving techniques from the proper domain to the new domain. However, this leads to discrepancies between problem-solving heuristics and methodological demands.

Evolution has not only shaped our brains but also our behavior and basic strategies in all aspects of life. Life history theory studies show that even basic decisions in our lives—which we perceive as our very own personal decisions—are deeply influenced by external factors. Surprisingly, number, gender, and birth interval of our children are among them, depending on environmental conditions (see, e.g., Bogin 2006; Mace 2000). It is well known for example that under bad economic conditions more female children are born (Trivers 1985; Catalano et al. 2005).

In the same vein, it is crucial how many siblings we have for our own "decision" about how many children we will have (Chasiotis et al. 2006). Another study demonstrates that women seem to be fitness-maximizing: In order to have four surviving children, over 80% of Dogon women gave birth to an average of ten children—which is optimal in terms of fitness (Strassmann and Gillespie 2002). Or consider our love as parents: for fathers, the resemblance of their children to themselves determines to a certain extent how much they care (Platek et al. 2004).

To conclude: In contrast to the belief in autonomy concerning decisions, sociobiology holds that we are substantially extrinsically determined by biological factors and do not have as much control over our decisions and actions as we think we do.

These are intriguing results, but the real impact of sociobiology has manifested itself in the field of altruism and morality. Sociobiology claims to have an explanation for altruistic behavior: It is—in some form or other—beneficial for the altruist (Voland 2009b), i.e. phenotypic altruism is in fact motivated by genetic egoism. Thus, the traditional view of morality in general and altruism in particular as entirely good is at least partly an illusion. This gene-centered view of modern biology caused an uproar (Hamilton 1964b, a; Dawkins 1976; Trivers 1985) due to the destruction of a certain concept of morality. Outside science it may also be due to misunderstandings and exaggerated fears. What is the debate about?

Modern sociobiology holds that what matters in the evolutionary process are genes—they are truly immortal, using their bearers (e.g. us) only as temporary "vehicles", as Dawkins famously put it. As a consequence, phenotypically altruistic motives have been revealed as genetically egoistic. Inclusive fitness theory (Hamilton) is the theoretical justification for kin selection explaining most forms of altruism. Other forms of cooperation can be explained by positive cost–benefit calculations for each participant (see Nowak 2006 for a short overview). Thus, if evolutionary theory is correct, then altruism without any benefit is impossible for altruists in the long run.

In this sense, underlying motives and reasons for our (moral) behavior that feel intuitively right are explained by these theories. *These reasons are not transparent to us.* Let us consider two examples. It is known from real incidents that humans run back into a burning house to save family members but most do not run back for friends (in Sime 1983 nobody does)! This is backed up by experiments: If we have to choose to save our own baby or instead *five* other babies, most of us still save our own baby. These moral behaviors can be explained by evolutionary theory (Hauser 2006).

A Darwinian interpretation of human morality is therefore quite contrary to the standard (traditional, philosophical) view (Voland 2005):

- Moral behavior is not the expression of free moral decisions; instead, it has to be interpreted as evolutionarily optimal, context-dependent behavior. Example: doing altruistic acts enhances the reputation in social groups and increases reproduction opportunities.
- Altruism and care for others is mostly confined to relatives and well-known individuals (due to kin selection and our history of hunters and gatherers in small groups).
- Our social mode of life in groups and extensive parental care are *the* essential forces shaping our morality.
- Moral intuitions change in the evolutionary process. They are heavily dependent on cultural input and context variables in sensitive periods. Hence, they are neither fixed, timeless, nor a priori principles.
- Neither group nor universal moral standards can be expected as a standard rule, because moral strategies are the result of individual selection, resulting in individual and diverging morals. This does not preclude their existence, as other factors may work against that tendency (and in fact do).
- Moral consensus may be no more than a tactical agreement, as moral behavior is the expression of strategies that can be switched to suit different needs.

This said, it should be clear that evolutionary informed theories do not at any time judge natural phenomena in a moral sense. Theories about morality are descriptive and operate outside moral considerations (Vogel 1989). In this sense, sociobiological theories are only *explanations* of behaviors—see the "Resistance" sections below for why there is so much resistance to this kind of explanation.

18.3.3 Why Do We Have These Illusions?

There are three reasons that contribute to these illusions.

Historically, sociobiology and gene-centered theories are very young—just about 40 years old. Before, there was no sound scientific base for exposing these illusions as illusions. In my opinion, the sociobiological revolution is still in its infancy, that is, not even the majority of scientists considers these theories as fundamental, let alone is aware of its far-reaching implications. The general public has been partially exposed to some of these ideas via the heated discussions about, for example, Dawkins's popular books (*The Selfish Gene, The God Delusion*). In general, however, these ideas are rejected, since they mostly do not fit into the traditional perception of the world and ourselves.

Psychologically, our perception of ourselves is modeled around a free personality responsible for its decisions, which is enhanced by Western education. Thus, we do not ascribe our decisions to extrinsic factors but to our own volition. In most cases this is a correct perception of the causal structure (e.g., I decide to learn for this test in order to pass it), but the underlying ultimate reasons for our decisions are mostly hidden to us or are not those we perceive them to be.

Moreover, everybody is necessarily their center point in the world. Thus, the world is perceived as centered around their own person. It follows that events acquire their relevance *only* in relation to their person, therefore everything seems to converge towards their person, in turn up-valuing all own decisions.

Intuitively, consciousness is not transparent to itself (Voland 2007b). A fortiori, there is no directly accessible link for us from our moral decisions to the genetic level. To give an example: Why I love my brother is intransparent to me, although, of course, I can give ostensible reasons for it.

18.3.4 Resistance

Although sociobiology is a very young academic discipline it has received an extensive amount of criticism—not surprising, if the argument about destroying illusions is correct. Of course, the criticism could be due to its theoretical shortcomings, but as sociobiology is deeply grounded in Darwinian theory and musters a wealth of empirical confirmations as well as convincing formal theoretical frameworks, this is assumed to be unlikely here.

The debate started with the "founding manifest" of sociobiology (Wilson 1975), which evoked vehement reactions. Not only was a jug of ice-water dumped over Wilson's head at an academic conference (3 years after the publication!), but his book was also attacked as "part of the continuing conspiracy by scientists in the ruling class". Moreover the fifteen co-signers (including his colleagues Lewontin and Gould) in the *New York Review of Books* linked Wilson's book to "sterilization laws and [. . .] the eugenics policies which led to the establishment of gas chambers in Nazi Germany" (Bethell 2001). (For an extensive and detailed account of the "Sociobiology Wars" see Segerstrale 2000.) Even in today's review processes some reviewers still see terms such as "evolution" or "adaptation" as red flags, resulting in immediate rejection of such manuscripts, indicating the heat of discussion.

Sociological circles mostly see sociobiology as conservative, denying that humans are moldable by education and culture, obstructing progress, and helping to establish forms of power permanently.

Criticism, however, has not been limited to academic discussions. Consider the following frequently heard, but exaggerated, and in these formulations false characterizations of sociobiological tenets (from Geiser 2005, who cites these examples as evoking resistance against sociobiology):

- Genocide means the victory of the fittest.
- Rape is a natural act.
- Killing children is healthy behavior.

Other examples are the countless hate-mails to Richard Dawkins, one of today's leading evolutionary biologists (see http://richarddawkins.net/theUgly).

To draw a conclusion: Resistance to the sociobiological revolution is fierce, but will probably be fiercer still once the encompassing and far-reaching consequences for our perception of ourselves (in law, social life, etc.) will have become more obvious.

18.3.5 Reasons for Resistance

The two negative consequences (Table 18.1) of the disillusionment of sociobiology are "extrinsic determination" and "little control over our decisions and actions".

There are basically two ways of having little control and being extrinsically determined. The first is being controlled by other humans. In this case, manipulations to the benefit of others and detrimental to oneself are a very real possibility. The second is being determined by uncontrollable forces (say being swept away by a torrent), which is almost always a sign of poor foresight or wrong judgment. It seems as if we are strongly biased to *avoid* such situations with no or only one option left. Therefore, it could be suggested that we feel emotionally repelled when contemplating such situations or concepts, even if these uncontrollable external forces are biological parameters that shape our decisions in life.

It is certainly highly counterproductive—in terms of fitness—to do what others, especially other species, want you to do. This ranges from friendly manipulation such as symbiosis, e.g., between bees and flowers, to hostile parasites killing the host species (examples: fungi and ants; wasps and caterpillars). Within a species, manipulations more often do have some positive effects for the manipulated agent, but this is certainly not a general rule. Consider harems or wars as examples for humans. Kings or, in general, leaders control and manipulate others qua status—up to millions of others—for the leader's purposes *and to the latter's harm*, be they reproductive (harem) or in regard to power (war). Control or manipulation by others can also result in a shorter life (e.g., hard work in a mine), bad health, or missed opportunities: in general, it is the forced abandonment of own goals for the ends of others.

Additionally, a major revision of the perception of ourselves and the world would be necessary on abandoning this illusion (as in the first illusion). This is not only costly, but risky in its outcome: It is not certain whether a coherent working perception of ourselves (personality) can be achieved without major contradictions. In any case, such a process could last years. However, there are some biases (e.g., hindsight bias) contributing to that end.

A third negative consequence linked to the teachings of sociobiology is the feeling of many that Darwinism justifies cruelty, brutality, and the law of the jungle with its cold logic of natural selection. Merciless forces are said to determine human nature—there is neither a secure and harmonious place in nature or among ourselves nor is an ethical system (morals) feasible.

This third consequence is emotionally repugnant. Why? If this (false) belief of merciless cruelty between humans were to be true, the most valuable and most fundamental backbone of daily life would lose its significance and its protective value: the social group as protective system. Its protection includes not only emotional comfort in times of distress, but also physical safety, opportunities to learn, and many more.

18.4 Illusion 5: Cultural Achievements Are Purely Social Constructions

18.4.1 Description of Illusion

Cultural constructions such as religion or ethics have been widely regarded as the antithesis to biology—they are achievements *freeing* us from biological constraints. Such a "standard social science model" has come under attack by evolutionary psychologists (Tooby and Cosmides 1992).

The following discussion is limited to religion. Religion can be regarded as a representative for other cultural constructions. Religion is a very old phenomenon (at least 28,000 years, Mithen 1996; up to 70,000 years, Rossano 2009; possibly even older); it is without doubt very differentiated and universal. For all its variety, there is no definition of religion that is commonly agreed on.

According to two World Values surveys in 1981 and 1990, in 20 of 22 countries, the exceptions being Sweden and France, more than 60% of the population believes in God. Some countries even reach 100% and most are well above the 80% mark (Inglehart 1997, p. 372). Thus, it may suffice to state the obvious: religion is ubiquitous, it is very real to the overwhelming majority of people worldwide, it is a guide in both moral and ethical questions and it has been there in enormous variations ever since the dawn of humanity. Therefore, almost nobody would tag religion as an illusion at all. An exception is Richard Dawkins, who even labels it a "delusion", a still stronger term. Instead, religion is an abstract, theoretical concept that is not to be explained by naturalistic means. This is above all true for faith.

Yet, various thinkers have seen religion as an illusion. Among the more famous are Xenophanes and Critias (600 BC) and Ludwig Feuerbach (1804–1872), who pointed out the weaknesses of anthropomorphic religious beliefs. More recent and more popular are Marx and Freud:

> The abolition of religion as the illusory happiness of the people is required for their real happiness. The demand to give up the illusion about its condition is the demand to give up a condition which needs illusions (Marx 1968, p. 208, translation).

Along the same lines Freud said:

> Coming back to religious teachings, we may say recapitulating: All of them are illusions, unprovable, and nobody should be forced to hold them for true and to believe in them. Some of them are so improbable [...] that they are comparable to delusions (Freud 1927/1993, p. 134, translated by the author).

The empirically based academic challenge to religion dates back only a little more than 20 years, although the idea of analyzing religion in evolutionary terms dates back to Gustav Jäger, a contemporary of Darwin (Jäger 1869). This challenge has gathered momentum—contrary to our intuitions—claiming that religiosity and religion are biologically grounded and have to be explained as adaptations or by-products (Voland 2009a; Atran 2002). This in turn means that the metaphysics of religion and religiosity are an illusion.

18.4.2 Why Is It an Illusion?

Religious practices are behaviors and mental states, e.g., going to church once a week, praying, etc. Behaviors are expressions of evolutionary strategies (see e.g. Buss 2004). Religious behavior as one form of cultural behavior, therefore, "[...] sits on top of a bed of biological constraints and dispositions." (Ruse 1995, p. 158). The huge variety of religious flavors is little else but different expressions of universal behavior patterns. These patterns can be explained by naturalistic, non-transcendent means: It has been shown

– how religiosity is predisposed in children (Carey 1987);
– what mechanisms are responsible for religious explanations (Barrett and Keil 1996);
– that religions can be reduced to four basic variations, common to the natural world around humans, biomorph, sociomorph, technomorph, and ecstatic-cathartic (Topitsch 1988);
– how cults, e.g., cargo cults, develop very fast, sometimes in decades, and similar in structure to world religions (Dawkins 2006);
– that all religions are anthropomorphic to a high degree (Topitsch 1988);
– that part of the success of religions lies in their remarkable utilization of existing cognitive patterns (Frey 2009; Barrett 2000; Boyer 2004);
– that part of the success of religions lies in their adjustment for worldly use by political leaders (Lahti 2009).

Taken together, this evidence suggests an alternative, naturalistic interpretation of why we believe. Faith is not due to a transcendent god but to our very own cognitive apparatus on which religiosity can piggyback.

A major dividing question in this area of research is whether religion is an adaptation (Voland 2009a; Bulbulia 2007) or a by-product (Atran 2002; Boyer 2004). This may be due to terminological fuzziness (Voland 2009a) or differing explanations for the various components (e.g., mysticism, myths, metaphysics, ethics, rituals, etc.). This is certainly true for individual behaviors (e.g., benefitting from placebo effects by practicing shamanism; using a tight religious system of regulations to foster group cohesion, etc.).

18.4.3 Why Do We Have This Illusion?

Evolutionary studies of religion have compiled a lot of facts why religious behavior is often useful and fitness-enhancing:

- religious people seem to be healthier than non-believers (McCullough et al. 2000; Newberg and Lee 2006; McClenon 1997; Kark et al. 1996)
- religious people seem to be more fertile (Blume 2009; Zhang 2008)
- the concept of an all-seeing God helps to solve the free-rider problem in larger groups (Roes and Raymond 2003)
- religious groups persist longer than secular ones (Sosis and Bressler 2003)
- religious practices such as rituals establish social identities and enhance group solidarity (Voland 2009a)
- existential crises can be overcome more easily by reverting to mysticism (Boyer 2004)

These associations with religion are positive, offering a ready explanation as to why we have this illusion.

Considering the advantages mentioned above it seems obvious why these illusions of believing in supernatural powers persist. First, most components of religion are adaptations—religion being the unifying concept. Therefore, believers do have real benefits, e.g., better health or help in crises. Second, religion gives authoritative answers to existential questions such as the meaning of life or life after death. This in turn means that religious people often feel that a naturalistic explanation of religiousness leads to the negative consequences shown in Table 18.1: "Life is meaningless." or "Life has no purpose". Although this may be reinterpreted in a positive way ("Everybody is free to give life whatever purpose he/she desires", Schmidt-Salomon 2002), a meaningless life may lead to negative effects concerning motivation, activities, long-term projects, probably even culminating in suicide (Cioran 1987). Lacking motivation could have been detrimental in comparison to religious competitors (see e.g. the irrigation of desert areas in Israel, the Promised Land, by Jews, an extremely laborious project, spanning decades and requiring multiple scientific agricultural breakthroughs.)

In short: Most components of religious behavior are adaptive. Thus, the unifying force of religion is beneficial to its adherents, its traditions learned anew in every generation. A contributing factor for the persistence of this illusion are many cognitive biases that help to establish and maintain religious beliefs (Frey 2009). Religious beliefs build upon such biases as a hypersensitive agency detector (a certain way to perceive causality), a confirmation bias for existing beliefs, and many others. This means that functional cognitive structures—in their proper domain—are used for a new domain, religion, and filled with religious content.

18.4.4 Resistance

The Christian Church has rejected Darwin's theory of evolution since its publication in 1859 for good reasons (see Desmond and Moore 1994 for a detailed account, including positive reactions): First, Darwin presented a viable (even superior!) alternative to creation, a central tenet of Christian belief. Second, Darwin held that humans were not "in the likeness of God" but instead were descended from apes. Third, evolution is a-teleological, that is without goal or direction—which contradicts the Christian belief of God's plan for everything. Fourth, if evolutionary theory were right, morality would have to be rethought completely. Many people within and without the Church were afraid that morality might break down completely. More recently, evolutionary explanations of religion have met the same resistance (e.g. Dawkins 2006).

18.4.5 Reasons for Resistance

Abandoning a complex belief system like religion (see sections 18.2.5 and 18.3.5) is associated with high costs—not a step to be taken lightly. Giving up fundamental beliefs, which possibly shape an entire life, often entails giving up social networks, friends, and sometimes even members of the family. Two quotes from believers having converted to atheism (reactions to *The God Delusion* by Richard Dawkins, see http://richarddawkins.net/theGood) may be apt examples:

> Although this change [from religious to rational beliefs, UF] has been the source of an unimaginable amount of pain and dislocation in my family and circle of friends (what friends?), I still wish to thank you from the bottom of my heart and my newly free mind.

And a second one:

> My de-conversion, 15 years later, was a long, exhausting and lonely experience. [. . .] and though this has had profound implications on my personal relationships with family and friends, I do not regret this personal renaissance for one moment.

Of course, this may be true also for adapting a new belief system—an example would be severing all social ties when joining a monastery for life.

Individual losses, however, may not be the only negative consequence that converts may have to face. Consider great schisms like the Reformation, the conflict in Northern Ireland, Sunni and Shia Islam or any religious secession; they are usually characterized by extreme violence and an exorbitant death toll. Still another negative consequence of switching religions is the risk of manipulation by others, which is not negligible if sects (like scientology) are considered.

A switch to a new perception of oneself and the world has always to take into account the problem mentioned above: the new belief system may be no more than a manipulation by others for their benefit. Therefore, reasons against the old and in favor of the new system have to be strong indeed.

Apart from these difficulties (losing social networks, facing suppression, fearing manipulation), it is difficult if not impossible to re-create a totally different self-perception. This new "personality" would be at odds with the "old" one, the latter possibly consisting of *decades* of contradicting decisions and experiences in life.

A last point may be that the alternative to religious belief is fairly bleak: Compare believing in a comforting God and a meaningful life (according to God's wishes) plus eternal bliss in heaven after death to the naturalistic alternative of just living and dying (and nothing more). Small wonder, most people opt for alternative number one. Studies even suggest health benefits for believers (Newberg and Lee 2006; McCullough et al. 2000). What is the benefit of the latter? Nothing but a more rational and disillusioned world-view and a few spared prayers. This just does not sound like a good alternative to most people—remember Pascal's wager?

18.5 Illusion 6: We Are Free in What We Want

18.5.1 Description of Illusion

One of our most cherished illusions, not only because it is intuitively very obvious, is the belief in free will. We are under the overwhelming conviction that we can want whatever we want (free will), within some given physical and other restrictions (freedom of action).

Since antiquity, free will has been analyzed by religion, philosophy, and natural sciences. It is free will in particular that is absolutely essential for ethical systems, whether philosophical or religious. Morality is the decision between good and evil—but without free will choices become meaningless (essential, e.g., for Kant (1788/1995) in his *Critique of Practical Reason*, one of the most influential treatises on the topic).

The range of answers is broad: From the position that freedom remains complete even in situations with minimal freedom of action (Sartre 1943/1991) to schools emphasizing the numerous constraints on "free" will, e.g., our genetic inheritance, our education, and our culture. According to this latter view, these factors account for the majority of our decisions in life. Even if there were something like free will, it would be of minor importance. Throughout human history—either religious or philosophical—the complete denial of free will has never had a great number of serious proponents (eminent exceptions are Hobbes, Locke, Mill, Nietzsche, and Luther).

Furthermore, within natural sciences there is a growing consensus that to posit a free will for humans is equivalent to abandoning the very fundament of science, as explaining natural phenomena means determining their causal relationships. Literally, gaps would appear in the world, if free decisions were possible. These gaps would not be explicable in terms of science, making in turn our most successful explanation system false and useless (for a detailed discussion, see e.g. Beckermann 2001). Finally, insights from the neurosciences have changed the debate about free will quite dramatically in recent decades (see next section).

18.5.2 Why Is It an Illusion?

One trend emerges: Traditional philosophy of mind advocating free will has come under pressure, particularly with (the ever) growing empirical research of neuroscience. There is neither a "control center" in the brain (Dennett 1991), nor evidence that a free will precedes decisions and actions. The famous experiments of Libet and follow-ups by Haggard and Eimer are evidence for that, but have received their share of criticism. Current research, however, makes these debates meaningless by demonstrating that even complex decisions can be predicted accurately (Haynes et al. 2007), that decisions can be influenced by technical means without disturbing the sense of free will (Brasil-Neto et al. 1992), and that actions are decided upon not only in milliseconds, but up to ten seconds before the "free" conscious decision to do so (Soon et al. 2008).

Evolutionary theories trace back our intuition of free will to our lives in highly complex social networks. The social brain hypothesis (Dunbar 1998) contends that brain development is mainly due to the problem of understanding social interactions. In particular, this applies to the *intentions* responsible for actions of other group members. Therefore, we infer from our own intentions to the intentions of other actors (looking for food when hungry, etc.). Another theory turns this idea on its head: We do not infer from us to others, but from others to ourselves (Voland 2007b).

As other living beings seem to act upon their (free) intentions, we calculate their actions as if they possessed a free will. This assumption is called intentionality stance by Dennett (Dennett 1991). This modeling is only the second-best way; the best would be to infer from our decisions and their reasons translating into actions to others. However, this way is barred: Our own brain is intransparent to us, thus, we are not able to understand its own workings from a first-person perspective. Owing to this intransparency, the assumption of "free will" is at the same time as much an eminently practical instrument in daily life as it is false according to physical and neurophysiological arguments.

18.5.3 Why Do We Have This Illusion?

The reasons for the illusion of a free will have already been pointed out in the section above. The most important point is that humans are equipped with a "social brain" trying to read the intentions of others as well as the intransparency of decision processes.

18.5.4 Resistance

Resistance to the idea that there is no free will is quite strong, mostly in philosophical circles. There may be several reasons for this. An obvious trend is the shift from philosophy (e.g., Pauen 2004) to neuroscience. In turn, philosophers are concerned about their area of expertise, which some feel is slipping away from them. The most commonly used argument of philosophers (e.g., Peter Bieri or Geert Keil) is that empirical evidence from neuroscience has nothing to contribute to the problem of free will—but denial has never been a particularly successful strategy against the progress of natural science.

18.5.5 Reasons for Resistance

Resistance comes from multiple levels and for different reasons. First, our decision processes are intransparent to us, therefore the idea of having no free will is blatantly against our intuitions. Second, the problem of free will has been a core problem for both philosophy and religions for centuries. *Any new* approach (here: naturalistic explanations of our brain) will be regarded skeptically. It may have too little scientific weight (just another unsuccessful approach) or it may have too much (it could dissolve the issue once and for all, making former theories obsolete).

Third, often having no free will is wrongly linked to the consequence of the meaninglessness of life—especially in popular media. If everything is determined, then we are no more than machines: we do not have a single choice in our lives, and we are not able to change anything at all.

The associated ultimate loss of control is important, too (see section 18.3.5, which describes the implications). Denying that my free will and my ability to make my own decisions are existent means taking away the core of self-control and self-determination. The fundamental uniting force of my personality is attacked. Thus, all ambitions and purposes are rendered nil, making me open for manipulations by others.

18.6 Conclusion

This chapter has argued that some scientific revolutions destroy popular beliefs, exposing them as illusions. In itself this is not particularly interesting—it happens all the time. But some of these beliefs are fundamental to our very perception of ourselves and the world (of modern Western people). Plus, in each case these beliefs are strongly supported by our intuition. Strong resistance is to be expected, if only for historical reasons (a new concept has to assert itself) or psychological reasons (it is counterintuitive). This is the proximate level of explanation.

But there is more. Why do we have these illusions in the first place? Why don't they just disappear as soon as the "truth" is known? This is the ultimate level of explanation—the evolutionary reason for their existence. We hold them, because they are beneficial for us, intuitively obvious, and they are the "default" option. We are dualists, naïve realists, pre-Newtonians, and believe in free will because these are simple, hence cognitively undemanding models of the world and almost always right. They don't disappear, because the consequences of giving them up *are negative* in several respects:

Self-perception has to be harmonious and without inner contradictions. Personality disorders such as schizophrenia attest to this claim. Accordingly, there are a host of mechanisms (biases) to ensure a "whole" personality (e.g., hindsight bias, confirmation bias, attributional biases; see Taylor 1989/1995; Gilovich 1991; Plous 1993). A "whole" personality is not only healthy, but constitutes also a precondition for successful long-term plans (goals are not switched every few days, months, years), for efficiency (plans can be based on other plans), and for the ability to stay capable of acting. If the perception of oneself is unbalanced and without self-efficacy (Bandura and Schunk 1981; Schwarzer 1999), negative consequences such as listlessness, equating to missing out important opportunities or chances, are sure to follow. Contributing to such a harmonious self-perception is, for example, the illusion of leading a meaningful life. Consider the opposite: A life perceived to be absolutely meaningless may lead—in extreme cases—to spiritlessness, depression, and even suicide. Compared to people believing this illusion this would be a serious disadvantage in a biological sense. Therefore, these illusions may serve the purpose of staying active, maintaining the capability to act, and avoiding contradictions in action plans throughout life.

Another positive effect of some illusions (2, 3, 4, 6) is to maintain autonomy *and* being in control. The evolutionary function seems clear enough: This serves as a guard against alien manipulations. It also allows one to keep on achieving one's *own* goals in a biological sense. To fend off exploitation is one of the most basic requirements of organisms in nature. In this line of reasoning, self-determination (in all creatures, and in sophisticated mechanisms in humans, too) is a general mechanism to avoid alien manipulations.

A third aspect besides the psychological implementation of evolutionary adaptations is worth noting. The mechanisms are within *constraints* of our past, including blueprint and ecological niche. Cognitive patterns and strategies provide the tracks for cultural and religious decisions. For example, the Christian creation myth is the *effect* of an agency-based cognitive pattern, not the *cause* of similar cultural figures of thought.

To conclude: these arguments suggest that the illusions are here to stay. There is a long road ahead before naturalization of our perception of ourselves and the world can be successful and generally accepted. But as shown in this chapter, there are downsides to these illusions, mostly in scientific respects. So the intuitive resistance mentioned should be strictly kept out of scientific debates about these questions—which I have tried to demonstrate has been nearly impossible up to now. It distorts our scientific view of what is *really the case* despite the apparent filters of our cognitive and emotional apparatus. Hopefully, eventually, humankind will be completely disillusioned.

References

Alexander RD (1987) The Biology of Moral Systems. Berlin: De Gruyter

Atran S (2002) *In Gods We Trust: The Evolutionary Landscape of Religion.* Oxford University Press, Oxford

Bandura A, Schunk DH (1981) Cultivating competence, self-efficacy, and intrinsic interest through proximal self-motivation. Journal of Personality and Social Psychology 41:586–598

Barrett JL (2000) Exploring the natural foundations of religion. Trends in Cognitive Sciences 4(1):29–34

Barrett JL, Keil FC (1996) Conceptualizing a nonnatural entity: anthropomorphism in God concepts. Cognitive Psychology 31:219–247

Beckermann A (2001) *Analytische Einführung in die Philosophie des Geistes.* De Gruyter, Berlin

Bethell T (2001) Against Sociobiology. http://www.firstthings.com/article/2007/01/against-sociobiology-12. Accessed 22 December 2009

Bird CD, Emery NJ (2009) Insightful problem solving and creative tool modification by captive nontool-using rooks. Proceedings of the National Academy of Sciences of the USA 106:10370–10375

Blume M (2009) The reproductive benefits of religious affiliation. In: Voland E, Schiefenhövel W (eds) *The Biological Evolution of Religious Mind and Behavior*. Springer, Heidelberg

Bogin B (2006) Modern human life history: the evolution of human childhood and fertility. In: Hawkes K, Paine RR (eds) *The Evolution of Human Life History*. School of American Research, Santa Fe, NM

Boyer P (2004) *Und Mensch schuf Gott*. Klett-Cotta, Stuttgart

Brague R (1994) Geozentrismus als Demütigung des Menschen. Internationale Zeitschrift für Philosophie 1:1–25

Brague R (2006) *Die Weisheit der Welt: Kosmos und Welterfahrung im westlichen Denken*. Beck, Munich

Brasil-Neto JP, Pascual-Leone A, Valls-Solé J, Cohen LG, Hallett M (1992) Focal transcranial magnetic stimulation and response bias in a forced-choice task. Journal of Neurology, Neurosurgery, and Psychiatry 55:964–966

Brosnan SF, de Waal FBM (2003) Monkeys reject unequal pay. Nature 425:297–299

Bulbulia JA (2007) The evolution of religion. In:.Dunbar RIM, Barrett L (eds) *Oxford Handbook of Evolutionary Psychology*. Oxford University Press, Oxford

Buss DM (2004) *Evolutionary Psychology: The New Science of the Mind*. Pearson, Boston, MA

Carey S (1987) *Conceptual Change in Childhood*. MIT Press, Cambridge, MA

Catalano R, Bruckner T, Anderson E, Gould JB (2005) Fetal death sex ratios: a test of the economic stress hypothesis. International Journal of Epidemiology 34:944–948

Chasiotis A, Hofer J, Campos D (2006) When does liking children lead to parenthood? Younger siblings, implicit prosocial power motivation, and explicit love for children predict parenthood across cultures. Journal of Cultural and Evolutionary Psychology 4(2):95–123

Choi J, Bowles S (2007) The coevolution of parochial altruism and war. Science 318:636–640

Cioran EM (1987) *Lehre vom Zerfall*. Klett-Cotta, Stuttgart

Cirkovic MM (2003) The temporal aspect of the Drake equation and SETI. ArXiv:astro-ph/0306186v1

Dawkins R (1976) *The Selfish Gene*. Oxford University Press, Oxford

Dawkins R (1990) *The Extended Phenotype: The Long Reach of the Gene*. Oxford University Press, Oxford

Dawkins R (2006) *The God Delusion*. Bantam, London

Dawkins R, Krebs JR (1978) Animal signals: information or manipulation? In: Krebs JR, Davies NB (eds) *Behavioral Ecology: An Evolutionary Approach*. Blackwell, Oxford

Dennett DC (1991) *Consciousness Explained*. Little, Brown and Co., Boston, MA

Desmond A, Moore J (1994) *Darwin*. Rowohlt, Reinbek

Dunbar K (1997) How scientists think: on-line creativity and conceptual change in science. In: Ward TB, Smith SM, Vaid S (eds) *Conceptual Structures and Processes: Emergence, Discovery and Change*. APA Press, Washington, DC

Dunbar RIM (1998) The social brain hypothesis. Evolutionary Anthropology 6(5):178–190

Flasch K (1987) *Das philosophische Denken im Mittelalter: Von Augustin zu Machiavelli*. Reclam, Stuttgart

Freud S (1927/1993) *Die Zukunft einer Illusion*. Fischer, Frankfurt

Freud S (1969) *Eine Schwierigkeit der Psychoanalyse*. Fischer, Frankfurt

Frey UJ (2007) *Der blinde Fleck—Kognitive Fehler in der Wissenschaft und ihre evolutionsbiologischen Grundlagen*. Ontos, Heusenstamm

Frey UJ (2009) Cognitive Foundations of religiosity. In: Voland E, Schiefenhövel W (eds) *The Biological Evolution of Religious Mind and Behavior*. Springer, Heidelberg

Gallagher MW, Lopez SJ (2009) Optimism is Universal: Exploring Demographic Predictors of Optimism in a Representative Sample of the World. Conference Paper

Geiser R (2005) Soziobiologie mit besonderer Berücksichtigung des Menschen. http://geocities.
ws/disputator2000/sociobiologia2005-06.html. Accessed 28.4.2010
Gigerenzer G, Todd PM (1999) Fast and frugal heuristics: the adaptive toolbox. In: Gigerenzer
G, Todd PM, ABC Research Group (eds) *Simple Heuristics that Make Us Smart*. Oxford
University Press, New York, NY
Gilovich T (1991) *How We Know What Isn't So: The Fallibility of Human Reason in Everyday
Life*. Macmillan, New York, NY
Gintis H, Bowles S, Boyd R, Fehr E (2003) Explaining altruistic behavior in humans. Evolution
and Human Behavior 24:153–172
Goldstein DG, Gigerenzer G (1999) The recognition heuristic: how ignorance makes us smart. In:
Gigerenzer G, Todd PM, ABC Research Group (eds) *Simple Heuristics that Make Us Smart*.
Oxford University Press, New York, NY
Güntürkün O (1996) Lernprozesse bei Tieren. In: Hoffmann J, Kintsch W (eds) *Enzyklopädie der
Psychologie*. Hogrefe, Göttingen
Hamilton WD (1964a) The genetical evolution of social behaviour. I. Journal of Theoretical
Biology 7:1–16
Hamilton WD (1964b) The genetical evolution of social behaviour. II. Journal of Theoretical
Biology 7:17–52
Hauser MD (2006) *Moral Minds: How Nature Designed Our Universal Sense of Right and Wrong*.
HarperCollins, New York, NY
Haynes J, Sakai K, Rees G, Gilbert S (2007) Reading hidden intentions in the human brain. Current
Biology 17(4):323–328
Inglehart R (1997) *Modernization and Postmodernization: Cultural, Economic, and Political
Change in 43 Societies*. Princeton University Press, Princeton, NJ
Inoue S, Matsuzawa T (2007) Working memory of numerals in chimpanzees. Current Biology
17:R1004–R1005
Jäger G (1869) *Die Darwin'sche Theorie und ihre Stellung zu Moral und Religion*. Hoffmann,
Stuttgart
Kant I (1788/1995) Kritik der praktischen Vernunft. In: Weischedel W (ed) *Die Kritiken/Immanuel
Kant*. Suhrkamp, Frankfurt
Kark JD, Shemi G, Friedlander Y, Martin O (1996) Does religious observance promote health?
Mortality in secular vs religious kibbutzim in Israel. American Journal of Public Health
86(3):341–346
Koestler A (1959) *Die Nachtwandler: Das Bild des Universums im Wandel der Zeit*. Emil Vollmer,
Wiesbaden
Kraiker C (1992) Die Geschichte von den drei Kränkungen. http://www.paed.uni-muenchen.de/
~kraiker/Kraenkungen.htm. Accessed 10 July 2009
Krützen M, Mann J, Heithaus MR, Connor RC, Bejder L, Sherwin WB (2005) Cultural transmis-
sion of tool use in bottlenose dolphins. Proceedings of the National Academy of Sciences of
the USA 102:8939–8943
Kuhn TS (1957/1981) *Die kopernikanische Revolution*. Suhrkamp, Frankfurt
Lahti DC (2009) The correlated history of social organization, morality and religion. In: Voland
E, Schiefenhövel W (eds) *The Biological Evolution of Religious Mind and Behavior*. Springer,
Heidelberg
Losito LJ (2002) Beetles. In: O'Toole C (ed) *Encyclopedia of Insects and Spiders*. Oxford
University Press, Oxford
Mace R (2000) Evolutionary ecology of human life history. Animal Behaviour 59:1–10
Marx K (1968) Zur Kritik der Hegelschen Rechtsphilosophie. In: Landshut S (ed) *Die
Frühschriften*. Kröner, Stuttgart
McClenon J (1997) Shamanic healing, human evolution, and the origin of religion. Journal for the
Scientific Study of Religion 36(3):345–354
McCullough ME, Larson DB, Hoyt WT, Koenig HG (2000) Religious involvement and mortality:
a meta-analytic review. Health Psychology 19(3):211–222
Mithen S (1996) *The Prehistory of the Mind: The Cognitive Origins of Art, Religion and Science*.
Thames and Hudson, New York, NY

Nesse RM (2001) The smoke detector principle: natural selection and the regulation of defensive responses. Annals of the New York Academy of Sciences 935:75–85

Newberg AB, Lee BY (2006) The relationship between religion and health. In: McNamara P (ed) *Where God and Science Meet: How Brain and Evolutionary Studies Alter Our Understanding of Religion, Vol. 3: The Psychology of Religious Experience.* Praeger, Westport, CT

Nietzsche F (1883/1999) Also sprach Zarathustra. In: Schlechta K (ed) *Friedrich Nietzsche: Werke in drei Bänden,* Band II. Hanser, Munich

Nowak MA (2006) Five rules for the evolution of cooperation. Science 314:1560–1563

Nowak M, Sigmund K (1993) A strategy of win-stay, lose-shift that outperforms tit-for-tat in the Prisoner's Dilemma game. Nature 364:56–58

Pauen M (2004) *Illusion Freiheit? Mögliche und unmögliche Konsequenzen der Hirnforschung.* Fischer, Frankfurt

Pauen M (2007) *Was ist der Mensch? Die Entdeckung der Natur des Geistes.* Deutsche Verlagsanstalt, Munich

Platek SM, Raines DM, Gallup JGG, Mohamed FB, Thomson JW, Myers TE, Panyavin IS, Levin SL, Davis JA, Fonteyn LCM, Arigo DR (2004) Reactions to children's faces: males are more affected by resemblance than females are, and so are their brains. Evolution and Human Behavior 25(6):394–405

Plous S (1993) *The Psychology of Judgment and Decision Making.* McGraw-Hill, New York, NY

Roes FL, Raymond M (2003): Belief in moralizing gods. Evolution and Human Behavior 24:126–135

Rossano M (2009) The African interregnum: the 'where', 'when', and 'why' of the evolution of religion. In: Voland E, Schiefenhövel W (eds) *The Biological Evolution of Religious Mind and Behavior.* Springer, Heidelberg

Ruse M (1995) *Evolutionary Naturalism: Selected Essays.* Routledge, London

Sartre J-P (1943/1991) *Das Sein und das Nichts: Versuch einer phänomenologischen Ontologie.* Rowohlt, Reinbek

Scheler M (1955) Vom Umsturz der Werte: Abhandlungen und Aufsätze. In: Scheler M (ed) *Max Scheler: Gesammelte Werke,* Band 3. Francke, Bern

Schmidt-Salomon M (2002) Hoffnung jenseits der Illusionen? Die Perspektive des evolutionären Humanismus. http://www.schmidt-salomon.de/homepage.htm. Accessed 21 September 2009

Schwarzer R (1999) Self-regulatory processes in the adoption and maintenance of health behaviors: the role of optimism, goals, and threats. Journal of Health Psychology 4(2):115–127

Segerstrale U (2000) *Defenders of the Truth: The Battle for Science in the Sociobiology Debate and Beyond.* Oxford University Press, Oxford

Sime JD (1983) Affiliative behaviour during escape to buildings exits. Journal of Environmental Psychology 3(1):21–41

Söling C (2002) *Der Gottesinstinkt: Bausteine für eine Evolutionäre Religionstheorie.* Dissertation, Universität Gießen. http://bibd.uni-giessen.de/ghtm/2002/uni/d020116.htm. Accessed 20 December 2009

Sommer V (1989) *Die Affen: Unsere wilde Verwandtschaft.* Gruner und Jahr, Hamburg

Soon CS, Brass M, Heinze H, Haynes J (2008) Unconscious determinants of free decisions in the human brain. Nature Neuroscience 11:543–545

Sosis R, Bressler ER (2003) Cooperation and commune longevity: a test of the costly signaling theory of religion. Cross-Cultural Research 37(2):211–239

Strassmann BI, Gillespie B (2002) Life-history theory, fertility and reproductive success in humans. Proceedings of the Royal Society of London B: Biological Sciences 269:553–562

Sukopp T (2006) *Naturalismus—Kritik und Verteidigung erkenntnistheoretischer Positionen.* Ontos, Heusenstamm

Taylor SE (1989/1995) *Mit Zuversicht: Warum positive Illusionen für uns so wichtig sind.* Rowohlt, Reinbek

Thornton A, McAuliffe K (2006) Teaching in wild Meerkats. Science 313:227–229

Tooby J, Cosmides L (1992) The Psychological Foundations of Culture. In: Barkow JH, Cosmides L, Tooby J (eds) *The Adapted Mind: Evolutionary Psychology and the Generation of Culture*. Oxford: Oxford University Press

Topitsch E (1988): *Erkenntnis und Illusion: Grundstrukturen unserer Weltauffassung*. Mohr Siebeck, Tübingen

Trivers R (1985) *Social Evolution*. Benjamin Cummings, San Francisco, CA

Tweney RD (2001) Scientific thinking: a cognitive-historical approach. In: Crowley KD, Schunn CD, Okada T (eds) *Designing for Science: Implications from Everyday, Classroom, and Professional Settings*. Lawrence Erlbaum, Hillsdale, NJ

Vogel C (1989) *Vom Töten zum Mord: Das wirklich Böse in der Evolutionsgeschichte*. Hanser, Munich

Voland E (2005) '... nur tierischer als jedes Tier zu sein'. Vom 'sogenannten' zum 'wirklich Bösen' in der Evolution. In: Elsner N, Lüer G (eds) '... *sind eben alles Menschen'—Verhalten zwischen Zwang, Freiheit und Verantwortung*. Wallstein, Göttingen

Voland E (2007a) Die Fortschrittsillusion. Spektrum der Wissenschaft 4:108–113

Voland E (2007b) We recognize ourselves as being similar to others: implications of the 'social brain hypothesis' for the biological evolution of the intuition of freedom. Evolutionary Psychology 5(3):442–552

Voland E (2009a) Evaluating the evolutionary status of religiosity and religiousness. In: Voland E, Schiefenhövel W (eds) *The Biological Evolution of Religious Mind and Behavior*. Springer, Heidelberg

Voland E (2009b) *Soziobiologie—Die Evolution von Kooperation und Konkurrenz*. Spektrum, Heidelberg

Vollmer G (1975/2002) *Evolutionäre Erkenntnistheorie: Angeborene Erkenntnisstrukturen im Kontext von Biologie, Psychologie, Linguistik, Philosophie und Wissenschaftstheorie*. Hirzel, Stuttgart

Vollmer G (1995) Die vierte bis siebte Kränkung des Menschen—Gehirn, Evolution und Menschenbild. In: Vollmer G (ed) *Auf der Suche nach der Ordnung: Beiträge zu einem naturalistischen Welt- und Menschenbild*. Hirzel, Stuttgart

de Waal FBM (1989) Food sharing and reciprocal obligations among chimpanzees. Journal of Human Evolution 18(5):433–459

Williams BJ (1968) Discussions. In: Lee RB, DeVore I (eds) *Man the Hunter*. Aldine, New York, NY

Wilson EO (1975) *Sociobiology: The New Synthesis*. Belknap Press of Harvard University Press, Cambridge, MA

Zhang L (2008) Religious affiliation, religiosity, and male and female fertility. Demographic Research 18(8):233–262

Index

U.J. Frey et al. (eds.), *Homo Novus – A Human Without Illusions*,
The Frontiers Collection, DOI 10.1007/978-3-642-12142-5,
© Springer-Verlag Berlin Heidelberg 2010

THE FRONTIERS COLLECTION

Series Editors:

A.C. Elitzur L. Mersini-Houghton M.A. Schlosshauer M.P. Silverman
J.A. Tuszynski R. Vaas H.D. Zeh